Connecting Territories

Connecting Territories

Exploring People and Nature, 1700–1850

Edited by

Simona Boscani Leoni
Sarah Baumgartner
Meike Knittel

BRILL

LEIDEN | BOSTON

This volume was funded by publication grants from the Swiss National Science Foundation, the UniBern Forschungsstiftung (Berne University Research Foundation) and the Friedrich-Emil-Welti-Fonds, Bern. The editors thank these foundations for their support.

FNSNF Swiss National Science Foundation

UniBern | Forschungsstiftung Berne University Research Foundation

Cover illustration: Detail of the *Teufelsbrücke* (bridge of the devil) over the river Reuss in the Swiss canton of Uri, from: Johann Jacob Scheuchzer. Nova Helvetiae tabula geographica illustrissimis et potentissimis cantonibus et rebus publicis reformatae religionis Tigurinae, Bernensi, Glaronensi, Basiliensi, Scaphusianae, Abbatis Cellanae / dominis suis clementissimis humillime dicata a Ioh. Iacobo Scheuchzero Tigurino; Ioh. Melch. Füsslin ornamenta pinx.; Ioh. Henr. Huber et Eman. Schalch sculps. Zurich, [Hofmeister], [1765?]. Zentralbibliothek Zürich, 5 Hb 04: 8-14. Wikimedia Commons.

The Library of Congress Cataloging-in-Publication Data is available online at https://catalog.loc.gov
LC record available at https://lccn.loc.gov/2021048739

Typeface for the Latin, Greek, and Cyrillic scripts: "Brill". See and download: brill.com/brill-typeface.

ISSN 2452-3283
ISBN 978-90-04-41246-0 (hardback)
ISBN 978-90-04-41247-7 (e-book)

This book is printed on acid-free paper and produced in a sustainable manner.

Contents

Figures and Tables

Figures

Tables

Notes on Contributors

Sarah Baumgartner
has recently completed a PhD at the University of Bern. She holds an MA in History and a BSc in Geography of the same University and now works as a freelance historian and archivist. Her research interests include the history of scientific societies in the 18th century and early modern agriculture.

Simona Boscani Leoni
is a Swiss National Science Foundation-Professor of Early Modern History at the University of Bern and senior researcher at the EPFL, Lausanne. She earned a PhD in History from the EHESS in Paris. Her main research interests are medieval social history, iconography, early modern history of knowledge and environmental history. Her current research project focuses on the main actors, networks and places of scientific communication from the 17th century to around 1830. Most recently, she published an annotated edition of the correspondence of the Swiss physician and naturalist Johann Jakob Scheuchzer: *"Unglaubliche Bergwunder". Johann Jakob Scheuchzer und Graubünden. Ausgewählte Briefe*, 1699–1707 (2019); an extended edition of this correspondence online on: https://hallerNet.org.

Stefanie Gänger
is Professor of Modern History at the University of Heidelberg. She holds an MPhil and a PhD in History from the University of Cambridge and completed her BA in History at the universities of Augsburg and Seville. Her work considers the histories of science and medicine in late-colonial and early Republican Spanish America, as well as the wider world. Her books and articles examine a variety of themes, ranging from the history of antiquarianism and collecting to that of vernacular medicine and plant trade.

Meike Knittel
is a postdoctoral researcher at the Museum für Naturkunde Berlin and an associated researcher at the University of Bern's Historical Institute, where she completed her PhD in 2018. She holds an MA in History from the University of Konstanz and has worked as a curator for an exhibition on the medieval coral trade. Her research interests include the history of collections and the history of knowledge with a focus on natural history museums and 18th-century botany.

Francesco Luzzini
(PhD, History of Science; BS/MS, Natural Sciences) is Affiliate Scholar at the Max Planck Institute for the History of Science in Berlin and Contributing Editor for the *Isis* Bibliography of the History of Science. His work focuses mainly on the earth and environmental sciences, natural philosophy, and medicine in early modern Europe. His new project *Sounding the Depths of Providence* focuses on the early modern debate on ore generation and on its influence on the human-environment system: in the upcoming years, this research will be carried out under an EU-sponsored Marie Sklodowska- Curie Global Fellowship (Ca' Foscari University Venice-Johns Hopkins University).

Jon Mathieu
is editor-in-chief of the journal *Histories* and professor emeritus of history at the University of Lucerne, Switzerland. He has widely published on the history of mountain regions. Currently he is working on a *Brief Global History of Sacred Mountains since 1500*.

Barbara Orland
is a senior lecturer (Privatdozentin) of history of science, technology, and medicine at the pharmacy museum of the University of Basel. The main field of her research and teaching lies in the history of the life sciences and biomedicine from the 17th to the 20th century. In 2011 she replaced the professor for history of science at the University of Konstanz. In 2007/2008 she has been awarded the Käthe-Leichter guest professorship at the University of Vienna. Her current research interests cover different fields of the history of life sciences and biomedicine, e.g. history of pharmaceutical materiality; scientific concepts of fertilization and pregnancy, nutrition and metabolism, biomaterials (like blood and milk).

Irina Podgorny
is a permanent research fellow at the Argentine National Council of Science (CONICET). She studied Archaeology at La Plata University, obtaining her PhD in 1994 with a dissertation on the history of archaeology and museums. She has been a research fellow at the MPIWG in Berlin and the Ibero-Amerikanisches Institut Berlin as well at MAST (Museu de Astronomia) in Rio de Janeiro. She has been a member of the Editorial Board of Science in Context since 2003 and History of Humanities since 2017 and has recently been elected to the Council of HSS (History of Science Society). Her current work includes the history of paleontology, museums of natural history and archaeological ruins.

Chetan Singh
former Professor of History at Himachal Pradesh University, Shimla (India), specialised in medieval north Indian history at the Centre for Historical Studies, Jawaharlal Nehru University, New Delhi. He has also been researching and writing on environmental change, history and culture of the western Himalaya for more than two decades. He was Director of the Indian Institute of Advanced Study, Shimla, from 2013 to 2016. His publications include *Himalayan Histories: Economy, Polity and Religious Tradition* (2018), *Natural Premises: Ecology and Peasant Life in the Western Himalaya, 1800–1950* (1998), and *Region and Empire: Panjab in the Seventeenth Century* (1991).

Martin Stuber
is senior researcher at the Historical Institute of the University of Bern, Switzerland. He published widely on the history of natural resources (forest, cultivated plants, wetlands, sustainability) and on the history of knowledge in the 18th century (Albrecht von Haller, Scientific Networks, Economic Enlightenment).

CHAPTER 1

Introduction: From Switzerland to the Indies

Simona Boscani Leoni, Sarah Baumgartner and Meike Knittel

In Zurich, in the early 18th century, the physician and naturalist Johann Jakob Scheuchzer (1672–1733) published a bibliography of all books on natural history that he knew about. His catalogue contained publications on Europe, Africa, Asia, and the Americas. Scheuchzer's idea was to bring together the different local natural histories into a natural history of the world. His own contribution to this "global natural history" was his research on the Alps and its fauna, flora, minerals and about "homo alpinus helveticus" ("Swiss Alpine Man").[1]

Half a century later, the Physical Society of Zurich, a patriotic society interested in natural sciences and agronomy, inspired by similar projects undertaken by other economic societies, conducted surveys of the city's subject territory.[2] Likewise, animated by reports on travels to hitherto unknown regions and news about the exploration of faraway territories, diverse actors – both academically and non-academically educated – all over Europe became increasingly interested in their surroundings. Drawing on travelogues and questionnaires of local and colonial actors as well as on their own observations, early modern natural historians collected information about both nature and people inhabiting these territories, which helped to turn them into distinct spaces.[3]

---

1 Johann Jakob Scheuchzer, *Bibliotheca scriptorum historiae naturalis omnium Terrae Regionum inservientium; historiae naturalis Helvetiae Prodromus; accessit Celeberrimi Viri Jacobi Le Long, Bibliothecarii Oratoriani de Scriptoribus historiae naturalis Galliae; collegit Johann Jakob Scheuchzer, Med. D. Math. Prof. Societ. Leopoldino-Carolinae, Regiae Angl. & Boruss.* (Tiguri: Typis Henrici Bodmeri, 1716). About Johann Jakob Scheuchzer, see: Simona Boscani Leoni, ed., *Wissenschaft – Berge – Ideologien: Johann Jakob Scheuchzer (1672–1733) und die frühneuzeitliche Naturforschung – Scienza – montagna – ideologie. Johann Jakob Scheuchzer (1672–1733) e la ricerca naturalistica in epoca moderna* (Basel: Schwabe, 2010); Urs B. Leu, ed., *Natura Sacra: Der Frühaufklärer Johann Jakob Scheuchzer (1672–1733)* (Zug: Achius Verlag, 2012); Michael Kempe, *Wissenschaft, Theologie, Aufklärung: Johann Jakob Scheuchzer (1672–1733) und die Sintfluttheorie* (Epfendorf: bibliotheca academica Verlag, 2003).

2 Sarah Baumgartner, "Das nützliche Wissen. Akteure, Tätigkeiten, Kommunikationspraxis und Themen der Naturforschenden Gesellschaft in Zürich, 1746 bis ca. 1830" (PhD diss., Universitity of Bern, 2019); Rolf Graber, *Bürgerliche Öffentlichkeit und spätabsolutistischer Staat: Sozietätenbewegung und Konfliktkonjunktur in Zürich 1746–1780* (Zurich: Chronos, 1993).

3 On travels, queries and instructions for travellers: Maurizio Bossi and Claudio Greppi, eds., *Viaggi e scienza: le istruzioni scientifiche per i viaggiatori nei secoli XVII–XIX*, Gabinetto scientifico letterario G. P. Vieusseux, Studi 13 (Florence: Olschki, 2005); Lorelai Kury, "Les

 | DOI:10.1163/9789004412477_002

FIGURE 1.1
Seal of the Physical Society Zurich (*Physikalische Gesellschaft Zürich or Naturforschende Gesellschaft Zürich*)
SOURCE: FERDINAND RUDIO, FESTSCHRIFT DER NATURFORSCHENDEN GESELLSCHAFT IN ZÜRICH, 1746–1896, ZURICH, IN KOMMISSION BEI FÄSI & BEER, 1896, 213

Since the early days of European presence in the Americas, information reaching Europe contributed to a critical re-reading of Aristotle and other classical texts: Renaissance scholars shared the view that the natural world was still largely unknown and that many *secrets* were waiting to be investigated.[4] Curiosity revived, not only for antiquities, but for "exotic" and local nature, especially for "exotic" and local flora and fauna, and for minerals, more particularly for local characteristics of the terrain, orology and hydrology. Civil and natural history, geography and antiquarianism are very long-standing "travelling companions" in the early modern age.

instructions de voyage dans les expéditions scientifiques françaises (1750–1830)," *Revue d'histoire des sciences* 51, no. 1 (1998); Justin Stagl, *A History of Curiosity: The Theory of Travel 1550–1800* (Chur: Harwood Academic Publishers, 1995); Daniel Carey, "Inquiries, Heads, and Directions: Orienting Early Modern Travel," in *Travel Narratives, the New Science and Literary Discourse: 1569–1750*, ed. Judy A. Hayden (Farnham: Ashgate, 2012); Lorelai Kury, *Histoire naturelle et voyages scientifiques (1780–1830)* (Paris: L'Harmattan, 2001); Joan-Pau Rubiés, "Instructions for Travellers: Teaching the Eye to See," *History and Anthropology* 9, no. 2–3 (1996); Alix Cooper, "Fragen ohne Antworten: Die Suche nach lokalen Informationen in der frühen Aufklärung," in *Vor Google: Eine Mediengeschichte der Suchmaschine im analogen Zeitalter*, ed. Thomas Brandstetter (Bielefeld: transcript, 2012).

4 On the complexity of natural history during the Renaissance, the "discovery" of the Americas and the development of a colonial botany, see: Brian W. Ogilvie, *The Science of Describing: Natural History in Renaissance Europe* (Chicago, London: The University of Chicago Press, 2006); Londa Schiebinger and Claudia Swan, eds., *Colonial Botany: Science, Commerce, and Politics in the Early Modern World* (Philadelphia: University of Pennsylvania Press, 2005); Samir Boumediene, *La colonisation du savoir: une histoire des plantes médicinales du "Nouveau Monde" (1492–1750)* (Vaulx-en-Velin: Les éditions des mondes à faire, 2016); Norbert Ortmayr, "Kulturpflanzentransfers 1492–1900," *Historische Sozialkunde: Geschichte, Fachdidaktik, politische Bildung* 1 (2002); Yota Batsaki, Sarah Burke Cahalan and Anatole Tchikine, eds., *The Botany of Empire in the Long Eighteenth Century* (Washington, DC: Dumbarton Oaks Research Library and Collection, 2016).

The final decades of the 18th century became a crucial period for the different fields of natural history and related disciplines of ethnology and archaeology, as academic subjects in light of the advancement of Linnaean systematic classification in botany and zoology.[5] The perception and systematization of the natural world went through a paradigmatic shift as a result. Collections and the practice of collecting itself played a major role in this process and influenced the global exchange of ideas, knowledge, specimens, and personnel. Material as well as intellectual exchange took place in diverse settings, including collections, media, lecture halls, and the natural world itself.

This publication brings together several researchers from around the globe, who have analysed the exploration of the different "local natures" (natural history) and the discovery of "local inhabitants and their history" (ethnography and antiquarianism) from a comparative perspective in the long 18th century.[6] The focus on this period, a time when scientific travels and expeditions around the world increased, is especially fruitful inasmuch as it re-visits debates on the periodization of this "discovery of the indigenous".[7] The articles collected here cast a transnational look at European science, at home and abroad, as well as at scientific practice globally, and at the involvement of a great variety of local actors, for example academies and learned societies. This comparative and *longue durée* approach is supported by the ability to transcend individual disciplines and to combine the histories of natural history, medicine, environmental history, ethnology and archaeology as well as their interconnections. Finally, the book questions the different contexts of knowledge creation, be they political (republics, monarchies, colonial rule, etc.), material or linguistic. Based on the findings of historians, with particular emphasis on the social and cultural practices of early-modern natural history, the book focuses especially on collecting and exchanging, measuring and classifying information on territories of different scope.

---

5 Lisbet Koerner, *Linnaeus: Nature and Nation*, 3rd ed. (Cambridge, MA: Harvard University Press, 2001).

6 The book was inspired by discussions and reflections developed during an international conference organized at the University of Bern by the editors of the volume in collaboration with the Lichtenberg-Kolleg of the Georg-August-University, Göttingen (Dominik Hünniger) under the title of "Mapping the Territory: Exploring People and Nature, 1700 until 1830" (21–23 September 2017). The conference was financed by the Swiss National Science Foundation (Project no. 174561), the University of Bern, the Centre for Global Studies of the University of Bern, the Swiss Academy of Humanities and Social Sciences, the Burgergemeinde Bern, and the Swiss Society for Eighteenth-Century Studies. The editors thank all speakers for their active participation in the success of the conference.

7 Alix Cooper, *Inventing the Indigenous: Local Knowledge and Natural History in Early Modern Europe* (Cambridge: Cambridge University Press, 2007).

Four main questions inspired this publication.[8] The first one concerns the actors of natural history. Over the last decades, the attention of historians has shifted from individual figures, more exactly learned men with an academic training, like physicians and professors, to the diverse actors doing natural history, i.e. amateurs (or grassroots), practitioners, family members, and women.[9] Through the proliferation of communication channels in the early modern age, university-trained scholars were able to establish contact with people of different social backgrounds, with or without academic training, with merchants, diplomats, (court and colonial) officials, clergymen (pastors of the Protestant and Reformed Church, priests or Jesuits) as well as with artisans, apothecaries, gardeners, hunters, and peasants.[10] Not only were the networks socially diverse, but they also ranged from local to global networks of research in people, fauna and flora.[11]

8 The book is linked to a Swiss National Science Foundation-sponsored Professorship project led by Simona Boscani Leoni (Project no.: 144731), which ran from 2013 (October) to 2017 (September) under the title: "Cultures of Natural History: Main Actors, Networks and Places of Scientific Communication in the Early-Modern Period". Sarah Baumgartner and Meike Knittel participated in the project as scientific collaborators and wrote their PhD dissertations during the period. Baumgartner, "Das nützliche Wissen"; Meike Knittel, "Netzwerke der Botanik. Johannes Gessner (1709–1790) und die botanische Forschung im 18. Jahrhundert" (PhD diss., University of Bern, 2018). An extension of the project under the title: "Cultures of natural history online. Main actors, networks, topics in the *longue durée* (1550 to around 1830)" is still continuing at the University of Bern (Project no.: 176978).

9 A pioneer work on women in the sciences was: Londa Schiebinger, *The Mind has no Sex? Women in the Origins of Modern Science* (Cambridge, MA: Harvard University Press, 1989).

10 About these (in)visible assistants/helpers: Steven Shapin, "The Invisible Technician," *American Scientist* 77, no. 6 (1989); Iwan R. Morus, "Invisible Technicians, Instrument-Makers and Artisans," in *A Companion to the History of Science*, ed. Bernard V. Lightman (New York: John Wiley and Sons, 2016); Dunja Bulinsky, *Nahbeziehungen eines europäischen Gelehrten: Johann Jakob Scheuchzer (1672–1733) und sein soziales Umfeld* (Zurich: Chronos Verlag, 2020); Sophie Ruppel, *Botanophilie: Mensch und Pflanze in der aufklärerisch-bürgerlichen Gesellschaft um 1800* (Vienna: Böhlau, 2019); Monika Mommertz, "Heroization in Science, Scholarship, and Knowledge-Production: The Intellectual Hero in Transdisciplinary and Trans-Epochal Perspective," *Helden. Heroes. Héros. Special Issue* 4 (2018); Simona Boscani Leoni, ed., *"Unglaubliche Bergwunder": Johann Jakob Scheuchzer und Graubünden. Ausgewählte Briefe 1699–1707* (Chur: Verlag Bündner Monatsblatt, 2019); Valentina Pugliano, "Natural History in the Apothecary's Shop," in *Worlds of Natural History*, ed. Helen A. Curry, Nicholas Jardine, James A. Secord, and Emma C. Spary (Cambridge: Cambridge University Press, 2018) and the special issue of *Berichte zur Wissenschaftsgeschichte* 44, no. 2 (2021): *Working at the Margins: Labor and the Politics of Participation in Natural History, 1700–1830*, ed. Patrick Anthony.

11 For this communicative network of naturalists/botanists, see for example: Bettina Dietz, *Das System der Natur: Die kollaborative Wissenskultur der Botanik im 18. Jahrhundert*

The second question focuses on the role of communication networks and their spread across local, transregional and global levels, beyond confessional and class differences. Thirdly, we wanted to examine the research practices of the different actors involved in "natural knowledge production" and their fieldworks. Finally, our fourth prompt was the definition of the different *lieux de savoir* and their function in this context.[12] By *lieux de savoir* we mean different spaces where practical-empirical knowledge was produced and exchanged, especially anatomy theatres, botanical gardens, natural history and antiquarian collections, but also academies or economic and patriotic societies.[13]

Considering these multiple networks, research practices and *lieux de savoir*, which enabled naturalists to collect and process natural history data and information during the early modern period, we can say that research in these fields was actually polyphonic and multifarious: a combination of different "cultures of natural history", learned as well as popular.[14] They came into contact through different research practices, such as questionnaires, and findings were eventually published in scholarly publications.[15] We can provide evidence of knowledge making as a multi-layered, non-linear process, a collective enterprise that

(Cologne: Böhlau, 2017); Regina Dauser et al., eds., *Wissen im Netz: Botanik und Pflanzentransfer in europäischen Korrespondenznetzen des 18. Jahrhunderts* (Berlin: Akademie-Verlag, 2008). For an example of local/transregional correspondence: Boscani Leoni, *"Unglaubliche Bergwunder"*.

12 Arthur MacGregor, ed., *Naturalists in the Field: Collecting, Recording and Preserving the Natural World from the Fifteenth to the Twenty-First Century* (Leiden: Brill, 2018).

13 On global scientific practices between 1750 and 1850: Patrick Manning and Daniel Rood, *Global Scientific Practice in an Age of Revolutions, 1750–1850* (Pittsburgh: University of Pittsburgh Press, 2016). For the concept of *lieu de savoir*: Christian Jacob, ed., *Lieux de savoir: espaces et communautés* (Paris: Albin Michel, 2007); Paula Findlen, "Anatomy Theaters, Botanical Gardens, and Natural History Collections," in *The Cambridge History of Science: Early Modern Science*, ed. Katharine Park and Lorraine Daston (Cambridge: Cambridge University Press, 2006); René Sigrist, Eric Widmer, and Wladimir Berelowitsch, "Les lieux des sciences dans l'Europe moderne," in *Lieux d'Europe: mythes et limites*, ed. Stella Ghervas and François Rosset (Paris: Ed. de la Maison des sciences de l'homme, 2008). See also: Bruno Latour, *Science in Action: How to Follow Scientists and Engineers Through Society* (Cambridge, MA: Harvard University Press, 1987).

14 For pioneer work on these topics, see: Nicholas Jardine, Emma Spary, and James A. Secord, eds., *Cultures of Natural History* (Cambridge: Cambridge University Press, 1996); Nicholas Jardine and Emma Spary, "Worlds of History," in Curry, Jardine, Secord, Spary, *Worlds of Natural History*.

15 Simona Boscani Leoni, "Queries and Questionnaires: Collecting Local and Popular Knowledge in 17th and 18th Century Europe," in *Wissenschaftsgeschichte und Geschichte des Wissens im Dialog – Connecting Science and Knowledge*, ed. Kaspar von Greyerz, Silvia Flubacher, and Philipp Senn (Göttingen: V&R unipress, 2013).

involved diverse institutions and actors. The notion of "nature" emerged from different projections and ideals.

Some topics we have been scrutinising during the last few years offer further entry points from which to reach a clearer understanding of the collaborative culture of naturalists worldwide in the long 18th century: in particular, the investigation of the role of botanical networks and of economic societies in the discussion and dissemination of natural knowledge. Looking at different types of sources, such as letters, plant lists and printed botanical works by different actors (naturalists, but also state officials, merchants, and priests), we have been examining the ways in which books, plants, and seeds circulated from one continent to another, between big cities, capitals, small towns and more isolated places, for example in mountain regions. From these sources we understand that the geography of knowledge was relative and complex indeed. It encompassed hub cities – which housed universities and a plethora of other *lieux de savoir* and were directly connected to Atlantic ports –, peripheral towns and villages on a global scale and the problem of acceptance of other (i.e. non-European) forms of knowledge (see Stefanie Gänger's contribution in this volume).[16] In that respect, we have to consider the existence of different levels in the relationship between centres of knowledge production and so-called "peripheries." The exchange of Alpine plants, for example, and the special therapeutic qualities credited to Alpine air, milk and plants shows that there was something exceptional about these supposedly barren regions in Europe. Their exceptionality, combined with the limited knowledge of their "wild" nature, turns them into a sort of intra-European counterpart to non-European exoticism, as well as other peripheral regions (for more on this, see Meike Knittel's, Simona Boscani Leoni's, Barbara Orland's and Stefanie Gänger's articles in this volume).[17]

---

16 On this topic: Kenneth Pomeranz, *The Great Divergence: China, Europe, and the Making of the Modern World Economy* (Princeton: Princeton University Press, 2000); Marie-Noëlle Bourguet, "Voyages lointains et mesure du monde: la projection du regard européen sur le monde (vers 1500-vers 1800)," in *L'Europe des sciences et des techniques XVe–XVIIIe siècles: un dialogue des savoirs*, ed. Liliane Hilaire-Pérez, Fabien Simon, and Marie Thébaud-Sorger (Rennes: Presses universitaires de Rennes, 2016); Bruno Latour, "Comment redistribuer le Grand Partage?," *Revue de synthèse* 110 (1983); Latour, "Visualisation and Cognition: Drawing Things Together," in *Knowledge and Society: Studies in the Sociology of Culture Past and Present*, vol. 6, ed. Henrika Kuklick and Elizabeth Long (Greenwich, CT: Jai Press, 1986), 6.

17 On the "exotic Alps": Bernhard C. Schär, "On the Tropical Origins of the Alps: Science and the Colonial Imagination of Switzerland, 1700–1900," in *Colonial Switzerland: Rethinking Colonialism from the Margins*, eds. Patricia Purtschert and Harald Fischer-Tiné (Basingstoke: Palgrave Macmillan, 2015); Simona Boscani Leoni, "The Discovery of the Alps:

Regarding the development of botanical and agronomic studies during the 18th century, it is worth mentioning the rise of economic-patriotic societies, which conducted surveys on territories (Sarah Baumgartner, Martin Stuber).[18] These institutions were very important local actors in the process of mapping territories: they were interested in conducting the most precise surveys possible of the agricultural, economic and human resources of a region, trying – somewhat paternalistically – to involve peasants in the process of "optimisation" of agriculture.[19] In this context, too, the reception and circulation of knowledge must be considered as an active process, strongly dependent on the particular circumstances and on the interests of the actors involved.[20] Measuring territories cartographically and cataloguing the flora, fauna, and minerals of an area also influenced (and still does) the perception of the

Between 'Science' and Exoticism," in *Exotic Switzerland? Looking Outward in the Age of Enlightenment*, ed. Noémie Etienne, Claire Brizon, Chonja Lee, and Etienne Wismer (Berlin: Diaphanes, 2020). For other remote regions and 18th century natural history, see e.g.: Linda Andersson Burnett, "An Eighteenth-Century Ecology of Knowledge: Patronage and Natural History," *Culture Unbound – Journal of Current Cultural Research* 6 (2014).

18 For example: Koen Stapelbroek and Jani Marjanen, eds., *The Rise of Economic Societies in the Eighteenth Century: Patriotic Reform in Europe and North America* (Basingstoke: Palgrave Macmillan, 2012); see also: Alix Cooper, "'The 'Possibilities of the Land': The Inventory of 'Natural Riches' in the Early Modern Territories," in *Oeconomies in the Age of Newton*, eds. Margaret Schabas and Neil de Marchi (Durham, NC: Duke University Press, 2003).

19 On the Economic Enlightenment and its "culture of innovation" (*Innovationskultur*): Marcus Popplow, "Die Ökonomische Aufklärung als Innovationskultur des 18. Jahrhunderts zur optimierten Nutzung natürlicher Ressourcen," in *Landschaften agrarisch-ökonomischen Wissens: Strategien innovativer Ressourcennutzung in Zeitschriften und Sozietäten des 18. Jahrhunderts*, ed. Marcus Popplow (Münster: Waxmann, 2010).

20 On the concepts of transfer/circulation of knowledge, for example: Claire Gantet and Markus Meumann, eds., *Les échanges savants franco-allemands au XVIIIe siècle: transferts, circulations et réseaux* (Rennes: Presses universitaires de Rennes, 2019); Mitchell G. Ash, "Wissens- und Wissenschaftstransfer – einführende Bemerkungen," *Berichte zur Wissenschaftsgeschichte* 29, no. 3 (2006); Silvia Capanema, Quentin Deluermoz, Michel Molin, and Marie Redon, eds., *Du transfert culturel au métissage: concepts, acteurs, pratiques* (Rennes: Presses universitaires de Rennes, 2015); Thomas DaCosta Kaufmann, "Interpreting Cultural Transfer and the Consequences of Markets and Exchange: Reconsidering *Fumi-E*," in *Artistic and Cultural Exchanges Between Europe and Asia, 1400–1900: Rethinking Markets, Workshops and Collections*, ed. Michael North (London: Routledge, 2010); Michel Espagne, "Der theoretische Stand der Kulturtransferforschung," in *Kulturtransfer: Kulturelle Praxis im 16. Jahrhundert*, ed. Wolfgang Schmale (Innsbruck: Studien-Verlag, 2003); Ortmayr, "Kulturpflanzentransfers 1492–1900".

different regions; in other words, such undertakings are an important component of "making landscapes."[21]

Natural history was at once "local" and "global." As we pointed out at the beginning of this Introduction, in 1716 Johann Jakob Scheuchzer published a bibliography of all books on natural history from the Renaissance to the 18th century, under the title of *Bibliotheca scriptorum historiae naturalis omnium Terrae Regionum inservientium*.[22]

His idea was to assemble the different "local" natural histories into a natural history of the world. To explain the intentions of his enterprise, he wrote in the Preface to his *Bibliotheca scriptorum*:

> I know that the natural history of a territory is related to the natural history of many others, which are sometimes very remote. We cannot investigate one territory without knowing the other ones. I sometimes needed to look for Switzerland in the Indies and for the Indies in Switzerland.[23]

The *Bibliotheca scriptorum* shows how Scheuchzer interpreted the natural history of Switzerland (and its "wild" landscapes, the Alps) from the perspective of an increased presence of Europeans in non-Europeans areas and in terms of a discovery of the "exotic."[24] In the chapter devoted to the Americas, Scheuchzer not only highlighted the vastness of the country and the fertility of the lands, but he also mentioned about forty authors of naturalistic texts devoted to the New Continent.[25] These included Johannes Fragoso, a Toledo doctor at the

---

21 We subscribe to W. J. Thomas Mitchell's definition of landscapes "as a process by which social and subjective identities are formed," W. J. Thomas Mitchell, "Introduction," in *Landscape and Power*, ed. W. J. Thomas Mitchell, 2nd ed. (Chicago, London: The University of Chicago Press, 2002), 1; Daniel Speich, "Mountains Made in Switzerland: Facts and Concerns in Nineteenth-Century Cartography," *Science in Context* 22, no. 3 (2009).

22 Scheuchzer, *Bibliotheca scriptorum historiae naturalis omnium Terrae Regionum inservientium; historiae naturalis Helvetiae Prodromus; accessit celeberrimi viri Jacobi Le Long, Bibliothecarii Oratoriani de Scriptoribus historiae naturalis Galliae; collegit Johann Jakob Scheuchzer, Med. D. Math. Prof. Societ. Leopoldino-Carolinae, Regiae Angl. & Boruss.*

23 Original: "Ita, nosti, innexa est unius Regionis Historia Naturalis cum aliorum, longissimè etiam dissitarum, Historia, ut vix unam plenè elaborare possis sine alia [...]. Helvetiam haud raró quaerere necesse habeo in Indiis; et Indiam in Helvetia," ibid., *Praefatio ad lectorem* (unpaginated).

24 This form of thought is at the basis of the development of colonial thought in the 18th century. On the topic: Bernhard C. Schär, "Bauern und Hirten reconsidered: Umrisse der 'erfundenen Schweiz' im imperialen Raum," in *Postkoloniale Schweiz: Formen und Folgen eines Kolonialismus ohne Kolonien*, ed. Patricia Purtschert (Bielefeld: transcript, 2012).

25 Scheuchzer, *Bibliotheca scriptorum historiae naturalis omnium Terrae Regionum inservientium; historiae naturalis Helvetiae Prodromus; accessit celeberrimi viri Jacobi Le Long,*

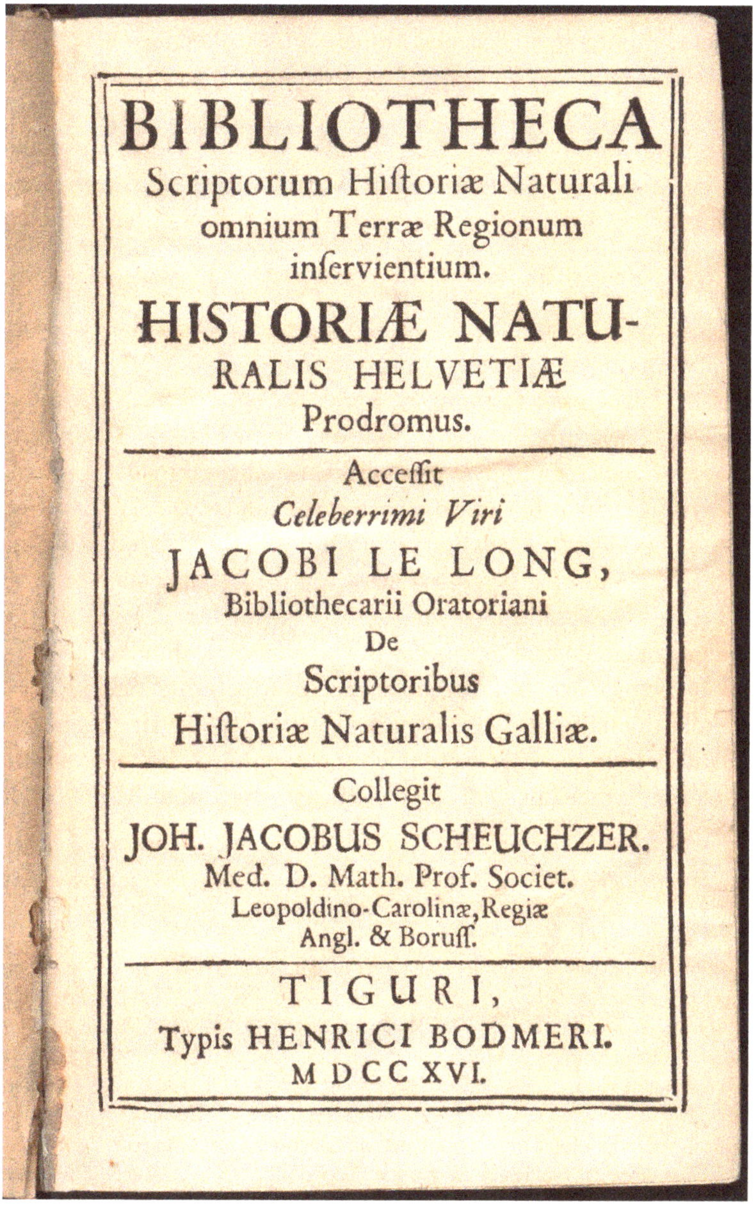

BIBLIOTHECA
Scriptorum Hiſtoriæ Naturali
omnium Terræ Regionum
inſervientium.
HISTORIÆ NATU-
RALIS HELVETIÆ
Prodromus.

Acceſſit
Celeberrimi Viri
JACOBI LE LONG,
Bibliothecarii Oratoriani
De
Scriptoribus
Hiſtoriæ Naturalis Galliæ.

Collegit
JOH. JACOBUS SCHEUCHZER.
Med. D. Math. Prof. Societ.
Leopoldino-Carolinæ, Regiæ
Angl. & Boruſſ.

TIGURI,
Typis HENRICI BODMERI.
M DCC XVI.

FIGURE 1.2 Frontispiece of Johann Jakob Scheuchzer, *Bibliotheca scriptorum historiae naturalis omnium Terrae Regionum inservientium; historiae naturalis Helvetiae Prodromus; accessit Celeberrimi Viri Jacobi Le Long, Bibliothecarii Oratoriani de Scriptoribus historiae naturalis Galliae; collegit Johann Jakob Scheuchzer, Med. D. Math. Prof. Societ. Leopoldino-Carolinae, Regiae Angl. & Boruss.* (Tiguri: typis Henrici Bodmeri, 1716). ETH-Bibliothek Zurich, Shelf mark: Rar 10537

court of Philip II, Francisco Hernández (in the 1651 edition), Nardo Antonio Recchi, Francisco Ximenez, and Nicolás Monardes (also cited in Scheuchzer's chapter on Asia) together with the Inca Garcilaso de la Vega. These observations about the *Bibliotheca scriptorum* are noteworthy because they underline once more a central topic of this book, i.e. the linkage between local and global natural history, as well as the importance of the American and Asian territories for an inventory of the natural history of the world (one that included human beings).[26]

The contributions assembled in this volume examine from a comparative perspective the exploration of the different "local natures" and the discovery of "local inhabitants and their history" in a period marked by an increase in scientific travels and expeditions around the world. They give crucial impetus to the debate on the periodization of this "discovery of the indigenous," and the different stages of this development, while bringing to light the social and cultural practices of natural history, the media, methods and tools used to collect, classify and communicate information on territories.[27] Through four chapters dealing with Switzerland the book also insists on the entanglement of local and global networks of natural and civil history in the modern age, placing the Old Confederacy, a landlocked territory with no colonies, within the developments of global knowledge (but also global trade). To return to Scheuchzer's quotation about the Indies in Switzerland and Switzerland in the Indies, this shows well the deep connections, of which the naturalists themselves were aware, during the European colonisation of the Americas and the import of

Bibliothecarii Oratoriani de Scriptoribus historiae naturalis Galliae; collegit Johann Jakob Scheuchzer, Med. D. Math. Prof. Societ. Leopoldino-Carolinae, Regiae Angl. & Boruss., Dedicatio (unpaginated), 184, 201, 211.

26 For a discussion on the development of forms of empirical research in the 16th century during the construction of the Spanish colonial empire in the Americas: Antonio Barrera-Osorio, *Experiencing Nature: The Spanish American Empire and the Early Scientific Revolution* (Austin: University of Texas Press, 2006); Kapil Raj, "Quand l'Amérique a inventé la science européenne," in *Europa: une autre histoire*, ed. Jakob Vogel, Thomas Serrier and Etienne François (Paris: Les Arènes, 2017); Raj, "Thinking Without the Scientific Revolution: Global Interactions and the Construction of Knowledge," *Journal of Early Modern History* 21 (2017); Eve Darian-Smith and Philip C. McCarty, eds., *The Global Turn: Theories, Research Designs, and Methods for Global Studies* (Oakland: University of California Press, 2017); Sebastian Conrad, *What Is Global History?* (Princeton: Princeton University Press, 2016); Romain Bertrand, "Histoire globale, histoires connectées. Un 'tournant' historiographique?," in *Le tournant global des sciences sociales*, ed. Alain Caillé and Stéphane Dufoix (Paris: Ed. La Découverte, 2013); Fa-ti Fan, "The Global Turn in the History of Science," *East Asian Science, Technology and Society: An International Journal* 6 (2012).

27 Cooper, *Inventing the Indigenous*.

new products from the West Indies as substitutes for products from the East Indies. The very definition of the quality of Swiss, respectively Alpine nature (i.e. Alpine products) must be read not only as a "discovery of indigenous nature," but also as an attempt to exploit on the market of knowledge (and new commercial products) plants, animals and minerals that could have had a therapeutic and commercial use. In this way, Scheuchzer and other Swiss naturalists like him, linked this Central European confederation of republican city-states to the global networks and placed the "Swiss/Alpine wonders" within the competition between the Atlantic World and Asia.[28]

The book is organised in three sections. Section 1, entitled *Naturalists' Methods*, concentrates on different aspects and methods of mapping a territory. In what ways was new knowledge of people and nature achieved based on data collected from travel and empirical research? Who were the actors involved? Who and what were the people and objects studied? What territories were surveyed and why? The first contribution (Simona Boscani Leoni) analyses the improvement of questionnaires for gathering information on people, on natural history and resources in the construction of colonial empires in early modern Europe (Spain and Britain), the relationship between knowledge and power, and the development of this empirical method to explore Europe's "wild" regions (the Alps and the Apennines). In the second contribution, Francesco Luzzini sheds light on the efforts of naturalists to define a systematic approach to investigations of nature and people at the beginning of the 18th century. His source is a recently rediscovered manuscript reporting the first scientific journey across the Apennines by the geologist Antonio Vallisneri (1661–1730). Luzzini shows how the Italian naturalist made ground-breaking contributions to many crucial issues in the history of early modern science. Vallisneri was also interested in the local communities and in the study of their relationship with the environment. Both papers (Boscani Leoni's and Luzzini's) consider different ways of mapping local nature and people (the resources of a country), collecting local knowledge ("local lore") and the role of the "inhabitants" in this context (whether Native Americans or European mountaineers).

The third paper of the section seeks to understand how an observer's perception of plants varies when they grow near or far away from them. Meike Knittel, in fact, examines how prints as well as specimens inspired and framed

28 Johann Jakob Scheuchzer entitled his questionnaire dedicated to collecting information on Swiss natural history *Letter of invitation to explore the natural wonders to be found in the Swiss countryside*: Johann Jakob Scheuchzer, *Einladungs-Brief, zu Erforschung natürlicher Wunderen, so sich im Schweizer-Land befinden* (Zürich: [D. Gessner], 1699), edited in Boscani Leoni, *"Unglaubliche Bergwunder"*, 35–49.

European naturalists' engagement with plants from different territories. The topos of the "secretive Indian knower" stands at the heart of Stefanie Gänger's contribution. Examining the development of this epistemic category until the Enlightenment, Gänger explores the role of the figure in the dynamic relationship between European and Creole science and its similarities with other "illiterate," "experiential," and "inscrutable" knowers such as European peasants and women.

Section II considers the *Authorities' and Societies' Strategies* in mapping procedures. The first paper of this section, by Irina Podgorny, examines the system of production and circulation of knowledge between Spanish America and Europe, and its links to bureaucracy and to Atlantic trade in the years between the end of the 18th and the beginning of the 19th century. Podgorny convincingly points out not only the continuity in the research practices after Spanish America's Revolution and Independence, but equally the importance of the circulation of products from the Indies, and the movements of experts frequently dispatched to these regions. Both observations signal the complexity and the long-term dimension of the history of knowledge in the Spanish and American world. This aspect was for a long time overlooked in the narrative of the history of science, which tended to emphasize or privilege the Protestant contribution to science by colonial powers, such as the Netherlands and Britain.

The bureaucratic roots of natural history research in the Americas find their counterpart in the mapping strategies developed during the second half of the 18th century especially by patriotic or economic societies. Supported by rulers driven by a growing interest in the use and benefits of knowing the natural resources of their subject territories, these societies took the initiative to survey different territories. In this context, the Physical Society of Zurich and the Economic Society of Bern offer two examples of the survey methods of the Economic Enlightenment. Sarah Baumgartner's paper analyses the first demographic surveys in the regional territory of Zurich. The author underlines the multifaceted solutions implemented by several members of the Economic Commission of the Physical Society to improve this kind of investigations. Their approaches mixed influences from different theoretical schools of statistical thought and combined innovative methods in a way that today would be seen to challenge a long-established system of people categorisation. In his contribution, Martin Stuber analyses these aspects in depth, showing how the Economic Society of Bern, through its investigations of the traditional uses of wetlands, developed a form of social anthropology *avant la lettre*. According to Stuber's thesis, the record of traditional economic practices not only was the prerogative of extra-European expeditions, but it was also widely practised in the context of the Economic Enlightenment in Europe ("Ökonomische Aufklärung").

Jon Mathieu's paper introduces the third and last section of the book: *Defining Territories*. The three articles focus on the definition and perception of mountain and desert landscapes from a global perspective (Mathieu, Orland, and Singh). On the basis of Alexander von Humboldt's *Views of Nature* in its different editions (from 1808 to 1849), Jon Mathieu analyses the historical perception of two major ecosystems – deserts and mountains – in the period between 1750 and 1850. Through Humboldt's work one can observe the emergence of a negative perception of the desert, against the background of the enlightened debates on deforestation and the presence of arid regions. According to a politically driven colonial assumption, "barren zones" are the result of traditional forms of land use. The debate on deforestation leads to a reaction against this practice (through "reforestation"), helping the development of a negative perception of landscapes without forests, i.e. arid landscapes such as the desert. As we saw in the case of the economic and patriotic associations (Baumgartner, Stuber) the watchword of the elite between 1750 and 1850 was "improvement": European style agriculture, or reforestation, must make deserts (like mountains) "flourish."

In her contribution, Barbara Orland reflects on the relationship between "geographical space," "production of knowledge," and the marginalization (or negative perception) of other forms of knowledge or traditional land use (as reflected in Gänger's and Stuber's articles). Orland's paper investigates the "birth of the Healthy Alps" and the relationship between space perception, land use and theories of a healthy lifestyle through a case study: the Swiss whey cures. During the second half of the 18th century, whey cures came to be a distinctive feature of the healthy environment of the Alps; later, in the 19th century, this cure became a means of prevention, wellbeing and sociability. In this process, good health became a benefit that could be purchased and the place where to buy the treatment became less important. In this way, the link between landscape and health became looser and the healthy qualities of mountain milk survive mostly as symbolic capital.

The last contribution, by Chetan Singh, deals with the creation of scientific knowledge about the Himalayan region between 1784 (foundation of the Asiatick Society) and 1850. As shown in the papers by Baumgartner and Stuber, scientific societies are interested in surveying regional territories in Europe, but also in the non-European world, using a European colonial lens. Singh's article shows how the members of the Asiatick Society explored the Himalayan regions, trying to categorize people, their diseases, to map the territory, its rivers, its mountains and its resources. Accompanying this mapping of the Himalayan territory step by step, Singh underlines the connection between this undertaking and European explorations of other territories, the Alps or Latin America, and the influence of naturalists such as Horace-Bénédict de Saussure (1740–1799) or Alexander von Humboldt (1769–1859).

With the present collective work, we contribute to a developing social history of knowledge of territories. Our ambition is to achieve a comparative approach on a global scale, highlighting the asymmetries of knowledge and power that have influenced the exploration of territories and their human and non-human inhabitants.

Acknowledgements

This contribution is based on research funded by the Swiss National Science Foundation and carried out in the context of the projects "Kulturen der Naturforschung" (Cultures of Natural History), Projects-Nr. 144731 and 176978, at the University of Bern (2014–ongoing).

We would also like to thank Simona Cain for her proofreading work, Dominik Hünniger (series editor of "Emergence of Natural History") and Rosanna Woensdregt at Brill for their cooperation in the publication of this volume.

Bibliography

Printed Sources

Boscani Leoni, Simona, ed. *"Unglaubliche Bergwunder": Johann Jakob Scheuchzer und Graubünden. Ausgewählte Briefe 1699–1707*. Cultura alpina 9. Chur: Verlag Bündner Monatsblatt, 2019.

Scheuchzer, Johann Jakob. *Bibliotheca scriptorum historiae naturalis omnium Terrae Regionum inservientium; historiae naturalis Helvetiae Prodromus; accessit Celeberrimi Viri Jacobi Le Long, Bibliothecarii Oratoriani de Scriptoribus historiae naturalis Galliae; collegit Johann Jakob Scheuchzer, Med. D. Math. Prof. Societ. Leopoldino-Carolinae, Regiae Angl. & Boruss.* Tiguri: Typis Henrici Bodmeri, 1716.

Scheuchzer, Johann Jakob *Einladungs-Brief, zu Erforschung natürlicher Wunderen, so sich im Schweizer-Land befinden* (Zurich: [D. Gessner], 1699). In *"Unglaubliche Bergwunder": Johann Jakob Scheuchzer und Graubünden. Ausgewählte Briefe 1699–1707*. Edited by Simona Boscani Leoni, 33–49. Cultura alpina 9. Chur: Verlag Bündner Monatsblatt, 2019.

Literature

Ash, Mitchell G. "Wissens- und Wissenschaftstransfer – Einführende Bemerkungen." *Berichte zur Wissenschaftsgeschichte* 29, no. 3 (2006): 181–89.

Andersson Burnett, Linda "An Eighteenth-Century Ecology of Knowledge: Patronage and Natural History," *Culture Unbound – Journal of Current Cultural Research* 6 (2014): 1275–1297.

Anthony, Patrick, ed. *Working at the Margins: Labor and the Politics of Participation in Natural History, 1700–1830*. Special Issue of: *Berichte zur Wissenschaftsgeschichte* 44, no. 2 (2021).

Barrera-Osorio, Antonio. *Experiencing Nature: The Spanish American Empire and the Early Scientific Revolution*. Austin: University of Texas Press, 2006.

Batsaki, Yota, Sarah Burke Cahalan, and Anatole Tchikine, eds. *The Botany of Empire in the Long Eighteenth Century*. Dumbarton Oaks Symposia and Colloquia. Washington, DC: Dumbarton Oaks Research Library and Collection, 2016.

Baumgartner, Sarah. "Das Nützliche Wissen. Akteure, Tätigkeiten, Kommunikationspraxis und Themen der Naturforschenden Gesellschaft in Zürich, 1746 bis ca. 1830." PhD diss., University of Bern, 2019.

Bertrand, Romain. "Histoire globale, histoires connectées. Un 'tournant' historiographique?" In *Le tournant global des sciences sociales*, edited by Alain Caillé and Stéphane Dufoix, 44–66. Collection Bibliothèque du MAUSS. Paris: Ed. La Découverte, 2013.

Boscani Leoni, Simona, ed. *Wissenschaft – Berge – Ideologien: Johann Jakob Scheuchzer (1672–1733) und die frühneuzeitliche Naturforschung – Scienza – montagna – ideologie. Johann Jakob Scheuchzer (1672–1733) e la ricerca naturalistica in epoca moderna*. Basel: Schwabe, 2010.

Boscani Leoni, Simona. "Queries and Questionnaires: Collecting Local and Popular Knowledge in 17th and 18th Century Europe." In *Wissenschaftsgeschichte und Geschichte des Wissens im Dialog – Connecting Science and Knowledge*, edited by Kaspar von Greyerz, Silvia Flubacher, and Philipp Senn, 187–207. Göttingen: V&R unipress, 2013.

Boscani Leoni, Simona. "The Discovery of the Alps: Between 'Science' and Exoticism." In *Exotic Switzerland? Looking Outward in the Age of Enlightenment*, edited by Noémie Etienne, Claire Brizon, Chonja Lee, and Etienne Wismer, 285–97. Berlin: Diaphanes, 2020.

Bossi, Maurizio and Claudio Greppi, eds. *Viaggi e scienza: le istruzioni scientifiche per i viaggiatori nei secoli XVII–XIX*. Gabinetto scientifico letterario G. P. Vieusseux, Studi 13. Florence: Olschki, 2005.

Boumediene, Samir. *La colonisation du savoir une histoire des plantes médicinales du "Nouveau Monde" (1492–1750)*. Vaulx-en-Velin: Les éditions des mondes à faire, 2016.

Bourguet, Marie-Noëlle. "Voyages lointains et mesure du monde: la projection du regard européen sur le monde (vers 1500-vers 1800)." In *L'Europe des sciences et des techniques XVe–XVIIIe siècle: un dialogue des savoirs*, edited by Liliane Hilaire-Pérez, Fabien Simon, and Marie Thébaud-Sorger, 193–201. Collection "Histoire." Rennes: Presses universitaires de Rennes, 2016.

Bulinsky, Dunja. *Nahbeziehungen eines europäischen Gelehrten: Johann Jakob Scheuchzer (1672–1733) und sein soziales Umfeld.* Zurich: Chronos Verlag, 2020.

Capanema, Silvia, Quentin Deluermoz, Michel Molin, and Marie Redon, eds. *Du transfert culturel au métissage: concepts, acteurs, pratiques.* Collection "Histoire." Rennes: Presses universitaires de Rennes, 2015.

Carey, Daniel. "Inquiries, Heads, and Directions: Orienting Early Modern Travel." In *Travel Narratives, the New Science and Literary Discourse: 1569–1750*, edited by Judy A. Hayden, 25–52. Farnham: Ashgate, 2012.

Conrad, Sebastian. *What Is Global History?* Princeton, NJ: Princeton University Press, 2016.

Cooper, Alix. "The 'Possibilities of the Land': The Inventory of 'Natural Riches' in the Early Modern Territories." In *Oeconomies in the Age of Newton*, edited by Margaret Schabas and Neil de Marchi, 129–53. Durham: Duke University Press, 2003.

Cooper, Alix. *Inventing the Indigenous: Local Knowledge and Natural History in Early Modern Europe.* Cambridge, New York: Cambridge University Press, 2007.

Cooper, Alix. "Fragen ohne Antworten: Die Suche nach lokalen Informationen in der frühen Aufklärung." In *Vor Google: Eine Mediengeschichte der Suchmaschine im Analogen Zeitalter*, edited by Thomas Brandstetter, 73–84. Kultur- und Medientheorie. Bielefeld: transcript, 2012.

Curry, Helen A., Nicholas Jardine, James A. Secord, and Emma C. Spary, eds. *Worlds of Natural History.* Cambridge, New York: Cambridge University Press, 2018.

DaCosta Kaufmann, Thomas. "Interpreting Cultural Transfer and the Consequences of Markets and Exchange: Reconsidering *Fumi-e*." In *Artistic and Cultural Exchanges Between Europe and Asia, 1400–1900: Rethinking Markets, Workshops and Collections*, edited by Michael North, 136–61. London: Routledge, 2010.

Darian-Smith, Eve, and Philip C. McCarty, eds. *The Global Turn: Theories, Research Designs, and Methods for Global Studies.* 1st ed. Oakland: University of California Press, 2017.

Dauser, Regina, Stefan Hächler, Michael Kempe, Franz Mauelshagen, and Martin Stuber, eds. *Wissen im Netz: Botanik und Pflanzentransfer in europäischen Korrespondenznetzen des 18. Jahrhunderts.* Colloquia Augustana 24. Berlin: Akademie-Verlag, 2008.

Dietz, Bettina. *Das System der Natur: Die kollaborative Wissenskultur der Botanik im 18. Jahrhundert.* Cologne: Böhlau, 2017.

Espagne, Michel. "Der theoretische Stand der Kulturtransferforschung." In *Kulturtransfer: Kulturelle Praxis im 16. Jahrhundert*, edited by Wolfgang Schmale, 63–76. Wiener Schriften zur Geschichte der Neuzeit 2. Innsbruck: Studien-Verlag, 2003.

Fan, Fa-ti. "The Global Turn in the History of Science." *East Asian Science, Technology and Society: An International Journal* 6 (2012): 249–58.

Findlen, Paula. "Anatomy Theaters, Botanical Gardens, and Natural History Collections." In *The Cambridge History of Science: Early Modern Science*, edited by Katharine Park and Lorraine Daston, 272–89. Cambridge History of Science 3. Cambridge, New York: Cambridge University Press, 2006.

Gantet, Claire, and Markus Meumann, eds. *Les échanges savants franco-allemands au XVIII*[e] *siècle: transferts, circulations et réseaux*. Collection "Histoire". Rennes: Presses universitaires de Rennes, 2019.

Graber, Rolf. *Bürgerliche Öffentlichkeit und spätabsolutistischer Staat: Sozietätenbewegung und Konfliktkonjunktur in Zürich 1746–1780*. Zurich: Chronos, 1993.

Jacob, Christian, ed. *Lieux de savoir: espaces et communautés*. Paris: Albin Michel, 2007.

Jardine, Nicholas, and Emma Spary. "Worlds of History." In *Worlds of Natural History*, edited by Curry, Jardine, Secord, Spary, 3–14.

Jardine, Nicholas, Emma C. Spary, and James A. Secord, eds. *Cultures of Natural History*. Cambridge, New York: Cambridge University Press, 1996.

Kempe, Michael. *Wissenschaft, Theologie, Aufklärung: Johann Jakob Scheuchzer (1672–1733) und die Sintfluttheorie*. Frühneuzeit-Forschungen 10. Epfendorf: bibliotheca academica Verlag, 2003.

Knittel, Meike. "Netzwerke der Botanik. Johannes Gessner (1709–1790) und die Botanische Forschung im 18. Jahrhundert." PhD diss., University of Bern, 2018.

Koerner, Lisbet. *Linnaeus: Nature and Nation*. 3. ed. Cambridge, MA: Harvard University Press, 2001.

Kury, Lorelai. "Les instructions de voyage dans les expéditions scientifiques françaises (1750–1830)," *Revue d'histoire des sciences* 51, no. 1 (1998): 65–92.

Kury, Lorelai. *Histoire naturelle et voyages scientifiques (1780–1830)*. Paris: L'Harmattan, 2001.

Latour, Bruno. "Comment redistribuer le Grand Partage?" *Revue de synthèse* 110 (1983): 202–36.

Latour, Bruno. "Visualisation and Cognition: Drawing Things Together." In *Knowledge and Society: Studies in the Sociology of Culture Past and Present*. Vol. 6, edited by Henrika Kuklick and Elizabeth Long, 1–40. Greenwich CT, London: Jai Press, 1986.

Latour, Bruno. *Science in Action: How to Follow Scientists and Engineers Through Society*. Cambridge, MA: Harvard University Press, 1987.

Leu, Urs B., ed. *Natura Sacra: Der Frühaufklärer Johann Jakob Scheuchzer (1672–1733)*. Zug: Achius Verlag, 2012.

MacGregor, Arthur, ed. *Naturalists in the Field: Collecting, Recording and Preserving the Natural World from the Fifteenth to the Twenty-First Century*. Emergence of Natural History, vol. 2. Leiden: Brill, 2018.

Manning, Patrick, and Daniel Rood. *Global Scientific Practice in an Age of Revolutions, 1750–1850*. Pittsburgh: University of Pittsburgh Press, 2016.

Mitchell, W. J. Thomas. "Introduction." In *Landscape and Power*, 2nd ed., edited by W. J. Thomas Mitchell. Chicago, London: The University of Chicago Press, 2002: 1–5.

Mommertz, Monika. "Heroization in Science, Scholarship, and Knowledge-Production: The Intellectual Hero in Transdisciplinary and Trans-Epochal Perspective." *Helden. Heroes. Héros. Special Issue* 4 (2018): 3–12.

Morus, Iwan R. "Invisible Technicians, Instrument-Makers and Artisans." In *A Companion to the History of Science*, edited by Bernard V. Lightman, 97–110. New York: John Wiley and Sons, 2016.

Ogilvie, Brian W. *The Science of Describing. Natural History in Renaissance Europe*. Chicago, London: The University of Chicago Press, 2006.

Ortmayr, Norbert. "Kulturpflanzentransfers 1492–1900." *Historische Sozialkunde: Geschichte, Fachdidaktik, politische Bildung* 1 (2002): 22–30.

Pomeranz, Kenneth. *The Great Divergence: China, Europe, and the Making of the Modern World Economy*. The Princeton Economic History of the Western World. Princeton NJ: Princeton University Press, 2000.

Popplow, Marcus. "Die Ökonomische Aufklärung als Innovationskultur des 18. Jahrhunderts zur optimierten Nutzung natürlicher Ressourcen." In *Landschaften agrarisch-ökonomischen Wissens: Strategien innovativer Ressourcennutzung in Zeitschriften und Sozietäten des 18. Jahrhunderts*, edited by Marcus Popplow, 2–48. Cottbuser Studien zur Geschichte von Technik, Arbeit und Umwelt 30. Münster: Waxmann, 2010.

Pugliano, Valentina. "Natural History in the Apothecary's Shop." In *Worlds of Natural History*, edited by Curry, Jardine, Secord, Spary, 44–60.

Raj, Kapil. "Quand l'Amérique a inventé la science Européenne." In *Europa: une autre histoire*, edited by Jakob Vogel, Thomas Serrier, and Etienne François, 1089–1102. Paris: Les Arènes, 2017.

Raj, Kapil. "Thinking Without the Scientific Revolution: Global Interactions and the Construction of Knowledge." *Journal of Early Modern History* 21 (2017): 1–14.

Rubiés, Joan-Pau. "Instructions for Travellers: Teaching the Eye to See," *History and Anthropology* 9, no. 2–3 (1996): 139–90.

Ruppel, Sophie. *Botanophilie: Mensch und Pflanze in der aufklärerisch-bürgerlichen Gesellschaft um 1800*. Vienna: Böhlau, 2019.

Schär, Bernhard C. "Bauern und Hirten reconsidered: Umrisse der 'erfundenen Schweiz' im imperialen Raum." In *Postkoloniale Schweiz: Formen und Folgen eines*

Kolonialismus ohne Kolonien, edited by Patricia Purtschert, 315–31. Postcolonial Studies 10. Bielefeld: transcript, 2012.

Schär, Bernhard C. "On the Tropical Origins of the Alps. Science and the Colonial Imagination of Switzerland, 1700–1900." In *Colonial Switzerland: Rethinking Colonialism from the Margins*, edited by Patricia Purtschert and Harald Fischer-Tiné, 29–49. Cambridge Imperial and Post-Colonial Studies Series. Basingstoke: Palgrave Macmillan, 2015.

Schiebinger, Londa, and Claudia Swan, eds. *Colonial Botany: Science, Commerce, and Politics in the Early Modern World*. Philadelphia: University of Pennsylvania Press, 2005.

Schiebinger, Londa. *The Mind Has No Sex? Women in the Origins of Modern Science*. Cambridge, MA: Harvard University Press, 1989.

Schmale, Wolfgang, ed. *Kulturtransfer: Kulturelle Praxis im 16. Jahrhundert*. Wiener Schriften zur Geschichte der Neuzeit 2. Innsbruck: Studien-Verlag, 2003.

Shapin, Steven. "The Invisible Technician." *American Scientist* 77, no. 6 (1989): 554–63.

Sigrist, René, Eric Widmer, and Wladimir Berelowitsch, "Les lieux des sciences dans l'Europe moderne." In *Lieux d'Europe: mythes et limites*, edited by Stella Ghervas and François Rosset, 45–64. Paris: Ed. de la Maison des sciences de l'homme, 2008.

Speich, Daniel. "Mountains Made in Switzerland: Facts and Concerns in Nineteenth-Century Cartography." *Science in Context* 22, no. 3 (2009): 387–408.

Stagl, Justin. *A History of Curiosity: The Theory of Travel 1550–1800*. Studies in Anthropology and History (Chur: Harwood Academic Publishers, 1995).

Stapelbroek, Koen, and Jani Marjanen, eds. *The Rise of Economic Societies in the Eighteenth Century: Patriotic Reform in Europe and North America*. Basingstoke: Palgrave Macmillan, 2012.

PART 1

Naturalists' Methods

CHAPTER 2

Between the Americas and Europe: Mapping Territories through Questionnaires, 16th–18th Centuries

Simona Boscani Leoni

In the last decades, the Internet and electronic media have revolutionized our capacity to acquire and disseminate knowledge. These circumstances have drawn scholarly attention to the origins, development, and consequences of new information and communication technologies and to the problem of the collection of data and information over the centuries. As Paul Slack writes in a study about government and information in 17th-century England: "Studies of the information revolution brought about by modern electronic media have naturally prompted interest in how information was created, diffused and manipulated in the past and with what effects."[1] The focus in this respect has very much been on the early-modern period in Europe in which developments such as the creation of a market for printed matter and the expanding bureaucracy of both State and Church contributed to a huge elaboration of ways in which ideas, information, and instructions were communicated and received.

Precisely at this time, queries and questionnaires first made their debut as a research tool – a tool so beloved today of government agencies, political analysts or market researchers – for data collection used by natural historians, antiquarians, and geographers, as well as by governments and public officials ("Beamte").

In speaking about queries and questionnaires in the early modern age, it is worth considering their genesis as a result of different processes. First, the process of state building and empire building and the need, for several countries, to extend their political, economic and commercial control on their colonies and to improve knowledge of natural resources, which could be exploited. On the one hand, the Spanish *Cuestionarios para la formación de las Relaciones geográficas de Indias*, and on the other hand pastoral visitations organized by bishops are similar instruments designed to achieve better administrative

1 Paul Slack, "Government and Information in Seventeenth-Century England," *Past and Present* 184 (2004): 33.

communication between centres (of the Spanish Empire or of a diocese) and their peripheries. We may even say that the origin of the European bureaucracy was strongly influenced by the Church, an institution very concerned with the retrieval of information from official registers. Royal chanceries were slower than the papacy in developing techniques for administration. After the Council of Trent (1545–1563), each parish priest of the Catholic Church was required to record births, deaths and marriages in registers. At the same time, Pastoral visits became regular good practice for bishops wanting to control the spiritual life of parishioners.[2] For ecclesiastical and secular rulers an improved knowledge in geography, natural sciences, and anthropology of a region – through a systematic collection of information – underpinned a more effective organization and centralization of their power.[3] Secondly, there was an intensification of contacts between the Old and the New World through travels and a growing flow of data and information. Travels and travel narratives were often the only way to gather information about yet unknown and little-explored territories: "travel across the globe was a major means for extending

---

2 Peter Burke, *A Social History of Knowledge: From Gutenberg to Diderot* (Cambridge: Blackwell Publishers, 2000), 120–23; Burke, *The Historical Anthropology of Early Modern Italy: Essays on Perception and Communication* (Cambridge: Cambridge University Press, 1987).

3 About pastoral visits, for example: Umberto Mazzone and Angelo Turchini, eds., *Le visite pastorali: analisi di una fonte* (Bologna: Il Mulino, 1991); on the Spanish *Cuestionarios*: Francisco de Solano, ed., *Cuestionarios para la formación de las relaciones geográficas de Indias: siglos XVI/XIX* (Madrid: Consejo Superior de Investigaciones Científicas, 1988); Howard F. Cline, "The Relaciones Geográficas of the Spanish Indies, 1577–8," *Hispanic American Historical Review* 44 (1964); Cline, "The Relaciones Geográficas of Spain, New Spain, and the Spanish Indies: An Annotated Bibliography," in *Handbook of Middle American Indians*, ed. Howard F. Cline and Robert Wauchope (Austin: University of Texas Press, 1972); Arndt Brendecke, *Imperium und Empirie: Funktionen des Wissens in der Spanischen Kolonialherrschaft* (Cologne: Böhlau, 2009); Brendecke, "Informing the Council. Central Institutions and Local Knowledge in the Spanish Empire," in *Empowering Interactions. Political Cultures and the Emergence of the State in Europe 1300–1900*, ed. Willem Pieter Blockmans, André Holenstein, and Jon Mathieu (Farnham: Ashgate, 2009); Antonio Barrera, "Empire and Knowledge: Reporting from the New World," *Colonial Latin American Review* 15, no. 1, June (2006); Antonio Barrera-Osorio, *Experiencing Nature: The Spanish American Empire and the Early Scientific Revolution*, 1st ed. (Austin: University of Texas Press, 2006); Barrera-Osorio, "Nature and Experience in the New World: Spain and England in the Making of the New Science," in *Más allá de la Leyenda Negra: España y la revolución científica = Beyond the Black Legend; Spain and the Scientific Revolution*, ed. Víctor Navarro Brotons and William Eamon (Valencia: Universitat de València, 2007); Barrera-Osorio, "Empiricism in the Spanish Atlantic World," in *Science and Empire in the Atlantic World*, ed. James Delbourgo and Nicholas Dew (New York: Routledge, 2008); Barrera-Osorio, "Knowledge and Empiricism in the Sixteenth-Century Spanish Atlantic World," in *Science in the Spanish and Portuguese Empires, 1500–1800*, ed. Daniela Bleichmar, Paula De Vos, Kristin Huffine, and Kevin Sheehan (Stanford: Stanford University Press, 2009).

experience and for testing the adequacy of scientific generalizations."[4] In this context, questionnaires or "general heads" as instructions for travellers played an important role and became a useful tool for pursuing a systematic collection of empirical data in a region, in order to improve knowledge of it and to enable broader comparisons between different territories.[5] Thirdly, especially after the "discovery" of the Americas, there was a rekindling of curiosity not only for "exotic" but also for local nature. An extensive knowledge of local flora and fauna, but also of minerals, local characteristics of the terrain, orology and hydrology called for new forms of empirical research methods.[6] Besides "official" queries such as the Spanish *Cuestionarios*, and pastoral visits or queries for travellers, scholars developed questionnaires designed to collect material in view of writing a comprehensive *civil* and *natural history* of *local* territories.[7]

In my paper, I would like to highlight the common elements underlying the Spanish questionnaires, and the queries for collecting information about nature and people, which were also developed in the context of the Spanish America, and their appropriation in Britain and on the Continent. I deliberately

4 Steven Shapin, *A Social History of Truth: Civility and Science in Seventeenth-Century England* (Chicago, London: The University of Chicago Press, 1994), 245.

5 Michael Hunter, "Robert Boyle's 'Heads' and 'Inquiries'," *Robert Boyle Project, Occasional Papers*, no. 1 (2005); Hunter, "Robert Boyle and the Early Royal Society: A Reciprocal Exchange in the Making of Baconian Science," *The British Journal for the History of Science* 40 (March 2007); Barbara J. Shapiro, *A Culture of Fact: England, 1550–1720* (Ithaca, London: Cornell University Press, 2000), 72–76; Daniel Carey, "Inquiries, Heads, and Directions: Orienting Early Modern Travel," in *Travel Narratives, the New Science and Literary Discourse: 1569–1750*, ed. Judy A. Hayden (Farnham: Ashgate, 2012); Justin Stagl, "Vom Dialog zum Fragebogen: Miszellen zur Geschichte der Umfrage," *Kölner Zeitschrift für Soziologie und Sozialpsychologie* 31 (1979); Maurizio Bossi and Claudio Greppi, eds., *Viaggi e scienza: le istruzioni scientifiche per i viaggiatori nei secoli XVII–XIX* (Florence: Olschki, 2005); Joan-Pau Rubiés, "Instructions for Travellers: Teaching the Eye to See," *History and Anthropology* 9, no. 2–3 (1996). In their works, Carey, Rubiés and Stagl analyse the origin of this type of questionnaires for travellers in the context of the *ars apodemica*; this topic is for my analysis of secondary importance. An edition in French of different questionnaires (Boyle, Woodward, Vallisneri) in: Silvia Collini and Antonella Vannoni, eds., *Les instructions scientifiques pour les voyageurs (XVIIe–XIXe siècle)* (Paris: L'Harmattan, 2005).

6 Alix Cooper, *Inventing the Indigenous: Local Knowledge and Natural History in Early Modern Europe* (Cambridge: Cambridge University Press, 2007); Brian W. Ogilvie, *The Science of Describing: Natural History in Renaissance Europe* (Chicago, London: The University of Chicago Press, 2006).

7 Travel instructions experienced renewed success with the ethnological questionnaires, especially during the 19th century. See: Claude Blanckaert, ed., *Le terrain des sciences humaines: instructions et enquêtes (XVIIIe–XXe siècle)* (Paris: L'Harmattan, 1996); Adam Fox, "Printed Questionnaires, Research Networks, and the Discovery of the British Isles, 1650–1800," *The Historical Journal* 53, no. 3 (2010).

use the term "appropriation" following Kosta Gavroglu's definition as "a form of the fusion of aspects of science and technology with local traditions," namely to denote an active and creative form of *translation* of these tools into other cultural contexts.[8]

Questionnaires are a good example of how to approach the topic "mapping people and territories" because they show the common origin of investigations into geography as well as *civil* and *natural history* of a territory.[9] They also allow us to scrutinize the geographies of the development of empirical practices in the early modern age. Two aspects deserve particular attention, here. First, the role of the Spanish American empire as a place where new forms of empirical research emerged and were institutionalized.[10] Secondly, the need to understand their development in a *longue durée* perspective, which can normalize the interpretation of the role of Britain and Bacon's empiricism and integrate other European peripheral territories (the Old Swiss Confederacy and the Italian Peninsula in the 18th century) in this process. In a wider sense, my paper wants to contribute to the discussion of provincializing Europe and of a pluralization of spaces in the geography of the history of knowledge, in which the different religious denominations can find a place.[11]

---

8 Kostas Gavroglu et al., "Science and Technology in the European Periphery: Some Historiographical Reflections," *History of Science* 46, no. 2 (2008): 159–60.

9 See, for example, Charles W. J. Withers, *Geography, Science, and National Identity: Scotland Since 1520*, (Cambridge: Cambridge University Press, 2001); Withers, *Placing the Enlightenment: Thinking Geographically About the Age of Reason* (Chicago, London: The University of Chicago Press, 2007); David N. Livingstone, *Putting Science in Its Place: Geographies of Scientific Knowledge* (Chicago, London: The University of Chicago Press, 2003); Lissa Roberts, "Situating Science in Global History: Local Exchanges and Networks of Circulation," *Itinerario* 33, no. 1 (2009); Fa-ti Fan, "The Global Turn in the History of Science," *East Asian Science, Technology and Society: An International Journal* 6 (2012); René Sigrist, Eric Widmer, and Wladimir Berelowitsch, "Les lieux des sciences dans l'Europe moderne," in *Lieux d'Europe: mythes et limites*, ed. Stella Ghervas and François Rosset (Paris: Ed. de la Maison des sciences de l'homme, 2008).

10 For example: Brendecke, *Imperium und Empirie*; Jorge Cañizares-Esguerra, "Spanish America: From Baroque to Modern Colonial Science," in *The Cambridge History of Science: Eighteenth-Century Science*, ed. Roy Porter (Cambridge: Cambridge University Press, 2003); Cañizares-Esguerra, *Nature, Empire, and Nation: Explorations of the History of Science in the Iberian World* (Stanford: Stanford University Press, 2006); Cañizares-Esguerra, "Iberian Science in the Renaissance: Ignored How Much Longer?," *Perspectives on Science* 12, no. 1 (2004).

11 For more on the problem (and the Merton-Thesis), see: Dipesh Chakrabarty, *Provincializing Europe: Postcolonial Thought and Historical Difference* (Princeton: Princeton University Press, 2000); Steven Shapin, "The House of Experiment in Seventeenth-Century England," *Isis* 79 (1988); Ierome B. Cohen, ed., *Puritanism and the Rise of Modern Science: The Merton Thesis* (New Brunswick: Rutgers University Press, 1990); Edward B. Davis and

1 Queries in the Spanish American Empire

From the earliest contacts with the New World, Spanish monarchs set up institutions to govern their territories, for instance the *Consejo de Indias* (1524), and demanded descriptions of them by sending official requests for reports on their geography, human settlements, and demography. Between 1550 and 1570, one may notice an institutionalization of information-gathering practices through questionnaires. This phase began in the mid-1540s with the questionnaire (1546) and *Memorial* of Alonso de Santa Cruz (c. 1500–1572). Santa Cruz's questionnaire was a pioneer for its format focused on the collection of natural history information, compared with other questionnaires with tax-related queries.[12] It consisted of seventeen questions with a preliminary note ("advertencia"); it mentioned many relevant aspects to an understanding of the newly discovered territories: the geographical location of the villages, their names in the local languages, the determination of ports and places based on longitude and latitude, the hydrography of an area and its inundation risks. Other topics concerned the description of the lands and their geographical characteristics, rivers, lakes, springs, mountains and mines, minerals, and rocks. Several other queries examined flora, fruit, spices, and herbs, but also the indigenous population, their kingdoms and costumes.[13] Between 1569 and 1577, we find similar questions about people and natural history in the *Cuestionarios* prepared by Juan de Ovando y Godoy (1530–1575), appointed Visitor to the Council of the Indies, and by Juan López de Velasco (1530–1598), *Cronista-Cosmógrafo* of the Spanish crown. In 1569, Juan de Ovando y Godoy sent a questionnaire with 31 chapters, asking for information about people, demographics, and ecclesiastical institutions to several jurisdictions in Mexico. A couple of years later, and helped by López de Velasco, his former secretary, he produced a new version of

Michael P. Winship, "Early Modern Protestantism," in *Science and Religion: A Historical Introduction*, ed. Gary B. Ferngren (Baltimore: Johns Hopkins University Press, 2002); Kapil Raj, "Relocations," in *Science and Technology Studies: Critical Concepts in the Social Sciences*, ed. Michael Lynch (London: Routledge, 2012), IV; Raj, "Thinking Without the Scientific Revolution: Global Interactions and the Construction of Knowledge," *Journal of Early Modern History* 21 (2017); Raj, "Quand l'amérique a inventé la science européenne," in *Europa: une autre histoire*, ed. Jakob Vogel, Thomas Serrier, and Etienne François (Paris: Les Arènes, 2017).

12 Carmen Millán de Benavides, *Epítome de la Conquista del Nuevo Reino de Granada: la cosmogonía española del Siglo XVI y el conocimiento por cuestionario* (Bogotá: Pontificia Universidad Javeriana, Instituto de Estudios Sociales y Culturales Pensar, 2001).

13 Raquel Álvarez Peláez, *La conquista de la naturaleza americana* (Madrid: Consejo Superior de Investigaciones Científicas, 1993).

the document, the *Ordenanzas para la formación del libro de las Descripciones de las Indias* (1573) with 135 questions, and sent it to Peru.[14]

These questionnaires were a part of an ambitious reform program that Ovando introduced in the 1560s: visiting and reforming the University of Alcalá (1564–1565), the reformation of the *Consejo de Indias* (1566–1571), and the scientific expedition to America of the medical attendant ("protomedico") of Philip II, Francisco Hernández (1517–1584) between 1570 and 1577.[15] In May 1577, the same López de Velasco was responsible for the publication of the key document in this tradition, the *Instrucción y Memoria de las Relaciones que se han de hacer para la descripcion de la Indias*.[16] If we look more closely at the 50 questions of this *Cuestionario*, we will find almost all the topics already present in Santa Cruz's questionnaire. Several questions concerned the

---

14 Solano, *Cuestionarios para la formación de las relaciones geográficas de Indias*, 11–15, 16–74; Pilar Ponce Leiva, "Las Ordenanzas sobre descripciones (1573). Su aplicación en la Real Audiencia de Quito," in Solano, *Cuestionarios para la formación de las relaciones geográficas de Indias*, ed. Solano, *Cuestionarios para la formación de las relaciones geográficas de Indias*, LXXIX–XCI. Raquel Álvarez Peláez, "El Cuestionario de 1577. La 'Instruccion y memoria de las relaciones que se han de hacer para la descripcion de las Indias' de 1577," in Solano, *Cuestionarios para la formación de las relaciones geográficas de Indias*; Barbara E. Mundy, *The Mapping of New Spain: Indigenous Cartography and the Maps of the Relaciones Geográficas* (Chicago, London: The University of Chicago Press, 1996); Burke, *A Social History of Knowledge*, 116–49.

15 Hernández brought sixteen volumes with drawings and observations back to Spain, including the descriptions of 3000 new species of plants and 500 new species of animals. His work was published posthumously: Francisco Ximenez, *Quatros libros de la naturaleza, y virtudes de las plantas, y animales que estan receuidos en el vso de medicina en la Nueva España, y la methodo, y correcion y preparacion, que para administrallas se requiere con lo que el Doctor Francisco Hernandez escrivio en lengua latina* (Mexico: Vid. Daualos, 1615). The Latin translation of the book was made by Federico Cesi, the founder of the Accademia dei Lincei in Rome: Francisco Hernández, *Nova plantarum, animalium et mineralium Mexicanorum historia a Francisco Hernandez medico in Indiis praestantissimo primum compilata, dein a Nardo Antonio Reccho in volumen digesta* (Romae: sumptibus Blasii Deversini, & Zanobii Masotti, typis Vitalis Mascardi, 1651); Jesús Bustamante García, "Francisco Hernández, Plinio del Nuevo Mundo: tradición clásica, teoría nominal y sistema terminológico indígena en una obra renacentista," in *Entre dos mundos: fronteras culturales y agentes mediadores*, ed. Berta Ares Queija and Serge Gruzinski (Sevilla: Escuela de Estudios Hispano-Americanos, 1997); Antonio Clericuzio, "Le accademie scientifiche del Seicento," in *Il contributo italiano alla storia del pensiero. Scienze*, ed. Antonio Clericuzio and Saverio Ricci (Rome: Istituto della Enciclopedia Italiana fondata da Giovanni Treccani, 2013); Luigi Guerrini, "Piante e animali del Nuovo Mondo. Federico Cesi e il *Tesoro messicano*," in *Le scienze*, vol. 5, ed. Antonio Clericuzio and Germana Ernst (Treviso: Colla, 2008).

16 The Spanish crown will continue using questionnaire as an administrative tool until 1812. Solano, *Cuestionarios para la formación de las relaciones geográficas de Indias*, 79–87.

geographical coordinates of places, their climate, and the quality of the soil. The queries also asked about the presence of mountains, springs, lakes, rivers, and about the vegetation, medicinal plants, and mines. There were also a series of questions examining the customs of the inhabitants.

At the same time, questionnaires were implemented and used by different actors involved in the evangelization, exploration and/or colonization of New Spain. The Franciscan friar Bernardino de Sahagún (c. 1499–1590), a missionary priest who participated in the Catholic evangelization of colonial Mexico, provides a very interesting example of this research practice. He worked on his *Historia general de las cosas de Nueva España* (first version 1576, second, revised version, 1585) by means of a "formulario de preguntas" (a questionnaire), which could be reconstructed from his book. Sahagún sent it to Indian villages. In his research practice, questionnaires are very valuable instruments, designed not only to investigate the religious or political organization of local communities but also to understand the "potential" of the land, i.e. local flora, fauna and minerals, involving local informants.[17] In this sense, we observe a simultaneous rise of questionnaires as information-gathering tools on an official level (in Spain's colonial state building) and on a more informal level in the context of early modern ethnological or natural history research by scholars, travellers, clergymen or friars. Like Sahagún, other naturalists – for example the British Robert Plot (1640–1696), the Swiss Johann Jakob Scheuchzer

17 For more on Sahagún: Florencio Vicente Castro, *Bernardino de Sahagún primer antropólogo en Nueva España (siglo XVI)* (Salamanca: Ed. Universidad de Salamanca, Institucíon "Fray Bernardino de Sahagún", Excma. Diputación provincial de León, 1986). Miguel León Portilla, *Bernardino de Sahagún*, Protagonistas de América (Madrid: Historia 16, 1987); Alfredo López Austin, "Estudio acerca del método de investigación de fray Bernardino de Sahagún," in *La investigación social de campo en México*, ed. Jorge Martínez Ríos (México: Universidad Nacional Autónoma, 1976); López Austin, "Estudio acerca del método de investigación de fray Bernardino de Sahagún," *Estudios de Cultura Náhuatl* 42 (2011); José M. López Piñero, "Sahagún, Bernardino de," in *Diccionario histórico de la ciencia moderna en España*, vol. 2, ed. José M. López Piñero (Barcelona: Ed. Península, 1983); Bernardino de Sahagún, *Florentine Codex: General History of the Things of New Spain*, transl. from the Aztec into English, with notes and ill. by Arthur James Outram Anderson, and Charles E. Dibble, 1. paperback ed. (Salt Lake City: University of Utah Press, 2012), 1–13. English edition of Sahagún's *Historia general*: Bernardino de Sahagún, *The Conquest of New Spain, 1585 Revision*, ed. Howard F. Cline (Salt Lake City: University of Utah Press, 1989). For more on questionnaires and lists as "social and material technology," especially in the context of 17th-Century British natural history, see: Elisabeth Yale, "Making Lists: Social and Material Technologies in the Making of Seventeenth-Century British Natural History," in *Ways of Making and Knowing: The Material Culture of Empirical Knowledge*, ed. Pamela H. Smith, Amy R. W. Meyers, and Harold J. Cook (Ann Arbor: University of Michigan Press, 2014).

(1672–1733) or the Italian Antonio Vallisneri (1661–1730) – used questionnaires and tried to involve local informants in their projects.

The implementation of questionnaires to obtain information about the territory, nature, and people was also an important element for the constitution of the Spanish colonial empire. In an inspiring article (1998), Bustamante García defined this process as a collective enterprise during Philip II's reign, aimed at establishing the monarchy and underpinning its legitimacy. Bustamante García claims that four factors were crucial for this process. First, the mathematization of space and its resources as a new form of political science (the *Relaciones geográficas* are an aspect of this); secondly, control over nature and the development of natural history research and of inventories of resources; third, control over time and the constitution of an "official" historical narrative. Finally, and no less importantly, he indicates the creation of a proper Spanish juridical and intellectual tradition. The first two factors are particularly interesting for our purpose: the mathematization of space and of its resources and control over nature alongside the development of natural history research and of inventories of resources. As Bustamante García rightly underlined, these fields are exactly "the same that France and England – from the Académie Royale des Sciences and the Royal Society – assumed as state projects in the succeeding two centuries and that are usually associated with the culmination of the so-called Scientific Revolution."[18]

In this sense, Bustamante García as well as new studies by Cañizares-Esguerra, Barrera-Osorio and, in the German speaking area, by Brendecke, confirm the role of colonial Spain in the development of a "new" form of empiricism, and challenge – once more – the role of Britain in the "so-called Scientific Revolution."[19] Particularly, Barrera-Osorio stressed the development of empirical strategies by pilots, natural historians and entrepreneurs and the fact that the Spanish Crown incorporated these strategies "in the structure of the imperial bureaucracy for collecting information."[20]

---

18 Jesús Bustamante García, "La empresa naturalista de Felipe II y la primera expedición cientifica en suelo americano. La creación del modelo expedicionario renacentista," in *Felipe II (1527–1598): Europa y la monarquía católica*, ed. José Martinez Millán (Madrid: Editorial Parteluz, 1998), 40: "los mismos que Francia e Inglaterra – desde la *Académie Royale des Sciences* y la *Royal Society* – asumieron como proyectos de Estado en los dos siglos siguientes y que habitualmente se asocian a la culminación de la llamada Revolución Científica." My translation.

19 Barrera-Osorio, *Experiencing Nature*; Brendecke, *Imperium und Empirie*; Cañizares-Esguerra, "Iberian Science in the Renaissance: Ignored How Much Longer?".

20 Barrera, "Empire and Knowledge: Reporting from the New World", 43.

As I will show in the next subchapters, while questionnaires were initially implemented as official information gathering tools in the early-modern state-building process, they continued their career as scholars' field research tools within the Royal Society in Britain and, later, on the European Continent. Johann Jakob Scheuchzer, the first naturalist to import the questionnaire method into the German-speaking world, knew the contemporary literature about natural history research in the Americas. In his *Historiae helveticae naturalis prolegomena* (1700), Scheuchzer mentioned Francisco Hernández' work and in his *Bibliotheca scriptorum historiae naturali* (1716), an international bibliography of natural history organized in chapters depending of the different continents, he mentioned not only Hernández but also the publications of Juan Fragoso (c. 1530–1597), a personal physician of Philip II of Spain.[21]

2 Queries in Britain

Interest in the Spanish colonial enterprise and the role of travels and travel narratives as (almost) reliable sources for compiling a comprehensive natural history played a similar central role in Britain during the 17th century. We can see a "chronological coincidence between European expansion and the institutionalization of empirical practices in European kingdoms."[22] Translations

21 Johann Jakob Scheuchzer and Johann Rudolf Lavater, *Historiae helveticae naturalis prolegomena publicae Eruditorum Syzētēsei subiecta praeside I. Iacobo Scheuchzero* [...], *Respondente Iohanne Rodolfo Lavatero* [...] *In Auditorio Aestivo H.L.Q.S. Die Septembr.* (Tiguri: Gessner, 1700), 17–19; Johann Jakob Scheuchzer, *Bibliotheca scriptorum historiae naturalis omnium Terrae Regionum inservientium; historiae naturalis Helvetiae Prodromus; accessit Celeberrimi Viri Jacobi Le Long, Bibliothecarii Oratoriani de Scriptoribus historiae naturalis Galliae; collegit Johann Jakob Scheuchzer, Med. D. Math. Prof. Societ. Leopoldino-Carolinae, Regiae Angl. & Boruss.* (Tiguri: Typis Henrici Bodmeri, 1716), 200–02, 211. He mentions Hernández' 1615 edition; Ximenez, *Quatros libros de la naturaleza, y virtudes de las plantas, y animales que estan receuidos en el vso de medicina en la Nueva España, y la methodo, y correcion y preparacion, que para administrallas se requiere con lo que el Doctor Francisco Hernandez escrivio en lengua latina* and the Italian edition of 1649 and 1651; Francisco Hernández, *Rerum medicarum Novae Hispaniae thesaurus seu plantarum animalium mineralium Mexicanorum historia* (Rome: Officina carte valori dell'Istituto Poligrafico e Zecca dello Stato, 1992 [1649]); Francisco Hernández, *Nova plantarvm, animalivm et mineralivm Mexicanorvm historia: cui demum accessere aliqvot ex Principis Federici Cæsii Frontispiciis Theatri naturalis phytosophicæ tabulæ; vna cum quamplurimis iconibus, ad octingentas* [...] *exhibentur* (Romae et al.: Mascardus, 1651) and Juan Fragoso, *Discurso de las cosas aromaticas, arboles y frutales* [...] (Madrid: Francisco Sanchez, 1572), in its Latin translation of 1610.

22 Barrera-Osorio, *Experiencing Nature*, 11.

of several books relating to the East and West Indies by Richard Eden (c. 1520–1576) and by Samuel Purchas (1577–1626) as well as the publication of José de Acosta's work in an English version (*The Naturall and Morall Historie of the East and West Indies*, London, 1604) were circulating in Britain, facilitating the reception of this empirical culture.[23]

Francis Bacon (1561–1626) was conscious of the importance of travel and travel explorations as instruments for acquiring new forms of knowledge of the world. He was also aware of their importance in securing first-hand information about newly discovered territories and, more generally, about "exotic" countries.[24] In his text *Of Travel* (1601), he compiled a list of "things to be seen and observed" during a tour.[25] In the *Parasceve ad historiam naturalem et experimentalem* (1620), he not only put together a series of suggestions about how to write comprehensive *civil* and *natural history*, but also named a series of topics underlining their usefulness for achieving such a commitment. He also emphasized the role of "factors and merchants to go everywhere in search of them [the materials] and bring them in."[26] Bacon also pointed out a range

23 Richard Eden, *The Decades of the Newe Worlde or West India: Conteyning the Nauigations and Conquestes of the Spanyardes,* [...] *Translated into Englysshe by Richarde Eden* (London: In ædibus Guilhelmi Powell, Anno. 1555); Samuel Purchas, *Purchas His Pilgrimes: In Fiue Bookes.* (London: Printed by William Stansby for Henrie Fetherstone, 1625); the original text of Acosta is: José de Acosta, *Historia natural y moral de las Indias*, [Reproducción en facsimil], introd., apéndice y antología por Barbara G. Beddall, Hispaniae scientia (Valencia: Albatros, 1977 [1590]).

24 In his *Novum Organon*, he wrote: "Rursus vero homines a progressu in scientiis detinuit et fere incantavit reverentia antiquitatis, et virorum qui in philosophia magni habiti sunt authoritas, atque deinde consensus. [...] Neque pro nihilo aestimandum, quod per longiquas navigationes et peregrinationes (quae seculis nostris increbuerunt) plurima in natura patuerint, et reperta sint, quae novam philosophiae lucem immittere possint." Francis Bacon, *Novum Organum, sive indicia vera de interpretatione naturae*, Part 1, LXXXIV. In: Francis Bacon, *The Works of Francis Bacon*, vol. 1, ed. coll. and ed. James Spedding, Robert Leslie Ellis, and Douglas Denon Heath, 15 volls. (Boston: Houghton, Mifflin and Company, 1857–1882), 290–91. Translation, vol. VIII, 116–17: "Again, men have been kept back as by a kind of enchantment from progress in the sciences by reverence for antiquity, by the authority of men accounted great in philosophy, and then by general consent. [...] Nor must it go for nothing that by distant voyages and travels which have become frequent in our times, many things in nature have been laid open and discovered which may let in new light upon philosophy."

25 For example, he asked about the "courts of princes," "courts of justice," but also church and monasteries, antiquities and ruins, libraries, cabinets and rarities, havens and harbors, antiquities and ruins, shipping and navies: Francis Bacon, "Of Travel" in *The Philosophical Works of Francis Bacon*, ed. John M. Robertson, Repr. from the texts and transl., with the notes and prefaces, of Ellis and Spedding (London: Routledge, 1905), 756–57.

26 Bacon, *The Works of Francis Bacon*, vol. 2, 44: "At intellectus materialia tam late patent ut ea (tamquam per procuratores et mercatores) undiqui conquiri et importari debeant.

of queries (*topica particularia* or *Articuli Inquisitionis*) and in his *Historia naturalis et experimentalis ad condendam philosophiam* (London, 1622), published as Part III of his *Instauratio magna*, intended to define the state of research in the natural sciences and at the same time to promote new ones.[27] Once again, in his utopian novel *Nova Atlantis* (1627) or, in English, *New Atlantis* (1628), which begins with the description of an overseas expedition from Peru to China and Japan, he underlines several times the importance of voyages to collect books, information and "patterns of experiments" from all parts of the world. In this text, we can read: "[…] we have twelve [fellows] that sail into foreign countries, under the names of other nations, (for our own we conceal); who bring us the books, and abstracts, and patterns of experiments of all other parts."[28]

Bacon's scheme for collecting data and the influence of questionnaires as information-gathering tools are evident in the activity of the Hartlib circle and in the early years of the Royal Society. After 1645, the brothers Gerard

Accedit etiam illud, quod coeptis nostris vix dignum esse aestimemus ut in re tali quae fere omnium industriae pateat nos ipsi tempus teramus. Quod autem caput rei est ipsi nunc praestabimus; ut eijusmodi historiae modum et descriptionem, qualis intentioni nostrae satisfaciat diligente et exacte proponamus." For the list of topics: Bacon, *The works of Francis Bacon*, vol. 2, 61–69; Translation: Bacon, *The works of Francis Bacon*, vol. VIII (Translations of the philosophical Works, vol. 1), 354: "[…] but the materials on which the intellect has to work are so widely spread, that one must emply factors and merchants to go everywhere in search of them and bring them in. Besides I hold it to be somewhat beneath the dignity of an undertaking like mine that I should spend my own time in a matter which is open to almost every man's industry." List of topics, 373–81.

27 Francisci Baconi, *Historia naturalis et experimentalis ad condendam philosophiam siue, phænomena vniuersi: quæ est Instauratoinis magnæ pars tertia* Londini: In officina Io. Hauiland, impensis Matthæi Lownes & Guilielmi Barret, 1622; Hunter, "Robert Boyle and the Early Royal Society: A Reciprocal Exchange in the Making of Baconian Science"; Hunter, "Robert Boyle's 'Heads' and 'Inquiries'". The bibliography on this topic (travel and science) is huge, and I shall mention only a few: Rob Iliffe, "Science and Voyages of Discovery," in *The Cambridge History of Science: Eighteenth-Century Science*, ed. Roy Porter (Cambridge: Cambridge University Press, 2003); Marie-Noëlle Bourguet, "Voyages lointains et mesure du monde: la projection du regard européen sur le monde (vers 1500-vers 1800)," in *L'Europe des sciences et des techniques XV*e*–XVIII*e *siècle: un dialogue des savoirs*, ed. Liliane Hilaire-Pérez, Fabien Simon, and Marie Thébaud-Sorger (Rennes: Presses universitaires de Rennes, 2016).

28 Francis Bacon, *New Atlantis: A Work Unfinished* (London: Printed by Tho. Newcomb, 1658), 32; for the Latin text: Francisci Baconi, *Nova Atlantis* (Turnhout: Brepols Publishers, 2014); about this work, for example: Josef Bordat, "Bacons Atlantis-Mythos und das Selbstverständnis der modernen Wissenschaft," in *Mythos – Helden – Symbole Legitimation, Selbst- und Fremdwahrnehmung in der Geschichte der Naturwissenschaften der Medizin und der Technik*, ed. Siegfried Bodenmann and Susan Splinter (Munich: Martin Meidenbauer, 2009).

(1604–1650) and Arnold Boate (1606–1653) were working on a comprehensive natural history of Ireland, which Gerald published in 1652.[29] At the same time, Arnold, hoping to collect enough information to complete the project, composed the first British questionnaire, *An Interrogatory Relating more particularly to the Husbandry and Natural History of Ireland*, published as an appendix to the second edition of Hartlib's *Legacie*.[30]

The work of gaining information in a more systematic way in the context of the British colonial expansion also prompted the Royal Society to draw up a series of "articles of inquiry," which provided a set of topics around which to organize observations. In 1661, the Society created a committee (whose members included Sir Robert Moray, William Petty, Laurence Rooke, Thomas Povey, Robert Boyle, John Wilkins, John Evelyn and Henry Oldenburg, first secretary of the Society and editor of the *Philosophical Transactions*) with the purpose of considering "proper questions to be inquired of in the remotest parts of the world." It is worth noticing that this committee was created at the same time as the Council for Foreign Plantations and not by accident; both institutions shared some members, like Robert Boyle and Thomas Povey. British scholars, merchants and the British Monarchy had a strong scientific (medical), economic and political interest in the resources of the newly discovered territories and in extra-European natural history (especially in botany).[31]

29 Gerard Boate, *Irelands Naturall History: Being a True and Ample Description of Its Situation, Greatness, Shape, and Nature, of Its Hills, Woods, Heaths, Bogs* [...] *With Its Heads or Promontories, Harbours, Roades, and Bayes* [...] *Of Its Metalls, Mineralls* [...] *Turf* [...] *And Lastly, of the Nature and Temperature of Its Air and Season, and What Diseases It Is Free From, or Subject Unto* (London: For John Wright, 1652).

30 Fox, "Printed Questionnaires, Research Networks, and the Discovery of the British Isles, 1650–1800," 595–96; Charles Webster, *The Great Instauration: Science, Medicine and Reform, 1626–1660*, 2nd ed., 1st ed. 1975 (Oxford: Peter Lang, 2002), 428–31; Arnold Boate, "An Interrogatory Relating more particularly to the Husbandry and Natural History of Ireland," in *Samuel Hartlib His Legacie, Or, an Enlargement of the Discourse of Husbandry Used in Brabant & Flaunders: Wherein Are Bequeathed to the Common-Wealth of England More Outlandish and Domestick Experiments and Secrets in Reference to Universall* Husbandry, ed. Samuel Hartlib, Cressy Dymock, Robert Child, and Richard Weston, eds., The 2nd ed., augmented with an appendix (London: Printed by R. & W. Leybourn for Richard Wodenothe, 1652), Appendix, without pagination.

31 On the relationship between colonial empires and botany, for example: Lucile H. Brockway, *Science and Colonial Expansion: The Role of the British Royal Botanic Gardens*, 2nd ed. (1st ed. 1979) (New Haven, London: Yale University Press, 2002). Londa Schiebinger and Claudia Swan, eds., *Colonial Botany: Science, Commerce, and Politics in the Early Modern World* (Philadelphia: University of Pennsylvania Press, 2005). Samir Boumediene, *La colonisation du savoir une histoire des plantes médicinales du "Nouveau Monde" (1492–1750)* (Vaulx-en-Velin: Les éditions des mondes à faire, 2016).

The most famous questionnaire produced in the context of the Royal Society is Robert Boyle's *General Heads for a Natural History of a Country, Great or Small* of (first published in 1665–1666).[32] The English questionnaires are rather similar to those developed by the Spanish scholars.[33] Like the Spanish questionnaire of 1577 and the questionnaires of Alonso de Santa Cruz, Boyle's queries contain a group of questions relating to "Supraterraneous, Terrestrial, and Subterraneous" topics.[34] In addition, he proposed queries about the heavens and air (i.e. weather, climate, winds, and diseases related to the weather or the occurrence of meteors) as well as about water and springs. Another group of questions related to the Earth, its configuration (dimensions, orology, the nature of soil), and its production, such as minerals, stones and mines (and it is not by chance that, in his questionnaire, Boyle refers to the technique used to separate "the Nobler from the Baser Metals" in Peru).[35] The "terrestrial questions" also included the inhabitants, the vegetation and the animals

32 In 1692, these queries were published with a supplement of later questionnaires printed in the *Philosophical Transactions*. Robert Boyle, *General Heads for the Natural History of a Country Great or Small: Drawn Out for the Use of Travellers and Navigators / Imparted by* [...] *Robert Boyle* [...]*; to Which Is Added, Other Directions for Navigators, Etc. With Particular Observations of the Most Noted Countries in the World; by Another Hand* (London: Printed for John Taylor [...] and S. Hedford, 1692); Boyle, "General Heads for a Natural History of a Countrey, Great or Small, Imparted Likewise by Mr. Boyle," *Philosophical Transactions* 1, no. 11 (1665); Boyle, "Other Inquiries Concerning the Sea," *Philosophical Transactions* 1, no. 18 (1665); "Inquiries for Suratte, and Other Parts of the East-Indies; Inquiries for Persia. – Inquiries for Virginia and the Bermudas. For Guaiana and Brasil," *Philosophical Transactions* 2, no. 23 (1666); "Inquiries of Directions for the Ant-Iles, or Caribe-Islands," *Philosophical Transactions* 3, no. 33 (1668); Daniel Carey, "Compiling Nature's History: Travellers and Travel Narratives in the Early Royal Society," *Annals of Science* 54, no. 3 (1997), 274; Boyle's questionnaire was addressed to travellers by sea and overland. About the different queries produced by members of the Royal Society: Thomas Birch, *The History of the Royal Society of London for Improving of Natural Knowledge* [Facsimile], 4 vols. (Bruxelles: Culture et civilisation, 1967–1968), vol. 1 (original edition: London, 1756–1757), 15 (for the quotation on p. 34), 68, 69, 79, 119, 130, 144, 165–66, 180, 192, 199, 297–98, and 318–19.

33 Barbara J. Shapiro, "History and Natural History in Sixteenth and Seventeenth-Century England: An Essay on the Relationship Between Humanism and Science," in *English Scientific Virtuosi in the 16th and 17th Centuries: Papers Read at a Clark Library Seminar, 5 February 1977*, ed. Barbara J. Shapiro and Robert G. Frank (Los Angeles: University of California, William Andrews Clark Memorial Library, 1979), 7.

34 "The things, to be observ'd in such a [natural] History, may be variously (and almost at pleasure) divided: As, into Supraterraneous, Terrestrial, and Subterraneous; and otherwise: but we will at present distinguish them into those things, that respect the *Heavens* or concern the *Air*, the *Water*, or the *Earth*," Boyle, "General Heads for a Natural History of a Countrey, Great or Small, Imparted Likewise by Mr. Boyle," 186.

35 Boyle, *General Heads*, 41.

of a territory. Boyle asked for information on the correct determination of a geographical place (i. e. the coordinates of a place, "what fixt stars are and what not seen there").[36]

Following these examples, Robert Plot, first Professor of Chemistry at the University of Oxford, Fellow of the Royal Society and the first Keeper of the Ashmolean Museum, wrote and printed questionnaires while preparing his studies on the natural history of Oxfordshire (published in 1677) and of Staffordshire (printed in 1686).[37] In his first printed questionnaire *Questions to be asked to the most ingenious of each County in my Travels through England*, Robert Plot, like in the Spanish and Robert Boyle's queries, first asked questions regarding the quality of the air and the weather (heaven). Plot was looking for firsthand information from seamen ("Enquire of Seamen," he wrote in his questionnaire) concerning navigation (tides, ebbs, currents), the recurrence of storms, and the presence of echoes. With these questions, Plot was following the *Directions for Sea-men, bound for far Voyages* of Laurence Rooke, also published in the *Philosophical Transactions* in 1665–1666.[38]

Other questions concerned the saltiness of the sea, the typology of the different fountains (springs), orography ("mountains, caves, barrows"), the quality of the earth, the presence of stones, gems and metals. Other points of interests were "strange or prodigious trees," "other curiosity in husbandry," and the different species of insects (especially bees and silkworms). The final questions of this first questionnaire concerned the inhabitants of the different regions (the presence of monsters, giants, hermaphrodites) and British, Roman or Saxon antiquities. Plot printed a second questionnaire, similar to this first one, after finishing his *History of Oxford-Shire* ("Enquiries to be propounded to the most

---

36 Boyle, *General Heads*, 17.

37 Robert Plot, *The Natural History of Oxford-Shire: Being an Essay Toward the Natural History of England* (Oxford: At Mr. S. Millers, 1677); Plot, *The Natural History of Stafford-Shire* (Oxford: Printed at the Theater, 1686); for Plot's biography: A. J. Turner, "Plot, Robert (Bap. 1640, D. 1696)," in *The Oxford Dictionary of National Biography*, ed. H. C. G. Matthew and B. Harrison (Oxford: Oxford University Press, 2004), https://doi.org/10.1093/ref:odnb/22385, accessed on 23 September 2019; A. G. Keller, "Plot, Robert," in *Complete Dictionary of Scientific Biography*, ed. Charles Coulston Gillispie (Detroit: Charles Scribner's Sons, 2008); for the questionnaires: Archives of the Royal Society, London, Classified Papers (Cl.P.), vol. XIX "Inquiries and Answers, December 24, 1662 – May 25, 1692". See: Cl.P. XIX/19/93: "Questions to be asked to the most ingenious of each county in my travels through England" by R[obert] P[lot]; Cl. P. XIX/19/94: "Enquiries to be propounded to the most Ingenious of each County in my Travels through England and Wales, in order to their History of Nature and Arts" by R[obert] P[lot].

38 Laurence Rooke, "Directions for Sea-Men, Bound for Far Voyages," *Philosophical Transactions* 1, no. 8 (1665–1666), 141–42.

Ingenious of each County in my Travels through England and Wales, in order to their History of Nature and Arts"). This questionnaire is organized in ten chapters: "Heaven and Air," "Waters," "Earth and Minerals," "Metals," "Stones," "Plants," "Husbandry," "Animals," "Arts," and "Antiquities." At the end of his flyer, Plot refers to his previous work about Oxfordshire and asks for the help of anyone who might be interested: their task is to "answer as many of them [questions] as they can, distinctly, according to the best of their skill." Replies have to be written "in a Paper apart, to lie ready against the Undertaker of this Design [Plot himself] shall call for them, in case they should then be absent from home."[39]

3 Queries and Natural History on the Continent

Henry Oldenburg (1619–1677), first secretary of the Society and editor of the *Philosophical Transactions*, acted as the broker (or the *trait d'union*) between Britain and the Continent in these years. He tried to export the model of the empirical way to investigate local territories, especially by exploiting his widespread correspondence network. One strategy was to send his correspondents copies of Robert Plot's *Natural History of Oxford-Shire* (Oxford, 1677) as a model for local investigations.[40]

Johann Jakob Wagner (1641–1695), doctor of the foundling hospital in Zurich and curator of that city's public library and museum (*Bürgerbibliothek* and *Kunstkammer*), attests a very early appropriation of this empirical way to investigate natural history in Switzerland. Wagner was active at the same time as

39 At the end of the text, Plot writes: "If anybody desire to be further informed in the nature of this Design, let him consult the Natural History of Oxford-Shire already extant [...]"; Archives of the Royal Society, London, Classified Papers, Cl. P. XIX/19/94: "Enquiries to be propounded to the most Ingenious of each County in my Travels through England and Wales, in order to their History of Nature and Arts" by R[obert] P[lot], without pagination. Keller, "Plot, Robert".

40 See: Alix Cooper, "Fragen ohne Antworten: Die Suche nach lokalen Informationen in der frühen Aufklärung," in *Vor Google: Eine Mediengeschichte der Suchmaschine im analogen Zeitalter*, ed. Thomas Brandstetter (Bielefeld: transcript, 2012). In the edition of Oldenburg's correspondence we can find several letters to Boyle, Hevelius, and other scholars, about queries and questionnaires, with some answers received from several physicians and naturalists; Alfred Rupert Hall and Marie Boas Hall, eds., *The Correspondence of Henry Oldenburg*, 13 vols. (Philadelphia: Taylor & Francis, 1965–1986), vol. 3 (1666–1667), 47, 49, 58, 65, 729 (queries for Hevelius), 87, 248–57 (answers by Hevelius) 340–41, 553–61 (answers by Peter Möller), 562–71 (answers by Michael Behm) and vol. 4 (1667–1668), 133, 166–68, 275, 306, 419–24, vol. 5 (1668–1669), 82–83, 105–07, 200, 285, 303, 326, 422–26, 474–76.

Plot and is the author of the first natural history of the Old Swiss Confederacy, the *Historia naturalis Helvetiae curiosa* (1680). In his book, he programmatically expressed his intention to follow Bacon in his commitment to write a novel natural history, applying the inductive method. In the first lines of his *Dedicatio*, he mentions Bacon as the most important figure in this field and he acknowledges the influence of his *De dignitate et augmentis scientiarum* on his personal work. Three years later, Wagner's book was glowingly reviewed in the *Philosophical Transactions*; its Baconian leanings were recognized and appreciated by the reviewer.[41] Interestingly, the sequence of topics in some chapters of his book closely mirror the organization of topics that we could find in the English questionnaires and in Plot's natural histories.

His successor as a doctor of the city of Zurich's foundling hospital, as curator of the public library and museum and as supporter of an empirical method to study nature was Johann Jakob Scheuchzer (1672–1733). Following the model of the Royal Society, in 1699 Scheuchzer published a questionnaire in Latin and German with approximately 190 questions, to investigate the natural history of the Old Swiss Confederacy, the first text of this kind in the German-speaking area. Like Wagner's book, Scheuchzer's *Einladungsbrief zur Erforschung natürlichen Wunderen, so sich im Schweizer-Land befinden* (1699, in Latin *Charta invitatoria*) begins with a reference to Bacon and the English questionnaires (fig. 2.1).[42] Furthermore, the author mentions the successful commitment of the members of the Royal Society to guide their compatriots and foreigners in making observations of all phenomena at sea and on land, of air, animals, plants, and minerals. Scheuchzer cultivated a close intellectual exchange with several members of this academy, especially with John Woodward (1665–1728),

41 Johann Jacob Wagner, *Historia naturalis Helvetiae curiosa* (Tiguri: Joh. Henrici Lindinneri, 1680): *Dedicatio* (not paginated) [p. 1]: "Ad Scientiæ Naturalis exactam notitiam comparandam, multorum præclarorum Virorum, rerum naturalium maxime peritorum, judiciô, inter quos Magnus ille Angliæ Cancellarius, Franciscus Baconus, Baro de Verulamio (de Augmento Scientiarum Lib. II. C. 3) facile principem obtinet locum, ere maxime foret, si Historia Naturalis Inductiva cujuslibet regionis, quantum fieri poterit exactissima adornaretur, quo ex variis Observationibus & Experimentis inde collectis Regulæ tandem certae & indubitatæ elicit, Systemaque absolutum componi ac perfici queam." (Italic in the original text); for Wagner's review: "[An Account of Four Books]," *Philosophical Transactions* 13, no. 143–54 (1683).

42 Johann Jakob Scheuchzer, *Charta invitatoria, quaestionibus, quae historiam Helvetiae naturalem concernunt, praefixa* (Tiguri: [D. Gessner], 1699); Johann Jakob Scheuchzer, *Einladungs-Brief, zu Erforschung Natürlicher Wunderen, so sich im Schweizer-Land befinden* (Tiguri: [D. Gessner], 1699); a recent edition of the German version of the text in: Simona Boscani Leoni, ed., *"Unglaubliche Bergwunder": Johann Jakob Scheuchzer und Graubünden. Ausgewählte Briefe 1699–1707*, Cultura alpina Band 9 (Chur: Verlag Bündner Monatsblatt, 2019), 33–49.

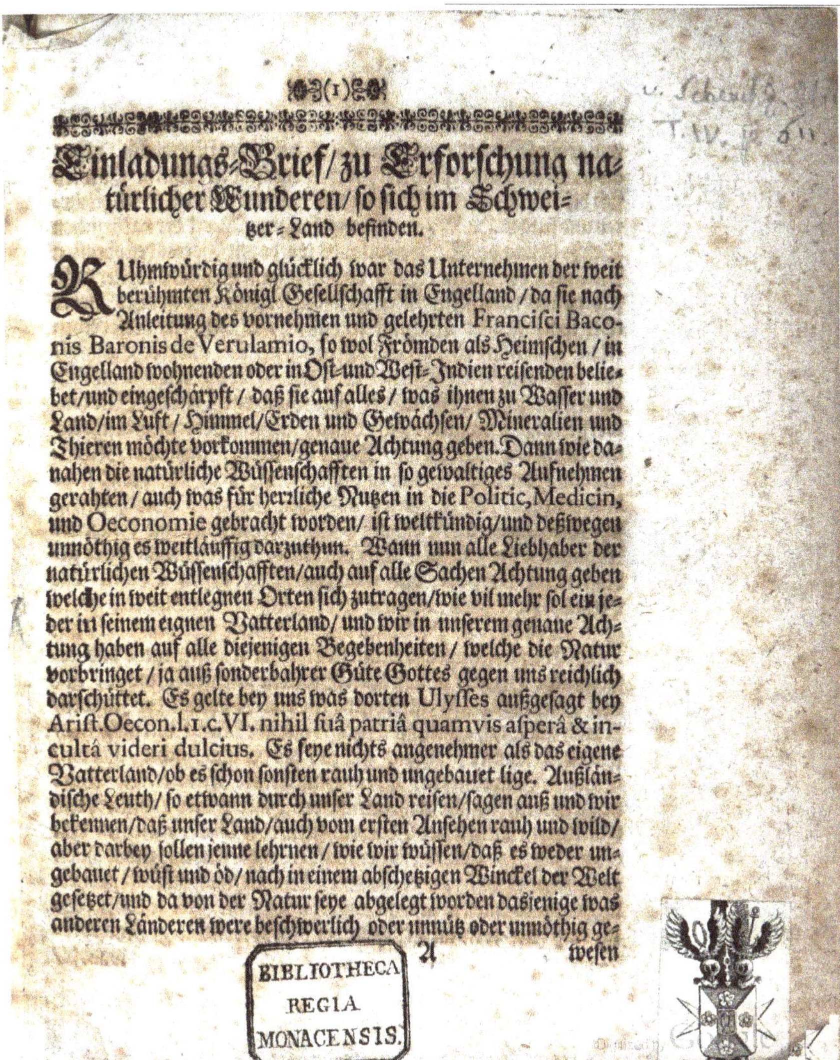

(1)

Einladungs-Brief/ zu Erforschung na-
türlicher Wunderen/ so sich im Schwei-
tzer-Land befinden.

Ruhmwürdig und glücklich war das Unternehmen der weit
berühmten Königl. Gesellschafft in Engelland / da sie nach
Anleitung des vornehmen und gelehrten Francisci Baco-
nis Baronis de Verulamio, so wol Frömden als Heimschen / in
Engelland wohnenden oder in Ost- und West-Indien reisenden belie-
bet/und eingeschärpft / daß sie auf alles / was ihnen zu Wasser und
Land/im Luft / Himmel/Erden und Gewächsen/ Mineralien und
Thieren möchte vorkommen/genaue Achtung geben. Dann wie da-
nahen die natürliche Wüssenschafften in so gewaltiges Aufnehmen
gerahten / auch was für herrliche Nutzen in die Politic, Medicin,
und Oeconomie gebracht worden/ ist weltkündig/und deßwegen
unnöthig es weitläuffig darzuthun. Wann nun alle Liebhaber der
natürlichen Wüssenschafften/auch auf alle Sachen Achtung geben
welche in weit entlegnen Orten sich zutragen/wie vil mehr sol ein je-
der in seinem eignen Vatterland/ und wir in unserem genaue Ach-
tung haben auf alle diejenigen Begebenheiten / welche die Natur
vorbringet / ja auß sonderbahrer Güte Gottes gegen uns reichlich
darschüttet. Es gelte bey uns was dorten Ulysses außgesagt bey
Arist. Oecon. l. 1. c. VI. nihil suâ patriâ quamvis asperâ & in-
cultâ videri dulcius. Es seye nichts angenehmer als das eigene
Vatterland/ob es schon sonsten rauh und ungebauet lige. Außlän-
dische Leuth/ so etwann durch unser Land reisen/sagen auß und wir
bekennen/daß unser Land/auch vom ersten Ansehen rauh und wild/
aber darbey sollen jenne lehrnen / wie wir wüssen/daß es weder un-
gebauet / wüst und öd/ nach in einem abschetzigen Winckel der Welt
gesetzet/und da von der Natur seye abgelegt worden dasjenige was
anderen Länderen were beschwerlich oder unnütz oder unnöthig ge-

A wesen

FIGURE 2.1 Johann Jakob Scheuchzer, *Einladungs-Brief, zu Erforschung natürlicher Wunderen, so sich im Schweizer-Land befinden*, [Zurich], 1699, 1. Bayerische Staatsbibliothek Digital, Signatur: 4 H.nat. 141 b

a physician and Professor of Physics at Gresham College in London.[43] Woodward was also the author of a questionnaire entitled *Brief Instructions for Making Observations in all parts of the World*.[44]

In Scheuchzer's questionnaire, we can find similarities with Boyle's and Woodward's texts. The Swiss scholar repeated Boyle's organization of the queries on "Supraterraneous, Terrestrial, and Subterraneous" topics, and the first groups of questions concerning the coordinates of a place, the quality of air and climate, are similar to Boyle's *General heads*. Other questions concerning observation of local inhabitants, weather forecast and the height of mountains, their changes and the presence of snow can be interpreted as examples of an intertextuality between Boyle's and Woodward's questionnaires and Scheuchzer's one. However, Scheuchzer adapted the questions to suit the peculiarity of his country and of its most unknown regions, the Alps. About fifteen queries referred to milk and dairy products, which Scheuchzer saw as very typical components of his compatriots' diet. Alps and mountains are mentioned almost thirty-five times in the *Einladungsbrief*, and fourteen questions concerned snow, glaciers, avalanches, or ice.[45]

Very interestingly, Scheuchzer received several long answers to his questionnaire, especially from pastors of the Reformed Church and from noblemen of an Alpine region, the Free State of the Three Leagues of Grisons (today Canton

43 Since 1704, Scheuchzer had been Fellow of the Royal Society. Woodward became one of the most important correspondents of Scheuchzer in London. The contacts between the two men remained very close from 1701 until 1726, because of Scheuchzer's keen interest in the diluvial theory advocated by the English doctor in his *Essay*: John Woodward, *An Essay Toward a Natural History of the Earth and Terrestrial Bodies, Especially Minerals As Also of the Sea, Rivers, and Springs With an Account of the Universal Deluge And of the Effects That It Had Upon the Earth* (London: Printed for Ric. Wilkin, 1695); about Woodward, Scheuchzer and diluvialism: Michael Kempe, *Wissenschaft, Theologie, Aufklärung: Johann Jakob Scheuchzer (1672–1733) und die Sintfluttheorie* (Epfendorf: bibliotheca academica Verlag, 2003).

44 The text has an appendix on queries about the "Natives of Guinea, Monomotapa, and other less known parts of Africa: of the East, and West Indies: Tartary, Greenland, or any other remote and uncivilized, or Pagan Countries". John Woodward, *Brief Instructions for Making Observations, and Collections in Order to the Promotion of Natural History, in All Parts of the World: As Also, for Collecting, Preserving, and Sending over Natural Things* (London: Printed for Richard Wilkin, 1696); Victor A. Eyles, "Woodward, John" in Gillispie, *Complete Dictionary of Scientific Biography*.

45 Boscani Leoni, "*Unglaubliche Bergwunder*," 37–49: questions no. 1–15 (air, weather), questions no. 97–99 (mountains, their height), questions no. 118–23 (inhabitants). See also: Boyle, "General heads for a natural history of a countrey, great or small, imparted likewise by Mr. Boyle", 186–87 (air, weather), 188 (inhabitants); Woodward, *Brief Instructions for making Observations, and Collections in order to the promotion of Natural History, in all parts of the World*, 3 (weather), 6 (weather forecast).

of Grisons). These informants often referred to the *Charta invitatoria* in their own letters, replying quite articulately, and enclosing sketches of mountain outlines, Alpine plants, minerals, crystals or again transcriptions of accounts about dragons encountered by shepherds, hunters, and valley dwellers in the Alps. Scheuchzer often re-used this information, sometimes mentioning his sources in his works.[46]

The last step in my investigation about the appropriation/adaptation of queries as information-gathering tools takes us to Italy. At the beginning of the 18th century, Scheuchzer was the centre of a scholars network interested in the question of the origin of the Earth and in the interpretation of fossils. In this network we can find scholars like John Woodward, Friedrich Leopold (1651–1722), Louis Bourguet (1678–1742), Karl Nikolaus Lang (1670–1741), Luigi Ferdinando Marsili (1658–1730), and Antonio Vallisneri (1661–1730). Contacts with Luigi Ferdinando Marsili and Antonio Vallisneri was intense, especially over the 1703–1705 period.[47] From 1705 onwards, Johann Jakob Scheuchzer's

46 Simona Boscani Leoni, "Centri e periferie. Alcune riflessioni sulla corrispondenza erudita tra Sei e Settecento," *Schweizerische Zeitschrift für Geschichte* 55, no. 4 (2005); "Queries and Questionnaires: Collecting Local and Popular Knowledge in 17th and 18th Century Europe," in *Wissenschaftsgeschichte und Geschichte des Wissens im Dialog – Connecting Science and Knowledge*, ed. Kaspar von Greyerz, Silvia Flubacher, and Philipp Senn (Göttingen: V&R unipress, 2013); "Men of Exchange: Creation and Circulation of Knowledge in the Swiss Republics of Eighteenth Century," in *Scholars in Action: The Practice of Knowledge and the Figure of the Savant in the 18th Century*, ed. André Holenstein, Hubert Steinke and Martin Stuber, 2 vols. (Leiden: Brill, 2013). Two long answers to Scheuchzer's *Einladungsbrief* came from the pastor of the reformed church Johann Leonhardi (1651–1725) and from the nobleman Rudolf von Rosenroll (1671–1730), Boscani Leoni, "*Unglaubliche Bergwunder*," 86–99 (Leonhardi to Scheuchzer, 23 November 1699 [Gregorian calendar 3 December 1699]), 117–23 (Rosenroll to Scheuchzer, 15 February.[1700] [Gregorian calendar 25 February 1700]); online: Boscani Leoni, *"Lettres des Grisons": Wissenschaft, Religion und Diplomatie in der Korrespondenz von Johann Jakob Scheuchzer. Eine Edition ausgewählter Schweizer Briefe (1695–1731), hallerNet.org, 2019: https://hallernet.org/edition/scheuchzer-korrespondenz*; for Leonhardi: https://hallernet.org/edition/scheuchzer-korrespondenz/letter/18055?view=tf&option1=yes&option2=yes&option3=yes&translation=inactive; for Rosenroll: https://hallernet.org/edition/scheuchzer-korrespondenz/letter/18174?view=tf&option1=yes&option2=yes&option3=yes&translation=inactive, accessed on 20 April 2021.

47 Marsili sent 62 letters to Zurich in the timespan between 1703 and 1730 and received 55 answers from Scheuchzer; in the period from 1701 to 1728 Scheuchzer wrote 16 letters to Vallisneri and received 90 answers. See: Rudolf Steiger, "Verzeichnis des Wissenschaftlichen Nachlasses von Johann Jakob Scheuchzer," *Beiblatt zur Vierteljahresschrift der Naturforschenden Gesellschaft in Zürich* 78, no. 21 (1933), 71. On the correspondence between Marsili and Vallisneri, see for example: Marta Cavazza, *Settecento inquieto: alle origini dell'Istituto delle Scienze di Bologna* (Bologna: Il Mulino, 1990); Dario Generali, *Antonio Vallisneri: gli anni della formazione e le prime ricerche* (Florence: Olschki, 2007); Francesco

brother, Johann (1684–1738), acted as secretary and scientific assistant to Marsili, particularly during his journeys through the Swiss Alps, when the latter collected very important information about the structure of mountains. Based on morphological criteria, Marsili classified mountains into three groups: as a whole, as a result of irregular ruins, and as a combination of strata. In the same period, Vallisneri – motivated by a desire to prove his new theories about the origin of springs and mountains – undertook a long explorative journey in the Garfagnana (a mountainous territory between the Apuan Alps and the Apennines) and in 1705 he wrote an account of his journey in his *Primi itineris per Montes specimen Physico-Medico*.[48] As we know, the *Philosophical Transactions* did not publish the text and only several years later, sometime between 1722 and 1726, Vallisneri printed it under the pseudonym of Giovanni Battista Perrucchini in the *Supplementi al Giornale de' Letterati d'Italia*. In the appendix to this account, there is a list of 26 queries.[49]

Vallisneri's queries focus on questions about springs, mountain strata, mountain configuration, and fossils, which happen to be the main topics of the correspondence with Johann Jakob Scheuchzer. The Italian scholar was not convinced of Woodward's and Scheuchzer's diluvial theory and interpreted mountain strata as a consequence of different inundations.[50] His and Scheuchzer's questionnaires have many similar questions in common. Without difficulty, we can interpret Vallisneri's queries as an "appropriation" of the genre following Scheuchzer's model. One can find similarities between the two texts in the questions on fossils and springs and on the quality of water (temperature, smell and colour, presence of metals like copper or materials like sulphur).[51]

Luzzini, *Il miracolo inutile: Antonio Vallisneri e le scienze della terra in Europa tra XVII e XVIII secolo* (Florence: Olschki, 2013).

48 Francesco Luzzini, ed. *Theory, Practice, and Nature In-between. Antonio Vallisneri's Primi Itineris Specimen* (Berlin: Edition Open Access/Max Planck Institute for the History of Science, 2018). http://www.edition-open-sources.org/sources/9/, accessed on 20 April 2021. See also the contribution of Francesco Luzzini in this volume.

49 Antonio Vallisneri, "Estratto d'alcune Notizie intorno alla Provincia della Garfagnana, cavate dal primo Viaggio Montano del Sig. Antonio Vallisnieri, [...] dal Sig. Dottore Giovanbatista Perrucchini," *Supplementi al Giornale de' Letterati d'Italia* II (1722, Articolo VII); Vallisneri, "Continuazione dell'estratto d'alcune Notizie intorno alla Garfagnana cavate dal primo Viaggio Montano del Sig. Antonio Vallisneri," *Supplemento al Giornale de' Letterati d'Italia* III (1726, Articolo VIII); for the questionnaire: 1726, 404–19. See also: Luzzini, *Il miracolo inutile*, 88–94.

50 See the manuscripts of the letters in the Zentralbibliothek Zürich, Ms. H 312, H 305.

51 Antonio Vallisneri, *Primi Itineris per Montes Specimen Physico-Medicum*. 1705, 52–53 of the manuscript; edition: Luzzini, *Theory, Practice and Nature In-between. Antonio Vallisneri's Primi Itineris Specimen*, 345–46, 124–25 and English translation, 198–99; online: http://www.edition-open-sources.org/sources/9/, accessed on 20 April 2021; the printed version

One can also find similarities in the queries on insects and plants, especially on mountain vegetation.[52] There is also a most remarkable similarity in the Scheuchzer's and Vallisneri's inquiries on the nature and customs of the local/ native inhabitants, on their body size and habits, on the techniques of preparation of dairy products by peasants, and on the measurement of the height of mountains with barometer and thermometer.[53] Mountains, hydrography as well fossils, mountains plants, animals and local economy (milk, cheese) figured very prominently in both questionnaires.

4 Conclusion

As we have seen, questionnaires played an important role in the process of state building in the early modern age, especially in the Spanish and British colonial empire building. They are also a very efficient instrument for the empirical investigation of civil and natural history. We have also stressed how important Spanish America was as a place for establishing new forms of empirical practices during the 16th century. One can notice a similar process in Britain, where the same types of information-gathering tools were "appropriated" and adapted to local needs. Here, one may observe the specialization of queries in two different types: queries for travellers that aimed to conduct and organize their scientific observations and questionnaires to obtain information about antiquities and natural history of a local territory. The questionnaires of Robert Plot for Oxfordshire and Staffordshire illustrate it very well. Scholars in Switzerland and in Italy, two countries on the sidelines of the colonial expansion, applied this tool as a "local" research instrument reorienting it to discovering less-well-known European territories, like the Alps, the Apuan Alps and the Apennines Mountains. Through Scheuchzer's and Vallisneri's

in: Vallisneri, "Continuazione dell'estratto d'alcune notizie intorno alla Garfagnana cavate dal primo Viaggio Montano del Sig. Antonio Vallisneri," 404, 406–09, Scheuchzer questions no. 71, 73, 170, 174 (questions no. 168 and 172 in the Latin questionnaire). Boscani Leoni, "*Unglaubliche Bergwunder*", 41–42, 47–48.

52 Vallisneri, "Continuazione dell'estratto d'alcune Notizie intorno alla Garfagnana cavate dal primo Viaggio Montano del Sig. Antonio Vallisneri," 409–10, 404; Scheuchzer, questions 162 (Latin no. 160) and 124–32; Boscani Leoni, "*Unglaubliche Bergwunder*", 44–47.

53 Vallisneri, "Continuazione dell'estratto d'alcune notizie intorno alla Garfagnana cavate dal primo Viaggio Montano del Sig. Antonio Vallisneri," 413–17; Scheuchzer, questions no. 97, 117–23, 175–89 (Latin questions no. 173–86). Scheuchzer did barometrical experiments and regular measurements of the temperature of the air during his journeys through the Swiss Alps.

information-gathering campaign, European peripheral mountain regions gradually became *centres* for investigating the origin of the Earth and studying natural history.

In addition, questionnaires are an extremely interesting tool if we want to follow the development of empiricism in the history of knowledge and in the modern history of science. They were an excellent device for improving knowledge of a territory, of its resources; and, as colonial history has shown, for controlling and exploiting territories and people. During the Enlightenment, they became again very valuable and widespread information-gathering tools in the form of economical, demographical, and agricultural surveys implemented by economic societies to improve "patriotic reforms" of the territories.[54]

For this reason, questionnaires force us to reflect on the close relationship between information and power, a very important issue not only for the past, but also for our present and future.

Acknowledgements

This contribution is based on research funded by the Swiss National Science Foundation and carried out in the context of the projects "Kulturen der Naturforschung" (Cultures of Natural History), Projects-Nr. 144731 and 176978, at the University of Bern (2014 – ongoing).

I would also like to thank the Royal Society Archives for providing me with copies of the documents used for this research.

Bibliography

Manuscripts

Royal Society, Archives of the Royal Society, London, Classified papers (Cl.P.):

Vol. XIX "Inquiries and Answers, December 24, 1662–May 25, 1692."

Vol. XIX/19/93: "Questions to be asked to the most ingenious of each County in my Travels through England" by R[obert] P[lot].

54 For example, see the surveys of the Economic society of Bern (*Oekonomische Gesellschaft Bern*): Gerrendina Gerber-Visser, *Die Ressourcen des Landes: Der ökonomisch-patriotische Blick in den Topographischen Beschreibungen der Oekonomischen Gesellschaft Bern (1759–1855)* (Baden: hier + jetzt, 2012).

Vol. XIX/19/94: "Enquiries to be propounded to the most Ingenious of each County in my Travels through England and Wales, in order to their History of Nature and Arts" by R[obert] P[lot].

Printed Sources

"[An Account of Four Books]." *Philosophical Transactions* 13, no. 143–154 (1683): 259–74.

Bacon, Francis. *New Atlantis: A Work Unfinished*. London: Printed by Tho. Newcomb, 1658.

Bacon, Francis. *The Works of Francis Bacon*, edited by James Spedding, Robert Leslie Ellis and Douglas Denon Heath. 14 vols. London: Longman [etc.], 1857–1874.

Bacon, Francis. *Nova Atlantis*. Library of Latin Texts. Turnhout: Brepols Publishers, 2014.

Bernardino de Sahagún, *Florentine Codex: General History of the Things of New Spain*, translated from the Aztec into English, with notes and ill. by Arthur James Outram Anderson, and Charles E. Dibble, 1. paperback ed., Monographs of the School of American Research 14. Salt Lake City: University of Utah Press, 2012.

Birch, Thomas. *The History of the Royal Society of London for Improving of Natural Knowledge: From Its First Rise, in Which the Most Considerable of Those Papers Communicated to the Society, Which Have Hitherto Not Been Published, are Inferted in Their Proper Order, as a Supplement to the Philosophical Transactions* [Facsimile]. Bruxelles: Culture et civilisation, 1967–1968, 4 vols. (original edition: London, 1756–1757).

Boate, Arnold. "An Interrogatory Relating more particularly to the Husbandry and Natural History of Ireland." In *Samuel Hartlib His Legacie, Or, an Enlargement of the Discourse of Husbandry Used in Brabant & Flaunders: Wherein Are Bequeathed to the Common-Wealth of England More Outlandish and Domestick Experiments and Secrets in Reference to Universall* Husbandry, edited by Samuel Hartlib, Cressy Dymock, Robert Child, and Richard Weston, The 2nd ed., augmented with an appendix (London: Printed by R. & W. Leybourn for Richard Wodenothe, 1652), Appendix, without pagination.

Boate, Gerard. *Irelands Naturall History: Being a True and Ample Description of Its Situation, Greatness, Shape, and Nature, of Its Hills, Woods, Heaths, Bogs* [...] *With Its Heads or Promontories, Harbours, Roades, and Bayes* [...] *Of Its Metalls, Mineralls* [...] *Turf* [...] *And Lastly, of the Nature and Temperature of Its Air and Season, and What Diseases It Is Free From, or Subject Unto*. London: For John Wright, 1652.

Boscani Leoni, Simona, ed. *"Lettres des Grisons": Wissenschaft, Religion und Diplomatie in der Korrespondenz von Johann Jakob Scheuchzer. Eine Edition ausgewählter Schweizer Briefe (1695–1731)*, https://hallernet.org/edition/scheuchzer-korrespondenz (accessed on 20 April 2021).

Boscani Leoni, Simona, ed. *"Unglaubliche Bergwunder": Johann Jakob Scheuchzer und Graubünden. Ausgewählte Briefe 1699–1707*. Cultura alpina Band 9. Chur: Verlag Bündner Monatsblatt, 2019.

Boyle, Robert. "General Heads for a Natural History of a Countrey, Great or Small, Imparted Likewise by Mr. Boyle." *Philosophical Transactions* 1, no. 11 (1665): 186–89.

Boyle, Robert. "Other Inquiries Concerning the Sea." *Philosophical Transactions* 1, no. 18 (1665): 315–16.

Boyle, Robert. *General Heads for the Natural History of a Country Great or Small: Drawn Out for the Use of Travellers and Navigators / Imparted by* [...] *Robert Boyle* [...]; *to Which Is Added, Other Directions for Navigators, Etc. With Particular Observations of the Most Noted Countries in the World; by Another Hand*. London: Printed for John Taylor [...] and S. Hedford, 1692.

Boyle, Robert. *The Works of Robert Boyle*. Edited by Hunter, Michael Cyril William and Edward B. Davis. London: Pickering & Chatto, 1999.

Collini, Silvia, and Antonella Vannoni, eds. *Les instructions scientifiques pour les voyageurs (XVII*e*–XIX*e *siècle)*. Paris: L'Harmattan, 2005.

Hall, Alfred Rupert, and Marie Boas Hall, eds. *The Correspondence of Henry Oldenburg*. 13 vols. Philadelphia, Pa.: Taylor & Francis, 1965–1986.

"Inquiries for Suratte, and Other Parts of the East-Indies; Inquiries for Persia. Inquiries for Virginia and the Bermudas. For Guaiana and Brasil." *Philosophical Transactions* 2, no. 23 (1666): 415–22.

"Inquiries of Directions for the Ant-Iles, or Caribe-Islands." *Philosophical Transactions* 3, no. 33 (1668): 634–39.

Plot, Robert. *The Natural History of Oxford-Shire: Being an Essay Toward the Natural History of England*. Oxford: At Mr. S. Millers, 1677.

Plot, Robert. *The Natural History of Stafford-Shire*. Oxford: Printed at the Theater, 1686.

Rooke, Laurence. "Directions for Sea-Men, Bound for Far Voyages." *Philosophical Transactions* 1, no. 8 (1665–1666): 140–43.

Solano, Francisco de, ed. *Cuestionarios para la formación de las Relaciones Geográficas de Indias: siglos XVI/XIX*. V centenario del descubrimiento de América 25. Madrid: Consejo Superior de Investigaciones Científicas, 1988.

Scheuchzer, Johann Jakob. *Einladungs-Brief, zu Erforschung natürlicher Wunderen, so sich im Schweizer-Land Befinden*. Zurich: [D. Gessner], 1699. In Boscani Leoni, *"Unglaubliche Bergwunder": Johann Jakob Scheuchzer und Graubünden. Ausgewählte Briefe 1699–1707*, 33–49.

Scheuchzer, Johann Jakob. *Charta invitatoria, quaestionibus, quae historiam Helvetiae naturalem concernunt, praefixa*. Tiguri: [D. Gessner], 1699.

Scheuchzer, Johann Jakob. *Bibliotheca scriptorum historiae naturalis omnium Terrae Regionum inservientium; historiae naturalis Helvetiae Prodromus; accessit Celeberrimi*

Viri Jacobi Le Long, Bibliothecarii Oratoriani de Scriptoribus historiae naturalis Galliae; collegit Johann Jakob Scheuchzer, Med. D. Math. Prof. Societ. Leopoldino-Carolinae, Regiae Angl. & Boruss. Tiguri: Typis Henrici Bodmeri, 1716.

Scheuchzer, Johann Jakob, and Johann Rudolf Lavater. *Historiae helveticae naturalis prolegomena publicae Eruditorum Syzētēsei subiecta praeside I. Iacobo Scheuchzero [...], Respondente Iohanne Rodolfo Lavatero [...] In Auditorio Aestivo H.L.Q.S. Die Septembr.* Tiguri: Gessner, 1700.

Luzzini, Francesco, ed. *Theory, Practice, and Nature In-between. Antonio Vallisneri's Primi Itineris Specimen.* Berlin: Edition Open Access/Max Planck Institute for the History of Science, 2018. http://www.edition-open-sources.org/sources/9/; accessed on 20 April 2021.

Vallisneri, Antonio. "Estratto d'alcune notizie intorno alla Provincia della Garfagnana, cavate dal primo Viaggio Montano del Sig. Antonio Vallisneri, [...] dal Sig. Dottore Giovanbatista Perrucchini, e da lui indirizzato in una lettera al Sig. Lodovico da Riva." *Supplementi al Giornale de' Letterati d'Italia* II (1722, Articolo VII): 270–312.

Vallisneri, Antonio. "Continuazione dell'estratto d'alcune notizie intorno alla Garfagnana cavate dal primo Viaggio Montano del Sig. Antonio Vallisneri." *Supplemento al Giornale de' Letterati d'Italia* III (1726, Articolo VIII): 376–428.

Wagner, Johann Jacob. *Historia naturalis Helvetiae curiosa.* Tiguri: Joh. Henrici Lindinneri, 1680.

Woodward, John. *An Essay Toward a Natural History of the Earth and Terrestrial Bodies, Especially Minerals As Also of the Sea, Rivers, and Springs With an Account of the Universal Deluge And of the Effects That It Had Upon the Earth.* Early. London: Printed for Richard Wilkin, 1695.

Woodward, John. *Brief Instructions for Making Observations, and Collections in Order to the Promotion of Natural History, in All Parts of the World: As Also, for Collecting, Preserving, and Sending over Natural Things.* London: Printed for Richard Wilkin, 1696.

Literature

Álvarez Peláez, Raquel. "El Cuestionario de 1577. La 'Instruccion y memoria de las relaciones que se han de hacer para la descripcion de las Indias' de 1577." In Solano, *Cuestionarios para la formación de las relaciones geográficas de Indias*, XCIII–CVI.

Álvarez Peláez, Raquel. *La conquista de la naturaleza americana.* Cuadernos Galileo de historia de la ciencia 14. Madrid: Consejo Superior de Investigaciones Científicas, 1993.

Barrera, Antonio. "Empire and Knowledge: Reporting from the New World." *Colonial Latin American Review* 15, no. 1 (June, 2006): 39–54.

Barrera-Osorio, Antonio. *Experiencing Nature: The Spanish American Empire and the Early Scientific Revolution.* 1st ed. Austin: University of Texas Press, 2006.

Barrera-Osorio, Antonio. "Nature and Experience in the New World: Spain and England in the Making of the New Science." In *Más allá de la Leyenda Negra: España y la revolución científica = Beyond the Black Legend; Spain and the Scientific Revolution*, edited by Víctor Navarro Brotons and William Eamon, 121–35. Valencia: Universitat de Valencia, 2007.

Barrera-Osorio, Antonio. "Empiricism in the Spanish Atlantic World." In *Science and Empire in the Atlantic World*, edited by James Delbourgo and Nicholas Dew, 177–202. New York: Routledge, 2008.

Barrera-Osorio, Antonio. "Knowledge and Empiricism in the Sixteenth-Century Spanish Atlantic World." In *Science in the Spanish and Portuguese Empires, 1500–1800*, edited by Daniela Bleichmar, Paula De Vos, Kristin Huffine, and Kevin Sheehan, 219–32. Stanford: Stanford University Press, 2009.

Benavides, Carmen Millán de. *Epítome de la Conquista del Nuevo Reino de Granada: la cosmogonía española del siglo XVI y el conocimiento por cuestionario.* Bogotá: Pontificia Universidad Javeriana, Instituto de Estudios Sociales y Culturales Pensar, 2001.

Blanckaert, Claude, ed. *Le terrain des sciences humaines: instructions et enquêtes (XVIIIe–XXe siècle).* Histoire des sciences humaines. Paris: L'Harmattan, 1996.

Bordat, Josef. "Bacons Atlantis-Mythos und das Selbstverständnis der modernen Wissenschaft." In *Mythos – Helden – Symbole Legitimation, Selbst- und Fremdwahrnehmung in der Geschichte der Naturwissenschaften der Medizin und der Technik*, edited by Siegfried Bodenmann and Susan Splinter, 121–30. Munich: Martin Meidenbauer, 2009.

Boscani Leoni, Simona. "Centri e periferie. Alcune riflessioni sulla corrispondenza erudita tra Sei e Settecento." *Schweizerische Zeitschrift für Geschichte* 55, no. 4 (2005): 441–47.

Boscani Leoni, Simona. "Men of Exchange: Creation and Circulation of Knowledge in the Swiss Republics of Eighteenth-Century." In *Scholars in Action: The Practice of Knowledge and the Figure of the Savant in the 18th Century*, edited by André Holenstein, Hubert Steinke, and Martin Stuber. 2 vols., 507–33. Leiden: Brill, 2013.

Boscani Leoni, Simona. "Queries and Questionnaires: Collecting Local and Popular Knowledge in 17th and 18th Century Europe." In *Wissenschaftsgeschichte und Geschichte des Wissens im Dialog – Connecting Science and Knowledge*, edited by Kaspar von Greyerz, Silvia Flubacher, and Philipp Senn, 187–207. Göttingen: V&R unipress, 2013.

Bossi, Maurizio, and Claudio Greppi, eds. *Viaggi e scienza: le istruzioni scientifiche per i viaggiatori nei secoli XVII–XIX.* Gabinetto scientifico letterario G.P. Vieusseux 13. Florence: Olschki, 2005.

Boumediene, Samir. *La colonisation du savoir une histoire des plantes médicinales du "Nouveau Monde" (1492–1750)*. Vaulx-en-Velin: Les éditions des mondes à faire, 2016.

Bourguet, Marie-Noëlle. "Voyages lointains et mesure du monde: la projection du regard européen sur le monde (vers 1500-vers 1800)." In *L'Europe des sciences et des techniques XV*e*–XVIII*e *siècle: un dialogue des savoirs*, edited by Liliane Hilaire-Pérez, Fabien Simon, and Marie Thébaud-Sorger, 193–201. Rennes: Presses universitaires de Rennes, 2016.

Brendecke, Arndt. *Imperium und Empirie: Funktionen des Wissens in der spanischen Kolonialherrschaft*. Cologne: Böhlau, 2009.

Brendecke, Arndt. "Informing the Council. Central Institutions and Local Knowledge in the Spanish Empire." In *Empowering Interactions: Political Cultures and the Emergence of the State in Europe 1300–1900*, edited by Willem Pieter Blockmans, André Holenstein, and Jon Mathieu, 235–52. Farnham: Ashgate, 2009.

Brockway, Lucile H. *Science and Colonial Expansion: The Role of the British Royal Botanic Gardens*. 2nd ed. (1st ed. 1979). New Haven, London: Yale University Press, 2002.

Burke, Peter. *The Historical Anthropology of Early Modern Italy: Essays on Perception and Communication*. Cambridge, New York: Cambridge University Press, 1987.

Burke, Peter. *A Social History of Knowledge: From Gutenberg to Diderot*. Cambridge: Blackwell Publishers, 2000.

Bustamante García, Jesús. "Francisco Hernández, Plinio del Nuevo Mundo: tradición clásica, teoría nominal y sistema terminológico indígena en una obra renacentista." In *Entre dos mundos. Fronteras culturales y agentes mediadores*, edited by Berta Ares Queija and Serge Gruzinski, 243–68. Sevilla: Escuela de Estudios Hispano-Americanos, 1997.

Bustamante García, Jesús. "La empresa naturalista de Felipe II y la primera expedición cientifica en suelo americano. La creación del modelo expedicionario renacentista." In *Felipe II (1527–1598): Europa y la monarquía católica*. Vol. 4, edited by José Martinez Millán, 39–59. Madrid: Editorial Parteluz, 1998.

Cañizares-Esguerra, Jorge. "Spanish America: From Baroque to Modern Colonial Science." In Porter, *The Cambridge History of Science: Eighteenth-Century Science*, 718–40.

Cañizares-Esguerra, Jorge. "Iberian Science in the Renaissance: Ignored How Much Longer?" *Perspectives on Science* 12, no. 1 (2004): 86–124.

Cañizares-Esguerra, Jorge. *Nature, Empire, and Nation: Explorations of the History of Science in the Iberian World*. Stanford, California: Stanford University Press, 2006.

Carey, Daniel. "Compiling Nature's History: Travellers and Travel Narratives in the Early Royal Society." *Annals of Science* 54, no. 3 (1997): 269–92.

Carey, Daniel. "Inquiries, Heads, and Directions: Orienting Early Modern Travel." In *Travel Narratives, the New Science and Literary Discourse: 1569–1750*, edited by Judy A. Hayden, 25–52. Farnham: Ashgate, 2012.

Cavazza, Marta. *Settecento inquieto: alle origini dell'Istituto delle Scienze di Bologna.* Bologna: Il Mulino, 1990.

Chakrabarty, Dipesh. *Provincializing Europe: Postcolonial Thought and Historical Difference.* Princeton studies in culture, power, history. Princeton: Princeton University Press, 2000.

Clericuzio, Antonio. "Le accademie scientifiche del Seicento." In *Il contributo italiano alla storia del pensiero. Scienze,* edited by Antonio Clericuzio and Saverio Ricci, 38–45. Roma: Istituto della Enciclopedia Italiana fondata da Giovanni Treccani, 2013.

Cline, Howard F. "The Relaciones Geográficas of the Spanish Indies, 1577–8." *Hispanic American Historical Review* 44 (1964): 341–74.

Cline, Howard F. "The Relaciones Geográficas of Spain, New Spain, and the Spanish Indies: An Annotated Bibliography." In *Handbook of Middle American Indians,* edited by Howard F. Cline and Robert Wauchope, 370–95. Austin: University of Texas Press, 1972.

Cohen, Ierome Bernard, ed. *Puritanism and the Rise of Modern Science: The Merton Thesis.* New Brunswick: Rutgers University Press, 1990.

Cooper, Alix. *Inventing the Indigenous: Local Knowledge and Natural History in Early Modern Europe.* Cambridge, London: Cambridge University Press, 2007.

Cooper, Alix. "Fragen ohne Antworten: Die Suche nach lokalen Informationen in der frühen Aufklärung." In *Vor Google: Eine Mediengeschichte der Suchmaschine im analogen Zeitalter,* edited by Thomas Brandstetter, 73–83. Kultur- und Medientheorie. Bielefeld: transcript, 2012.

Davis, Edward B., Winship Michael P. "Early Modern Protestantism." In *Science and Religion: A Historical Introduction,* edited by Gary B. Ferngren, 117–29. Baltimore: Johns Hopkins University Press, 2002.

Eyles, Victor Ambrose. "Woodward, John." In Gillispie, *Complete Dictionary of Scientific Biography,* vol. 14, 500–03.

Fan, Fa-ti. "The Global Turn in the History of Science." *East Asian Science, Technology and Society: An International Journal* 6 (2012): 249–58.

Fox, Adam. "Printed Questionnaires, Research Networks, and the Discovery of the British Isles, 1650–1800." *The Historical Journal* 53, no. 3 (2010): 593–621.

Gavroglu, Kostas, Manolis Patiniotis, Faidra Papanelopoulou, Ana Simões, Ana Carneiro, Maria Paula Diogo, José Ramón Bertomeu Sánchez, Antonio García Belmar, and Agustí Nieto-Galan. "Science and Technology in the European Periphery: Some Historiographical Reflections." *History of Science* 46, no. 2 (2008): 153–75.

Generali, Dario. *Antonio Vallisneri: gli anni della formazione e le prime ricerche.* Biblioteca dell'edizione nazionale delle opere di Antonio Vallisneri 2. Florence: Olschki, 2007.

Gerber-Visser, Gerrendina. *Die Ressourcen des Landes: Der ökonomisch-patriotische Blick in den Topographischen Beschreibungen der Oekonomischen Gesellschaft Bern (1759–1855)*. Archiv des Historischen Vereins des Kantons Bern 89. Baden: hier + jetzt, 2012.

Gillispie, Charles Coulston, ed. *Complete Dictionary of Scientific Biography*. 18 vols. Detroit: Charles Scribner's Sons, 2008.

Guerrini, Luigi. "Piante e animali del Nuovo Mondo. Federico Cesi e il *Tesoro messicano*." In *Le scienze*, edited by Antonio Clericuzio and Germana Ernst. 12 vols., 197–213. Il rinascimento italiano e l'Europa 5. Treviso: Colla, 2008.

Hunter, Michael. "Robert Boyle's 'Heads' and 'Inquiries'." *Robert Boyle Project, Occasional Papers* no. 1 (2005). URL: http://www.bbk.ac.uk/boyle, accessed on 20 April 2021.

Hunter, Michael. "Robert Boyle and the Early Royal Society: A Reciprocal Exchange in the Making of Baconian Science." *The British Journal for the History of Science*, 40 (2007): 1–23.

Iliffe, Rob. "Science and Voyages of Discovery." In Porter, *The Cambridge History of Science: Eighteenth-Century Science*, 618–46.

Keller, A. G. "Plot, Robert." In Gillispie, *Complete Dictionary of Scientific Biography*, vol. 11, 40–41.

Kempe, Michael. *Wissenschaft, Theologie, Aufklärung: Johann Jakob Scheuchzer (1672–1733) und die Sintfluttheorie*. Frühneuzeit-Forschungen 10, edited by Peter Blickle, Richard van Dülmen, Heinz Schilling, and Winfried Schulze. Epfendorf: bibliotheca academica Verlag, 2003.

León Portilla, Miguel. *Bernardino de Sahagún*. Protagonistas de América. Madrid: Historia 16, 1987.

Livingstone, David N. *Putting Science in Its Place: Geographies of Scientific Knowledge*. Chicago, London: The University of Chicago Press, 2003.

López Austin, Alfredo. "Estudio acerca del método de investigación de fray Bernardino de Sahagún." In *La investigación social de campo en México*, edited by Jorge Martínez Ríos, 9–56. México: Universidad Nacional Autónoma, 1976.

López Austin, Alfredo. "Estudio acerca del método de investigación de fray Bernardino de Sahagún." *Estudios de Cultura Náhuatl* 42 (2011): 353–400.

López Piñero, José M. "Sahagún, Bernardino de." In *Diccionario Histórico de la Ciencia Moderna en España*. Vol. 2, edited by José M. López Piñero, 281–82. Historia, ciencia, sociedad 181. Barcelona: Ed. Península, 1983.

Luzzini, Francesco. *Il miracolo inutile: Antonio Vallisneri e le scienze della terra in Europa tra XVII e XVIII secolo*. Biblioteca dell'edizione nazionale delle opere di Antonio Vallisneri 9. Florence: Olschki, 2013.

Mazzone, Umberto and Angelo Turchini, eds. *Le visite pastorali: analisi di una fonte*. Annali dell'Istituto storico italo-germanico. Quaderno. Bologna: Il Mulino, 1991.

Mundy, Barbara E. *The Mapping of New Spain: Indigenous Cartography and the Maps of the Relaciones Geográficas*. Chicago, London: The University of Chicago Press, 1996.

Ogilvie, Brian W. *The Science of Describing. Natural History in Renaissance Europe*. Chicago, London: The University of Chicago Press, 2006.

Ponce Leiva, Pilar. "Las Ordenanzas sobre Descripciones (1573). Su aplicación en la Real Audiencia de Quito." In Solano, *Cuestionarios para la formación de las relaciones geográficas de Indias*, LXXIX–XCI.

Porter, Roy, ed. *The Cambridge History of Science: Eighteenth-Century Science* 4. Cambridge, New York: Cambridge University Press, 2003.

Raj, Kapil. "Relocations." In *Science and Technology Studies: Critical Concepts in the Social Sciences*. Vol. 4, edited by Michael Lynch, 3–11. London: Routledge, 2012.

Raj, Kapil. "Quand l'Amérique a inventé la science européenne." In *Europa: une autre histoire*, edited by Jakob Vogel, Thomas Serrier, and Etienne François, 1089–1102. Paris: Les Arènes, 2017.

Raj, Kapil. "Thinking Without the Scientific Revolution: Global Interactions and the Construction of Knowledge." *Journal of Early Modern History* 21 (2017): 1–14.

Roberts, Lissa. "Situating Science in Global History: Local Exchanges and Networks of Circulation." *Itinerario* 33, no. 1 (2009): 9–30.

Rubiés, Joan-Pau. "Instructions for Travellers: Teaching the Eye to See." *History and Anthropology* 9, no. 2–3 (1996): 139–90.

Schiebinger, Londa, and Claudia Swan, eds. *Colonial Botany: Science, Commerce, and Politics in the Early Modern World*. Philadelphia: University of Pennsylvania Press, 2005.

Shapin, Steven. "The House of Experiment in Seventeenth-Century England." *Isis* 79, (1988): 373–404.

Shapin, Steven. *A Social History of Truth: Civility and Science in Seventeenth-Century England*. Chicago, London: The University of Chicago Press, 1994.

Shapiro, Barbara J. "History and Natural History in Sixteenth and Seventeenth-Century England: An Essay on the Relationship Between Humanism and Science." In *English Scientific Virtuosi in the 16th and 17th Centuries: Papers Read at a Clark Library Seminar, 5 February 1977*, edited by Barbara J. Shapiro and Robert G. Frank, 3–55. Los Angeles: University of California, William Andrews Clark Memorial Library, 1979.

Shapiro, Barbara J. *A Culture of Fact: England, 1550–1720*. Ithaca, London: Cornell University Press, 2000.

Sigrist, René, Eric Widmer Eric, and Wladimir Berelowitsch. "Les lieux des sciences dans l'Europe moderne." In *Lieux d'Europe: mythes et limites*, edited by Stella Ghervas and François Rosset, 45–64. Paris: Ed. de la Maison des sciences de l'homme, 2008.

Slack, Paul. "Government and Information in Seventeenth-Century England." *Past and Present* 184 (2004): 33–68.

Stagl, Justin. "Vom Dialog zum Fragebogen: Miszellen zur Geschichte der Umfrage." *Kölner Zeitschrift für Soziologie und Sozialpsychologie* 31 (1979): 611–38.

Steiger, Rudolf. "Verzeichnis des Wissenschaftlichen Nachlasses von Johann Jakob Scheuchzer." *Beiblatt zur Vierteljahresschrift der Naturforschenden Gesellschaft in Zürich* 78, no. 21 (1933): 1–75.

Turner, A. J. "Plot, Robert (Bap. 1640, D. 1696), Published Online: 23 September 2004." In *The Oxford Dictionary of National Biography*. Edited by H. C. G. Matthew and B. Harrison. Oxford: Oxford University Press, 2004. https://doi.org/10.1093/ref:odnb/22385, accessed on 23 September 2019.

Vicente Castro, Florencio. *Bernardino de Sahagún primer antropólogo en Nueva España, (siglo XVI)*. Acta Salmanticensia Filosofía y letras, edited by José-Luis Rodríguez Molinero. Salamanca: Ed. Universidad de Salamanca, Institucíon "Fray Bernardino de Sahagún", Excma. Diputación provincial de León, 1986.

Webster, Charles. *The Great Instauration: Science, Medicine and Reform, 1626–1660*. 2nd ed., 1st ed. 1975. Studies in the History of Medicine. Oxford: Peter Lang, 2002.

Withers, Charles W. J. *Geography, Science, and National Identity: Scotland Since 1520*. Cambridge Studies in Historical Geography 33. Cambridge, New York: Cambridge University Press, 2001.

Withers, Charles W. J. *Placing the Enlightenment: Thinking Geographically About the Age of Reason*. Chicago, London: The University of Chicago Press, 2007.

Yale, Elisabeth. "Making Lists: Social and Material Technologies in the Making of Seventeenth-Century British Natural History." In *Ways of Making and Knowing: The Material Culture of Empirical Knowledge*, edited by Pamela H. Smith, Amy R. W. Meyers, and Harold J. Cook, 280–301. Ann Arbor: University of Michigan Press, 2014.

CHAPTER 3

(Re-)Shaping a Method: Field Research and Experimental Legacy in Vallisneri's *Primi Itineris Specimen* (1705)

Francesco Luzzini

1 Chasing the Manuscript

So wrote an excited Antonio Vallisneri to a friend of his, the physician Flaminio Corghi (16?–17?), in a letter dated June 1705:

> My Alpine journey will probably come out this year and – for my part – I consider it most interesting, given the new medical and physical things which I have observed in those mountains. It will be [written] in Plinian Latin, and first will be submitted to the eyes of the [reviewer], and then it will be published. In light of the furious work I have done, you will see that my Latin style has changed completely.[1]

The work had been "furious" indeed, considering what we now know about this matter. And such was Vallisneri's enthusiasm for the content of his manuscript that, in early February of the same year, he had sent a copy of it to Sir Hans Sloane (1660–1753), secretary of the Royal Society of London and editor of the *Philosophical Transactions*: he was hoping that this report (entitled *Primi Itineris per Montes Specimen Physico-Medicum*, "Physico-Medical Example of a First Journey through the Mountains") would be published in that august journal, allowing him to thrust himself into the forefront of the European Republic of Letters.

Alas, events went differently. The manuscript was never published, and the reasons for this missed opportunity are shrouded in speculation – although both the considerable length of the text and the fact that it was written in Latin

1 Antonio Vallisneri, *Epistolario, vol. I (1679–1710)*, ed. Dario Generali (Milan: Franco Angeli, 1991), 322: "Uscirà forse quest'anno il mio viaggio alpino, che per me stimo curiosissimo, per le cose mediche e fisiche osservate di nuovo in quelle Alpi. Sarà in latino pliniano, e passerà prima sotto gli occhi del Davino, e poi uscirà. Vedrete mutata affatto la mia maniera latina, per lo studio rabbioso che vi ho fatto." My translation here and throughout.

are the most plausible culprits. In fact, by the early eighteenth century the Royal Society tended to favour the publication of English-language papers. It is perhaps not by chance that in a letter dated March 9, 1710, Vallisneri declared to Sloane that his works were for the most part "written in the Italian language."[2] Presumably, Sloane had asked him for an English (or, at least, a shorter Latin) version of the manuscript, which he could – or would – not provide.[3]

Whatever the causes, the consequences of this unfortunate setback turned out to be unmistakably negative. The missed publication doomed the manuscript to oblivion, and the document disappeared from sight. Supposedly, this precious paper still lingers in the hallways of the Archives of the Royal Society of London, although there is no trace of it in the collections database. Thus, for a long time the only known and most complete surviving records of Vallisneri's journey were two Italian summaries of the original version, the *Estratti* ("Extracts"), published in 1722 and 1726 in the *Supplementi al Giornale de' Letterati d'Italia*.[4] Even in this condensed form, however, the mass of information provided by the author was such that it was no surprise that the original text had been deemed by him worthy of international prestige. The amazingly eclectic content of the *Estratti* encompassed the whole range of natural sciences, including topics such as stratigraphy, mineralogy, petrography, geomorphology, palaeontology, hydrogeology, geography, technology, mining meteorology, medicine, chemistry, botany, and biology. Not to

2 Vallisneri, *Epistolario, vol. I (1679–1710)*, 503.

3 Vallisneri's strong advocacy of the use of the Italian language by all Italian scholars (even when addressing the international audience) may have played a role in his hesitation to comply with Sloane's request. This ideal was publicly upheld in: Antonio Vallisneri, "Che ogni Italiano debba scrivere in Lingua purgata Italiana, o Toscana," *Supplementi al Giornale de' Letterati d'Italia* I (1722). This work is now published in: Antonio Vallisneri, *Che ogni Italiano debba scrivere in Lingua purgata Italiana*, ed. Dario Generali (Florence: Olschki, 2013). On this topic, see: Dario Generali, "Il testo tra comunicazione scientifica e strategie editoriali. Il caso di Antonio Vallisneri," in *Edizioni e traduzioni di testi filosofici*, ed. Marialuisa Baldi and Barbara Faes de Mottoni (Milan: Franco Angeli, 2006); Generali, *Antonio Vallisneri: gli anni della formazione e le prime ricerche* (Florence: Olschki, 2007), 384–86; Generali, "La Repubblica delle Lettere e i carteggi scientifici: il caso di A. Vallisneri," in *La Repubblica delle Lettere, il Settecento italiano e la Scuola del secolo XXI*, ed. Andrea Battistini, Claudio Griggio, and Renzo Rabboni (Pisa, Rome: Fabrizio Serra Editore, 2011); Francesco Luzzini, *Il miracolo inutile: Antonio Vallisneri e le scienze della Terra in Europa tra XVII e XVIII secolo* (Florence: Olschki, 2013), 217–226; Rhoda Rappaport, *When Geologists were Historians, 1665–1760* (Ithaca: Cornell University Press, 1997), 218–219.

4 Antonio Vallisneri, "Estratto d'alcune Notizie intorno alla Provincia della Garfagnana, cavate dal primo Viaggio Montano del Sig. Antonio Vallisnieri," *Supplementi al Giornale de' Letterati d'Italia* II (1722); Vallisneri, "Continuazione dell'Estratto d'alcune Notizie intorno alla Garfagnana," *Supplementi al Giornale de' Letterati d'Italia* III (1726).

mention, of course, the many humanistic forays which were a typical feature in texts written by early modern savants, and which embraced philosophy, history, literature, archaeology, religion, anthropology, and even material culture and folklore. In brief, there were few doubts that finding the original document would allow us both to recreate Vallisneri's travel, and to appreciate its historical and scientific value, with an accuracy unrivalled in the history of early modern naturalistic explorations.

This premonition proved to be true in 2009, when the draft copy of the *Primi Itineris Specimen* was fortuitously discovered by me and my mentor, Dario Generali, in the State Archive of Reggio Emilia (where the vast and still largely unexplored bulk of the author's unpublished writings – the Archivio Vallisneri, or "Vallisneri Archive" – is held).[5] Needless to say, this finding exceeded my best expectations. And, in working my way through the content of those faded papers (fig. 3.1) and in comparing it with that from the *Estratti*, I realized that an exhaustive study of Vallisneri's adventure would take much longer than I had imagined. But the result was definitely worth the effort: I succeeded in recreating the itinerary almost in its entirety, and by the summer of 2010 I was also able to replicate in person Vallisneri's journey and many of his observations and explorations.[6] The crowning moment of this scholarly endeavour arrived in 2015, when the manuscript became the source text for the critical edition work that I carried out at the University of Oklahoma and at the Max Planck Institute in Berlin for the Edition Open Sources Project.[7]

5 Antonio Vallisneri, *Primi Itineris per Montes Specimen Physico-Medicum* (Manuscript, 1705).

6 See: Francesco Luzzini, "La Tana che urla: cenni di speleologia vallisneriana," in *Antonio Vallisneri: la figura, il contesto, le immagini storiografiche*, ed. Dario Generali (Florence: Olschki, 2008); Luzzini, "Cavità naturali ed artificiali in Garfagnana e Lunigiana. Le esplorazioni di Antonio Vallisneri," in *Antonio Vallisneri dalla Garfagnana alla Scienza*, ed. Pietro Rocchi (Lucca: Maria Pacini Fazzi Editore, 2010); Luzzini, "Multa curiosa. Vallisneri's early studies on Earth sciences," *Nuncius* 26, no. 2 (2011); Luzzini, *Il miracolo inutile*, 90–143; Luzzini, "An Uncomfortable, yet Wonderful Journey. Antonio Vallisneri and his Exploration of the Northern Apennines," in *Nel nome di Lazzaro. Saggi di storia della scienza e delle istituzioni scientifiche tra il XVII e il XVIII secolo*, ed. Centro Studi Lazzaro Spallanzani (Bologna: Edizioni Pendragon, 2014).

7 Francesco Luzzini, *Theory, Practice, and Nature In-between. Antonio Vallisneri's Primi Itineris Specimen* (Berlin: Edition Open Access/Max Planck Institute for the History of Science, 2018), http://www.edition-open-sources.org/sources/9/, accessed on 26 April 2021.

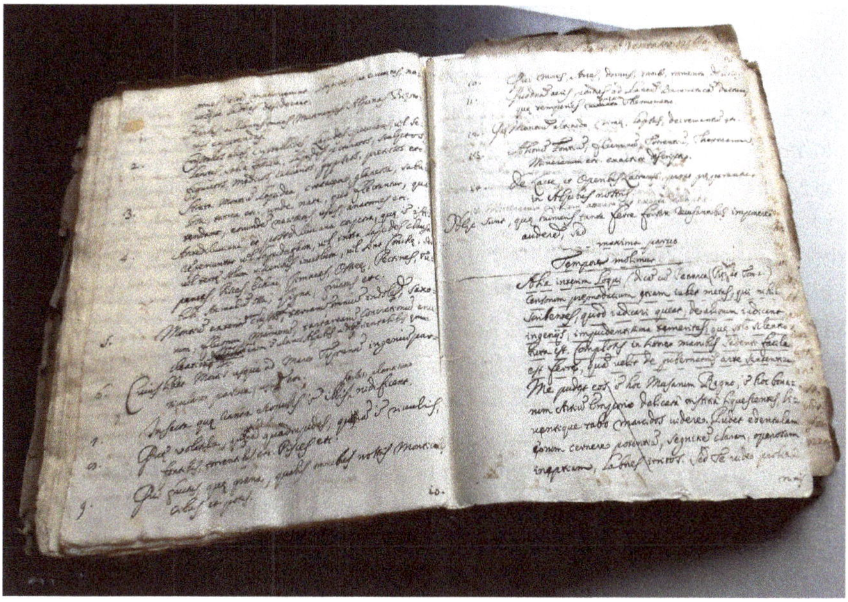

FIGURE 3.1 The draft copy of the "Primi Itineris Specimen". State Archive of Reggio Emilia, Italy, Archivio Vallisneri, 10, "Scritti, minute e appunti scientifici e letterari d'Antonio Vallisneri sr.," mazzo IV, 52–53

2 The Journey

As said above, Vallisneri's work had been "furious" – and not euphemistically: the zeal he had poured in the manuscript affected form and content alike. The seamlessly reworked draft (replete with deletions, slips of paper pasted here and there, marginal notes, endless corrections) is the result of an ornate and continuous stylistic effort; and the author's colourful and enthusiastic personality shines through in every inch of this rich text, his subtle and ubiquitous irony being particularly effective at catching the reader's attention.

The *Primi Itineris Specimen* is the report of a journey embarked on in the summer of 1704, when Vallisneri resolved to travel – with "daring soul," and "trembling foot" – across the "silent horrors" of the northern Apennines. And it is precisely in the description of these "horrors," and in the methods and tools used by the Paduan professor to penetrate their secrets, that the manuscript discloses its greatest value for the history of science. In fact, this document gives us crucial clues to understand a critical stage in the development of methodologies in field research: the moment when Italian experimentalism impacted early modern natural philosophy and spread across national and confessional boundaries, expanding its influence on different disciplines.

Vallisneri's route (fig. 3.2) stretched from north-northeast to south-southwest, covering a total distance of about 130 kilometres. From Scandiano, just south of the city of Reggio Emilia, he reached the Alpe di San Pellegrino ("Alp of Saint Peregrine"), one of the highest peaks in the northern Apennines. There he crossed the homonymous pass (which now links the Province of Modena with the Province of Lucca) and descended to Garfagnana. Then he headed west and climbed the valley of the Petrosciana Torrent (a tributary of the Serchio River), and finally reached the Apuan Alps, where he visited the iron mines of Fornovolasco and the cavern known as *Tana che urla* ("Screaming Cave"). It was in these remote places that he collected conclusive data to support a "great argument"[8] of his (which was also the main – though not the only – reason that had induced him to wander in the mountains): a theoretical system on the meteoric origin of springs, triumphantly and thoroughly explained in 1715 in the *Lezione Accademica intorno all'Origine delle Fontane* ("Academic Lecture on the Origin of Springs").[9] By upholding this theory with strong empirical evidence, in this treatise he confirmed measurements and observations

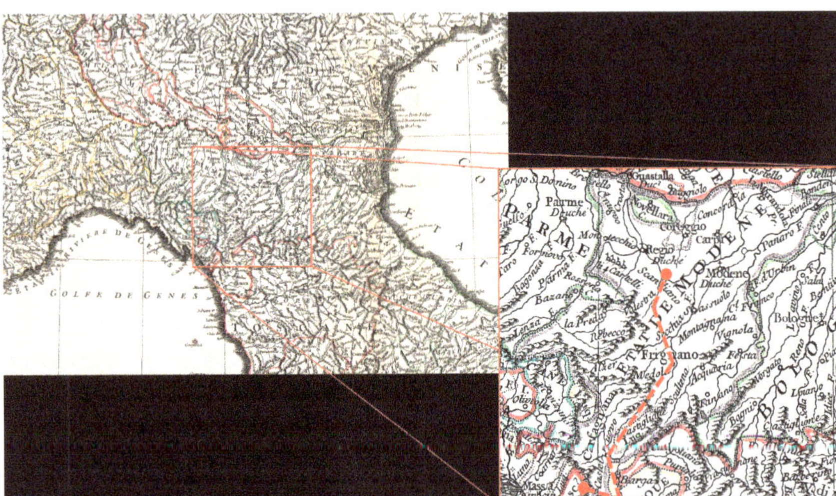

FIGURE 3.2 Northern Italy, the Duchy of Modena and Reggio, and Vallisneri's journey. Original map: "L'Italie: publiée sous les auspices de Monseigneur le Duc d'Orleans", 1743
COURTESY OF THE NORMAN B. LEVENTHAL MAP CENTER AT THE BOSTON PUBLIC LIBRARY, BOSTON (MA), USA

8 Vallisneri, *Epistolario, vol. I (1679–1710)*, 282.

9 Antonio Vallisneri, *Lezione Accademica intorno all'Origine delle Fontane* (Venezia: Appresso Gio. Gabbriello Ertz, 1715).

made previously by scholars, engineers, artisans and various experts all over Europe, and dealt a lethal blow to the competing and still far from unpopular theories of a compound origin of fresh water – like, for example, those which assumed the existence of hidden channels connecting the oceans to the earth and the exclusive or partial distillation and/or filtration of sea water through rock strata.[10]

Although unquestionably dominant, however, the origin of springs was not the only issue Vallisneri meant to discuss, given the countless number of natural phenomena that he had been studying over the years and during his journey. And from its very beginning, the *Primi Itineris Specimen* – with its astoundingly eclectic content – bears eloquent witness to the author's omnivorous curiosity.

The adventure begins among the gypsum layers and the close sulphur mines of Mount Gesso, a hill in the gypsum-sulphur formation of the northern Apennines:[11] in this "foul-smelling" cave, Vallisneri observes various kinds of sulphur and other minerals and rocks, and notes down the peculiar terms used by local miners ("canopi") to describe them.[12] Nor does he omit to discuss the therapeutic properties and uses of that place: indeed, it is an ascertained fact that the sulphur miners from Scandiano live "healthy, to the no small relief of the working people."[13]

After this first set of *physico-medical* observations, the route continues along the Tresinaro River, heads south and passes through the first slopes of the Emilian Apennines. Five miles past Scandiano, the author faces the barren landscape of the badlands: here, the "grim sight" of the eroded clay soil is enhanced by the "rude variety" of the many colours "wrapping and adorning all the slopes with black, reddish, ferruginous, sallow, white [hues]."[14] The

10 On this topic, see: Francesco Luzzini, *Il miracolo inutile*, 69–161; "Through Dark and Mysterious Paths. Early Modern Science and the Search for the Origin of Springs from the 16th to the 18th Centuries," *Earth Sciences History* 34, no. 2 (2015).

11 The gypsum layers of Mount Gesso are part of the gypsum-sulphur formation of the northern Apennines, whose evaporitic strata resulted from the Messinian salinity crisis which occurred in the late Miocene epoch (between 5.95 and 5.33 million years ago). During this epoch, a temporary closure of the Strait of Gibraltar made the Mediterranean Sea desiccate almost completely. This event originated the evaporitic rocks which are now visible in this region (http://www.edition-open-access.de/sources/9/8/index.html#16, accessed on 26 April 2021).

12 Luzzini, *Theory, Practice, and Nature In-between*, 152–53, http://www.edition-open-access.de/sources/9/8/index.html#20 (accessed 26 on April 2021).

13 Ibid., 153, http://www.edition-open-access.de/sources/9/8/index.html#21 (accessed on 26 April 2021).

14 Ibid., 162, http://www.edition-open-access.de/sources/9/8/index.html#54 (accessed on 26 April 2021).

gloominess of this picture is reinforced by another strange phenomenon on the other side of the river, the mysterious "salse": muddy mixtures of salt, water, carbon dioxide, clay, and hydrocarbons that leak from the ground as bubbling, smoking craters.[15] This "not unsightly source of pleasure" is studied with particular interest by Vallisneri, who is equally devoted to natural philosophy and medicine. He pays attention both to the oddities of these little volcanoes and to the medical properties of their mud, whose virtues he has been testing for years. Their waters, for example, are "a remedy for many diseases proceeding from viscous humours, especially from the cold ones"; and their "salty clay" dispel "tumors, [...] scabies, [...] and stagnant fluid," and is also "beneficial for nerves" and "oedematous legs."[16]

On the following day, the author reaches Mount Valestra and then arrives at the ancient and renowned baths of Quara (on the western side of the Dolo Creek). These mineral springs, whose fame dates back to the Roman age and in the fifteenth century were still widely and successfully used, now lie in ruins. Such a pity! Their waters – Vallisneri assures – can treat effectively a number of ailments such as "weak stomach, shortness of breath, flatulence, hypochondria, painful colics, sterility," and even "dizziness, the pain proceeding from slow, especially polypous blood, glutinous phlegm," and many others.[17] Sadly, none of these remarkable virtues are valued by moderns: and these ancient springs, once renowned all over Europe, are now clogged up with stones, mud, and sand.

The time has come to cross the Dolo Creek and head east. Vallisneri visits the village of Rubbiano, where he meets a family of surgeons known for healing viper bites. They all have a snake-like mark on their shoulders; and, especially in springtime (when the scar has "a brighter colour"), this mark portrays "the rough image of a nest of vipers."[18] After trying in vain to ascertain

15 A *salsa* is a phenomenon of secondary volcanism. It is a cold, muddy mixture composed of water, clay, carbon dioxide and hydrocarbons (usually methane and oil) leaking out from the ground. Once the mud reaches the surface, it dries and accumulates, forming little mud volcanoes a few meters tall. The gas leak from the surface is caused by slow and constant movements of the Earth's crust: these trigger the underground sacks in which the mixture is enclosed to open or to compress. The term "salsa" means "salty," as the mixture contains NaCl. Its salinity is equivalent to 1/2–1/3 of sea water (http://www.edition-open-access.de/sources/9/8/index.html#55, accessed on 26 April 2021).

16 Luzzini, *Theory, Practice, and Nature In-between*, 162–63, http://www.edition-open-access.de/sources/9/8/index.html#55 (accessed on 26 April 2021).

17 Ibid., 166–67, http://www.edition-open-access.de/sources/9/8/index.html#66 (accessed on 26 April 2021).

18 Ibid., 167–68, http://www.edition-open-access.de/sources/9/8/index.html#67 (accessed on 26 April 2021).

whether or not the image is artificial, the author resumes his journey and comes in sight of the springs of Vitriola, "provided by nature with gratuitous colouring properties."[19] These clear and tasteless waters colour the surrounding soil and the aquatic weeds with a "yellowish, ferruginous dye," and are used by the people of the countryside to "blacken linen clothes and wool" (but not without "some previous preparation" whose complex procedures are meticulously described in the manuscript).[20]

The path heads southwest, and finally – after "arduous efforts, and a rough journey" – the Alp of Saint Peregrine, with the homonymous Pass and the village of San Pellegrino in Alpe, comes into sight.[21] From this privileged viewpoint, a wealth of information is disclosed to Vallisneri's eyes. Although it is the "scorching month of August," winter still rages here "with snow and cold":[22] it is now that he reconsiders "the origin of springs and rivers from a higher [perspective]," and his mind follows the "immense mass of water [...] absorbed by underground streams through the darkness of [those] paths."[23] He realizes that the different nature and arrangement of rock layers is the key to understanding *where* and *how* springs emerge. And these considerations, in turn, serve as empirical premises to the next step: *why*. In this part of the manuscript, natural philosophy and field research come together in their most refined form, and the core of Vallisneri's theory on the hydrologic cycle is revealed and explained.

> From whence the mind is inclined to guess, why the waters hide themselves here, and appear there; why the perennial springs are uncommon here, and the course of rivers is more infrequent, while the both of them flow more abundantly in that [other place]. For this, I thought, is the only circulation of waters (in the bosom of these lands of ours, at least). From

19 Ibid., 168, http://www.edition-open-access.de/sources/9/8/index.html#68 (accessed on 26 April 2021).

20 Ibid., 168–69, http://www.edition-open-access.de/sources/9/8/index.html#68 (accessed on 26 April 2021). The village of Vitriola is located in an area delimited by the mountain ridge on which Montefiorino rests (on the west) and by the Dragone Creek (on the east). Arenaceous and calcareous rocks – typically turbidites – dating back to the Campanian, Maastrichtian, and Danian ages (Late Cretaceous-Early Paleocene, 83.5–61.6 Ma) dominate the lithology of this area. Most likely, the colouring properties of the springs described by Vallisneri result from high concentrations of iron oxides in the water.

21 Ibid., 170, http://www.edition-open-access.de/sources/9/8/index.html#73 (accessed on 26 April 2021).

22 Ibid., 170, http://www.edition-open-access.de/sources/9/8/index.html#73 (accessed on 26 April 2021).

23 Ibid., 171, http://www.edition-open-access.de/sources/9/8/index.html#75 (accessed on 26 April 2021).

> the sky to the earth, from the earth to the sea: and, in turn, from the sea to the sky, from the sky to the earth. That is to say, the cavernous mountains, and the thirsty lands absorb the waters pouring from the sky; [and these waters], flowing for the most part, and absorbed along the way, sink back to the sea through obscure paths. From there, they rise back to the clouds, which make them thin; and from the clouds they descend once more, in a perpetual circulation of the liquid element, whose operation never fails.[24]

After a description of many "crystals and crystal-like" minerals found in these mountains (all of them proving the existence of a "geometric design in nature, and of a somewhat indistinct vegetative power," caused "by an exhalation from the ground"),[25] the highest summit of the Apennines is crossed. Thereupon, brooks and torrents follow "an opposite course, as if the empire of the waters was divided,"[26] and descend to the Tyrrhenian Sea. At last, the Province of Garfagnana appears to the southwest.

Now natural philosophy merges with the "history of people,"[27] and an additional stack of nine papers in the manuscript features a long digression into history, literature, material culture, anthropology, and folklore. This province is bathed by many "perennial and clear torrents, rills, springs, and rivers," and abounds with "most excellent fishes" (trouts, it seems, are particularly delicious).[28] Its main stream, the Serchio River, runs into the sea about three miles from the estuary of the more famous Arno; and just like its noble neighbour, it is "swollen at times, and threatening."[29] The entire land is "rich in metals" and does not lack for "wheat, wine, hemp, fruits, vegetables, and fishes"; what is more, it flourishes with "plenty of meat, cheese, chestnuts," so that, it is 'sufficiently furnished with the former [goods]' and has "far more than enough

24 Ibid., 172, http://www.edition-open-access.de/sources/9/8/index.html#75 (accessed on 26 April 2021).

25 Ibid., 174, http://www.edition-open-access.de/sources/9/8/index.html#80 (accessed on 26 April 2021).

26 Ibid., 175, http://www.edition-open-access.de/sources/9/8/index.html#81 (accessed on 26 April 2021).

27 Ibid., 177, http://www.edition-open-access.de/sources/9/8/index.html#84 (accessed on 26 April 2021).

28 Ibid., 180, http://www.edition-open-access.de/sources/9/8/index.html#91 (accessed on 26 April 2021).

29 Ibid., 181, http://www.edition-open-access.de/sources/9/8/index.html#91 (accessed on 26 April 2021).

of the latter ones."[30] As to the inhabitants, the men are usually "short, and – for the most part – dark, muscular, strong, always ready to fight, easily inclined to anger, vengeful, mindful of injuries."[31] Yet they are also "smart, clever, friendly to strangers, lovers of hospitality, loyal to their lord, inclined to literature, naturally gifted with the most beautiful Tuscan language," and even "cheerful, lively, skilled in mechanics, and constantly engaged in commerce."[32] And although these people had suffered "under various lords" in the past, now they live happily "under the rule of the Most Serene House of Este."[33]

It is now time to return to the main path of natural philosophy. Vallisneri resumes his *physico-medical* account and describes his descent from the Alp of Saint Peregrine into Garfagnana, where he first reaches the town of Castiglione. After noticing "silvery pyrites from an underground copper and silver mine," and many other mineral and rock samples, he visits the thermal springs known as Bagno della Pieve[34] and comments on the "amazing properties" of these clear, salty, lukewarm, bituminous waters.[35] Then he descends to Camporgiano and moves on to Castelnuovo, the administrative centre of the province, and there he observes the ancient and half-ruined Torrite thermal baths.[36] The path continues south, following the course of the Serchio: once in Gallicano, Antonio heads west along a small tributary of this river (the Petrosciana Torrent) and climbs a steep valley, proceeding towards the western end of Garfagnana. There, among "such precipitous rocky ridges, and [...] high lands, and rough crags," he can see "strong and brawny men living long and happily," and "charming women" who surpass "even the urban Venuses in beauty and in gentle appearance"; and yet, they drink only "clearest water," and

30 Ibid., 182, http://www.edition-open-access.de/sources/9/8/index.html#98 (accessed on 26 April 2021).

31 Ibid., 182, http://www.edition-open-access.de/sources/9/8/index.html#99 (accessed on 26 April 2021).

32 Ibid.

33 Ibid., 184, http://www.edition-open-access.de/sources/9/8/index.html#107 (accessed on 26 April 2021).

34 Bagno della Pieve, a spa in the municipality of Pieve Fosciana (Province of Lucca) (http://www.edition-open-access.de/sources/9/8/index.html#138; accessed on 26 April 2021).

35 Luzzini, *Theory, Practice, and Nature In-between*, 186–188, http://www.edition-open-access.de/sources/9/8/index.html#137 (accessed on 26 April 2021).

36 Torrite thermal waters, an ancient spa in Castelnuovo di Garfagnana. The spring was located on the western side of the Serchio River, along the Turrite Secca Torrent. It disappeared in 1948, as a consequence of the construction of a nearby hydroelectric power plant (http://www.edition-open-access.de/sources/9/8/index.html#145; accessed on 26 April 2021).

fill the "growling stomach" with "rustic food."[37] The report lingers for a moment on the (quite intriguing) description of some peculiar habits of that people, who thrive in a land where "neither Minerva, nor Ceres, nor Bacchus" dispensed gifts.[38] Because of the scarcity of wheat, they prepare starch ("for stiffening linen clothes") from the "arum root,"[39] by removing its "external peel" and dissolving its "corrosive salts" with water; and the "shining wite substance" resulting from this process could not be distinguished from common starch.[40] In "times of famine" they even use it as food, all the "caustic power, and the corrosive strength [...] having been absorbed by the aqueous particles."[41]

At last, the Apuan Alps and the "raging Tyrrhenian Sea" come into view.[42] Vallisneri enters a small and poor village called Fornovolasco, home to a "hard and most warlike people."[43] This place is made up of just some huts and a few houses lying at the foot of the mountains; and not far from there are the renowned iron and vitriol mines, exploited for military purposes on behalf of the Dukes of Este since the second half of the fifteenth century. It seems that many of the workers are descendants of expert miners from the Lombard city of Brescia, whom the Duke Ercole I (1431–1505) hired to find and work the first ore veins: a "non-trivial proof of this" is that many dialect terms from Brescia can still be heard, combined by the "unaware people" with the "gracefulness" of the Tuscan language.[44]

Visiting the mines is not supposed to be an easy, nor a safe task. The village and the nearby tracks are infested with robbers and bandits; and even its inhabitants, according to rumours, have often been an unsafe company to strangers. But here comes an unexpected turn of events. A "certain sagacious youngster"

---

37 Luzzini, *Theory, Practice, and Nature In-between*, 191–92, http://www.edition-open-access.de/sources/9/8/index.html#149 (accessed on 26 April 2021).

38 Ibid., 192, http://www.edition-open-access.de/sources/9/8/index.html#151 (accessed on 26 April 2021).

39 Italian arum, or "gigaro chiaro" (*Arum italicum* Miller), Family Araceae. It is a herbaceous, perennial plant native to the Mediterranean region, growing 30 to 46 cm in height (12–18 in). Its tuberous rhizome is particularly rich in starch (http://www.edition-open-access.de/sources/9/8/index.html#151, accessed on 26 April 2021).

40 Luzzini, *Theory, Practice, and Nature In-between*, 192, http://www.edition-open-access.de/sources/9/8/index.html#151 (accessed on 26 April 2021).

41 Ibid., 192, http://www.edition-open-access.de/sources/9/8/index.html#151 (accessed on 26 April 2021).

42 Ibid., 192, http://www.edition-open-access.de/sources/9/8/index.html#152 (accessed on 26 April 2021).

43 Ibid., 192, http://www.edition-open-access.de/sources/9/8/index.html#152 (accessed on 26 April 2021).

44 Ibid., 192, http://www.edition-open-access.de/sources/9/8/index.html#154 (accessed on 26 April 2021).

steps into the "small inn" where Vallisneri has found (a rather uncomfortable) shelter, covering him with "devoted and trustworthy embraces" and "showing clear signs of joy with his voice and face":

> I was amazed at such kindness in such a rude place; and when I asked where so much courtesy [...] could live among crags and caves, he openly revealed that he, too, was a foreigner, and that his name was Domenico de' Corradi d'Austria, superintendent of mines [...]. Since, by an unexpected gift of fate, I was not unknown to him, he invited me to share dinner with him; nor did he want me to spend the night in [that] desolate tavern, which was often unsafe for strangers. As soon as I heard [that] name (which was equally familiar to me), and since it seemed to me that I was imprisoned in [that] cruel inn in the [company] of a deadly gang, or as if I was in a jail, I did not refuse the loyal hospitality and the friendly services of [my] host; and, with the promise of a safe shelter, and with the most pleasant conversation, I restored my energies, drained by the difficult journey.[45]

Domenico de' Corradi d'Austria (1677–1756) was a very expert miner himself, and his knowledge of the Apuan Alps would prove crucial for the success of Vallisneri's underground investigations. In the days that followed he would provide his guest with advice, helpers, equipment and direct assistance, making possible for him to explore the iron mines and the *Tana che urla*, which still today is one of the most famous caverns in Garfagnana.[46] There, Vallisneri collected new data for the development of his hydrogeological theory on the origin of springs (and even for other theories, like the controversial one about the existence of seminal and vegetative principles in minerals).[47] Thus, thanks

---

45 Ibid., 193, http://www.edition-open-access.de/sources/9/8/index.html#154 (accessed on 26 April 2021).

46 Ibid., 194–97, http://www.edition-open-access.de/sources/9/8/index.html#155 (accessed on 26 April 2021).

47 With respect to the debated issue of mineral genesis and growth, Vallisneri's thought (which was strongly influenced by Robert Boyle's corpuscularianism) was not exempt from fluctuations. Though he supposed and – somehow – admitted the existence in minerals of such biological features as seeds (or 'matrices') and nourishment, by the last decade of his life he did not seem to persist in supporting the view of a vegetative power in minerals. On this topic (and, more generally, on the development of the early modern debate on the generation of mineral ores), see: Francesco Luzzini, "Matrices, not Seeds. Vallisneri's Research on Mines: Between Empiricism and Philosophy," in *History of Research in Mineral Resources*, ed. José Eugenio Ortiz, Octavio Puche, Isabel Rábano, and Luis F. Mazadiego (Madrid: Instituto Geológico y Minero de España, 2011), 109–10;

to this final adventure in the deepest bowels of the mountains, he felt ready to work on a model of the water cycle that would deal a final blow to the rival theories of alembics and rock filters: for (as he would repeatedly assert ten years later in his *Lezione Accademica intorno all'Origine delle Fontane*), any origin of springs other than the meteoric was proved wrong by experience, all the "imaginary pipes or supposed gutters"[48] being nothing more than the product of vain speculations.

3 Rise of a Method

Interdisciplinarity can be considered as a dominant feature of the *Primi Itineris Specimen*. It is no surprise, then, that not even Vallisneri's attempt to define a methodology of field research was exempt from an interdisciplinary approach. In fact, it is in this aspect of his work that lies the key to understanding the methodological and theoretical problems the author faced in this effort – and, therefore, it is here that we can best understand his contribution to the development of Earth sciences and natural history.

Of course, background matters. Vallisneri had studied medicine at the University of Bologna: there he became a proud supporter of Marcello Malpighi (1628–1694) who, together with Francesco Redi (1626–1697), was the leading proponent of the experimental method in those years. Malpighi and his followers had enthusiastically merged experimentalism with the theoretical frameworks of Baconian philosophy and Cartesian mechanism and corpuscularism, extending the influence of Galilean school from physics to medicine and biology. However, this transition did not occur without hurdles. As important studies have explained, in the last decades of the seventeenth century the dispute between experimental and empirical medicine was far from clearly resolved; indeed, the undeniable progress of anatomical knowledge still lacked proven positive effect on people's health. Hence the strong opposition from the advocates of empiricism, such as the Galenic physicians Giovanni Girolamo Sbaraglia (1641–1710) and Paolo Mini (1642–1693) in Bologna; and hence the persistence of traditional therapeutics, essentially based on statistical and empirical criteria.

Luzzini, *Il miracolo inutile*, 132–37; Luzzini, "Sounding the Depths of Providence: Mineral (Re)Generation and Human-Environment Interaction in the Early Modern Period," *Earth Sciences History* 39, no. 2 (2020).

48 Luzzini, *Theory, Practice, and Nature In-between*, 197, http://www.edition-open-access.de/sources/9/8/index.html#157 (accessed on 26 April 2021).

Quite significantly, these methods were still used by even those physicians who rejected the theoretical principles of Hippocratism and Galenism – and both Malpighi and Vallisneri were no exception.[49] Nor did Vallisneri uncritically reject empirical medicine: during his studies in Bologna, he followed both Sbaraglia's and Malpighi's classes and tried to learn and merge the best of both approaches, although on an ideological level he took the side of experimentalism. However, in carrying on Malpighi's legacy he dared go further, attempting to extend the use of experimental method from medicine to the even more heterogeneous field of natural sciences. This was a transition whose challenge did not lie merely in the different nature of the subject matter: for – and not surprisingly – the very *size* of this subject matter played a role, too. In fact, for however incredibly complex a human or an animal body could be, it was possible to examine them in laboratories, anatomical theatres, or in other environments where controlled tests could be performed. But this was obviously impossible when studying mountains, caverns, lakes, rivers, and other geological contexts. In these cases, nature could not enter laboratories: therefore, laboratories had to enter – literally – nature.

Following this inverted approach, any savant venturing on the field must necessarily adopt an inverted perspective which, in turn, entailed an interpretive strain that significantly increased the tension between theory and practice. Consequently, both the notion and the method of "experimentalism" had to be reconsidered and reshaped, and its meaning stretched to encompass activities like the exploration and the observation of geological contexts, which were more suitably described as sensorial experiences than as contrived tests. It is not by chance that both the legacy of Galilean experimentalism and traditional empiricism were equally and ubiquitously present in Vallisneri's field research, and they confronted and supported each other in a constant and fruitful dialogue. However, this interaction – which permeates every page of the *Primi Itineris Specimen* – proved to be as fertile as laden with methodological and epistemological issues; and, in light of this dichotomy, we can also better understand the meaning of a cautious disclaimer at the end of the manuscript, one inviting the reader to

49 On this topic, see: Marta Cavazza, *Settecento inquieto: alle origini dell'Istituto delle Scienze di Bologna* (Bologna: Il Mulino, 1990), 185–201; Cavazza, "The Uselessness of Anatomy: Mini and Sbaraglia versus Malpighi," in *Marcello Malpighi Anatomist and Physician*, ed. Domenico Bertoloni Meli (Florence: Olschki, 1997); Claire Crignon, Carsten Zelle, and Nunzio Allocca, eds., *Medical Empiricism and Philosophy of Human Nature in the 17th and 18th Century* (Leiden/Boston: Brill, 2014); Roger French, *Medicine before Science* (Cambridge: Cambridge University Press, 2003), 215–21; Generali, *Antonio Vallisneri: gli anni della formazione e le prime ricerche*, 30–47.

> […] accept without blame of rashness what I will declare about the fountains, brooks, and rivers of our Alps: that, for the most part, carry their load downstream to the Po [River], all of them owing everything to the rains, and to the melted snows. As to my opinion on the Danube, Rhine, and Rhône Rivers, I am unacquainted [with them]. I am truly astounded at such great names, and can [only] think great things about [their] origin, having not been familiar with their springs. Should we observe with our own eyes [those] enormous mountains, [those] immense regions, and [those] greatest wildernesses, with [their] almost eternal masses of snow, […] perhaps the astonishment will give way to laughter, and we will not mix the mountains with the sea, nor the seas with the mountains.[50]

Vallisneri was well aware that his quest required "longer journeys and new works,"[51] for his explorations (and, therefore, his experiences) had been chronologically and geographically limited. This was all but a small hurdle, given the kind of audience he was addressing: an international context whose knowledge of Italian science and Galilean experimentalism was far from obvious, and where the notion of uniformity of natural laws across time and space was far from unanimously accepted – being, on the contrary, a matter of ferocious and still unresolved debate. And yet, by the first half of the nineteenth century this concept of uniformitarianism would assert itself as an essential element of science, thanks to the mass of data gathered by countless natural philosophers, scholars, practitioners, and technicians during their experiences on the field.[52]

50 Luzzini, *Theory, Practice, and Nature In-between*, 197, http://www.edition-open-access.de/sources/9/8/index.html#157 (accessed on 26 April 2021).

51 Ibid., 198, http://www.edition-open-access.de/sources/9/8/index.html#162 (accessed on 26 April 2021).

52 On this topic, see: Luzzini, "Through Dark and Mysterious Paths," 184–85. For an in-depth discussion of the notion of uniformitarianism, see: Claude C. Albritton Jr., *The Abyss of Time: Changing Conceptions of the Earth's Antiquity after the Sixteenth Century* (San Francisco: Freeman Cooper, 1980), 87–88, 100, 132, 139–47, 186–89, 215; Henry Faul and Carol Faul, *It Began with a Stone: A History of Geology from Stone Age to the Age of Plate Tectonics* (New York: John Wiley & Sons, 1983), 89, 90–91, 138, 184, 218; Stephen J. Gould, "Is Uniformitarianism Necessary?" *American Journal of Science* 263, no. 3 (1965): 223–28; David R. Oldroyd, *Thinking About the Earth: A History of Ideas in Geology* (London: Athlone, 1996), 82, 134, 136–39, 142–43, 177, 202, 301, 330–31; Rappaport, *When Geologists were Historians*, 95, 162, 164, 167, 205–06, 238–42, 246; Martin J. S. Rudwick, *The Meaning of Fossils: Episodes in the History of Paleontology* (Chicago, London: The University of Chicago Press, 1972), 110, 130–32, 169–72, 179, 185–88, 231, 235, 259; Davis A. Young, *The Biblical Flood: A Case Study of the Church's Response to Extrabiblical Evidence* (Grand Rapids: William B. Eerdmans Publishing Company; Carlisle: The Paternoster Press, 1995), 176, 178, 182, 209, 246, 248, 260, 262–63, 270, 285, 290, 292–93, 296.

In upholding field research among the members of the European Republic of Letters, Vallisneri clearly aligned himself with the enthusiastic supporters of this still widely neglected way to knowledge, and disclosed a view of immutable harmony of natural laws that would be explained in even greater detail in 1715 with the publication of the seminal *Lezione Accademica intorno all'Origine delle Fontane* – which, in fact, resumed and extended many concepts and data already featured in the *Primi itineris Specimen*. In doing so, Vallisneri made a significant contribution to the early development of the debate on uniformitarianism, as well as to the epistemological change produced by this notion in the years to follow.

In accordance with this view, Vallisneri never gave up the ambition to provide his field research with a tool that was as essential and long sought-after as still largely imperfect: a well-defined and univocal methodology to be applied in case of further travels in the mountains. The most eloquent expression of this effort can be found at the end of the manuscript in the form of a detailed list of instructions (or "Index of observations"). According to this prototypical field handbook, a worthy natural philosopher was expected to study and describe a wide array of subjects and objects: from every plant and animal living in the mountains, to the countless and different crystals, fossils, and stones found along the way; from the "stony, chalky, gravelly, sandy layers," to the "outer surface of the mountains" and to the "quality of every soil"; from the "air's weight" and the climate (measured, respectively, with "a barometric device" and with "a thermometer"), to the "height of mountains" and the many different "springs, rivers, torrents, thermal waters, mines."[53] Nor was the history of people to be passed over in silence, being neither useless, nor unnecessary to describe their different cultures and habits. Thus, for example, it was important to study the use they made of mountains and pastures; which "fruits and grains" were "used as food and drink"; how they prepared "milk and dairy products"; which were their "customs, arts, buildings," and even their "diseases, torments, and delights."[54]

The thematic variety of the "Index of observations" is matched by a crowning display of eclecticism in the last papers of the manuscript. Here, natural and human history merge once more in what could be described as a visual celebration of Vallisneri's interdisciplinary attitude: two perfectly preserved,

53 Luzzini, *Theory, Practice, and Nature In-between*, 198–99, http://www.edition-open-access.de/sources/9/8/index.html#163 (accessed on 26 April 2021).

54 Ibid., 199, http://www.edition-open-access.de/sources/9/8/index.html#168 (accessed on 26 April 2021). In the next two decades Vallisneri enriched and refined his "index," whose new version – published in 1726 – listed up to 26 points: Vallisneri, "Continuazione dell'Estratto d'alcune Notizie intorno alla Garfagnana," 404–17.

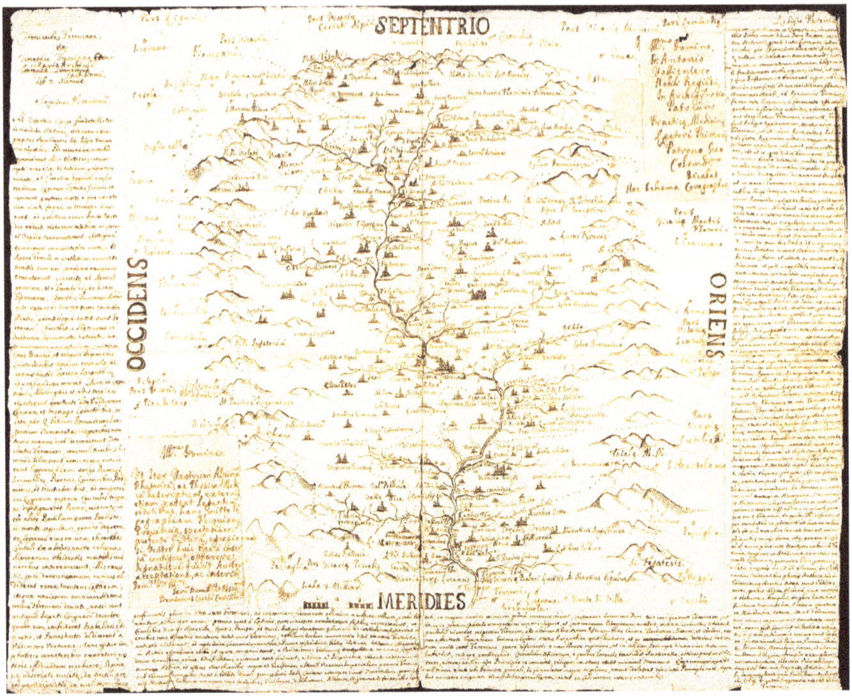

FIGURE 3.3 *Primi Itineris Specimen*, the map of Garfagnana by Domenico Cecchi
IMAGE COURTESY OF THE STATE ARCHIVE OF REGGIO EMILIA, ITALY

hand-drawn maps of Garfagnana. The first one, which is particularly interesting for us (fig. 3.3), bears an autographed dedication by the cartographer Domenico Cecchi (1678–1745). This beautiful document is much more than a simple map, being framed on three sides by a dense text. According to the title, the two quoted passages ("Chronology of Garfagnana" and "Religion of Garfagnana") are from Book 2 of the *Silvae Feronianae*: a (now lost) manuscript composed by Timoteo Tramonti (ca. 16th–17th century), an antiquarian from Castiglione di Garfagnana[55] and "Chancellor of the Archive" of this town.[56] The charming mixture of facts and legends featured in the text provides many literary, religious, folkloristic, mythological anecdotes about the history of the province: from the foundation of the first "fortified huts" and the arrival of Greek heroes and demigods, to the (then) recent dominion of the Princes of Este – passing through the Etruscans kingdoms, the Celts, the Gauls,

55 Currently, this town is part of the Province of Lucca (Tuscany).

56 Luzzini, *Theory, Practice, and Nature In-between*, 209, http://www.edition-open-sources.org/sources/9/10/index.html#7 (accessed on 26 April 2021).

the Romans, the advent of Christianity, the barbarian invasions, the bloody and infamous mediaeval struggles between the Guelphs and the Ghibellines (who "almost destroyed each other and their belongings"), and the exhausting battles for independence against the Republic of Lucca.[57]

4 As a Conclusion

In this contribution I have insisted on highlighting Vallisneri's proud support of a unitary view of knowledge, and described this approach as a distinctive feature of his work. It could be argued, reasonably, that interdisciplinarity was all but a peculiar aptitude in the late seventeenth and early eighteenth centuries: at a time when disciplinary boundaries were as fuzzy as enthusiastically crossed, it was relatively common to see doctors, clerics, engineers, alchemists, chief miners, artisans, noblemen, and other (variously) learned scholars and practicians expanding their inquiries beyond the flexible limits of their competencies. Still, as many studies have explained, it is undeniable that Vallisneri's omnivorous curiosity and his ability to recognize and make use of the links between medicine and the many fields of natural philosophy made him an uncommonly versatile scholar even for the standards of his time: to the point that, I believe, it would be reductive to interpret his ambitious work as a mere case of "early modern eclecticism."[58]

What is beyond any doubt is that the *Primi Itineris Specimen* is a worthy expression of the wide-ranging talent of its author. Indeed, it was unfortunate that unknown events doomed the manuscript to such a long oblivion, preventing it from being disclosed to the European Republic of Letters at the proper time. Even more unfortunate, because of the historical and scientific

57 Ibid., 209–215, http://www.edition-open-sources.org/sources/9/10/index.html (accessed on 26 April 2021).

58 See, for example: Generali, *Antonio Vallisneri: gli anni della formazione e le prime ricerche*; Generali, *Antonio Vallisneri: la figura, il contesto, le immagini storiografiche*; Generali, "La Repubblica delle Lettere e i carteggi scientifici: il caso di A. Vallisneri"; Luzzini, *Il miracolo inutile*; Luzzini, *Theory, Practice, and Nature In-between*, 9–51, http://www.edition-open-sources.org/sources/9/2/index.html (accessed on 26 April 2021); Antonio Vallisneri, *Saggio d'Istoria medica, e naturale, colla spiegazione de' nomi, alla medesima spettanti, posti per alfabeto*, ed. Massimo Rinaldi (Florence: Olschki, 2012); and, more generally, see the many studies and critical editions produced since 2000 by the National Edition of Antonio Vallisneri's Works (https://www.olschki.it/la-casa-editrice/collane-Olschki/edizioni-nazionali/edizione-nazionale-Vallisneri, accessed on 26 April 2021; www.vallisneri.it, accessed on 26 April 2021), one of the most lively and renowned cultural institutions of this kind in Europe.

significance of this document, which lies not just in the quantity and quality of information it contains, nor in its exceptional thematic variety, but also in its very flaws. It is thanks to these limitations – the many descriptive and lexical inconsistencies, the constant tension between theoretical interpretation and field data, the (unavoidable) methodological awkwardness – that we can fully understand the difficulties faced by the Galilean experimental tradition when it turned its attention from medicine to mountains, venturing in the *mare magnum* of natural philosophy and setting the basis for an epistemological revolution which, years later, would affect both notions of experimental "verification" and "observation."

As we have seen, this move did not come at no cost. The doubts of the many who questioned the uniformity of natural laws across time and space and the validity of field research for the advancement of human knowledge were not groundless speculations, as the mere concept of "field research" was still far from well defined in the early eighteenth century. Yet, as historians, we can't help but applauding Vallisneri (and the intellectual network he was part of) for his effort to define a methodology of naturalistic experimentalism. For this attempt, in being equally interdisciplinary and unifying, reflects an uncommon awareness of the challenges and ideals that would shape the path of science in the years to come.

Acknowledgement

This contribution has been supported by the MSCA-IF project SOUNDEPTH, funded from the European Union's Horizon 2020 Research and Innovation Programme under grant agreement no. 101019781.

Bibliography

Printed Sources

Luzzini, Francesco, ed. *Theory, Practice, and Nature In-between. Antonio Vallisneri's Primi Itineris Specimen*. Berlin: Edition Open Access/Max Planck Institute for the History of Science, 2018. http://www.edition-open-sources.org/sources/9/, accessed on 26 April 2021.

Vallisneri, Antonio. *Lezione Accademica intorno all'Origine delle Fontane*. Venezia: Appresso Gio. Gabbriello Ertz, 1715.

Vallisneri, Antonio. "Che ogni Italiano debba scrivere in Lingua purgata Italiana, o Toscana." *Supplementi al Giornale de' Letterati d'Italia* 1 (1722): 252–330.

Vallisneri, Antonio. "Estratto d'alcune Notizie intorno alla Provincia della Garfagnana, cavate dal primo Viaggio Montano del Sig. Antonio Vallisnieri." *Supplementi al Giornale de' Letterati d'Italia* II (1722): 270–312.

Vallisneri, Antonio. "Continuazione dell'Estratto d'alcune Notizie intorno alla Garfagnana." *Supplementi al Giornale de' Letterati d'Italia* III (1726): 376–428.

Vallisneri, Antonio. *Epistolario, vol. I (1679–1710)*, edited by Dario Generali. Milan: Franco Angeli, 1991.

Vallisneri, Antonio. *Saggio d'Istoria medica, e naturale, colla spiegazione de' nomi, alla medesima spettanti, posti per alfabeto*, edited by Massimo Rinaldi. Florence: Olschki, 2012.

Vallisneri, Antonio. *Che ogni Italiano debba scrivere in Lingua purgata Italiana*, edited by Dario Generali. Florence: Olschki, 2013.

Literature

Albritton, Claude C. Jr. *The Abyss of Time: Changing Conceptions of the Earth's Antiquity after the Sixteenth Century*. San Francisco: Freeman Cooper, 1980.

Cavazza, Marta. *Settecento inquieto: alle origini dell'Istituto delle Scienze di Bologna*. Bologna: Il Mulino, 1990.

Cavazza, Marta. "The Uselessness of Anatomy: Mini and Sbaraglia versus Malpighi." In *Marcello Malpighi Anatomist and Physician*, edited by Domenico Bertoloni Meli, 129–45. Florence: Olschki, 1997.

Crignon, Claire, Carsten Zelle, and Nunzio Allocca, eds. *Medical Empiricism and Philosophy of Human Nature in the 17th and 18th Century*. Leiden, Boston: Brill, 2014.

Faul, Henry and Carol Faul. *It Began with a Stone: A History of Geology from Stone Age to the Age of Plate Tectonics*. New York: John Wiley & Sons, 1983.

French, Roger. *Medicine before Science*. Cambridge, New York: Cambridge University Press, 2003.

Generali, Dario. "Il testo tra comunicazione scientifica e strategie editoriali. Il caso di Antonio Vallisneri." In *Edizioni e traduzioni di testi filosofici. Esperienze di lavoro e riflessioni*, edited by Marialuisa Baldi and Barbara Faes de Mottoni, 183–205. Milan: Franco Angeli, 2006.

Generali, Dario. *Antonio Vallisneri: gli anni della formazione e le prime ricerche*. Florence: Olschki, 2007.

Generali, Dario, ed. *Antonio Vallisneri: la figura, il contesto, le immagini storiografiche*. Florence: Olschki, 2008.

Generali, Dario. "La Repubblica delle Lettere e i carteggi scientifici: il caso di A. Vallisneri." In *La Repubblica delle Lettere, il Settecento italiano e la Scuola del secolo XXI*, edited by Andrea Battistini, Claudio Griggio, and Renzo Rabboni, 43–50. Pisa, Rome: Fabrizio Serra Editore, 2011.

Gould, Stephen J. "Is Uniformitarianism Necessary?" *American Journal of Science* 263, no. 3 (1965): 223–28.

Luzzini, Francesco. "La Tana che urla: cenni di speleologia vallisneriana." In *Antonio Vallisneri: la figura, il contesto, le immagini storiografiche*, edited by Dario Generali, 349–369. Florence: Olschki, 2008.

Luzzini, Francesco. "Cavità naturali ed artificiali in Garfagnana e Lunigiana. Le esplorazioni di Antonio Vallisneri." In *Antonio Vallisneri dalla Garfagnana alla Scienza*, edited by Pietro Rocchi, 93–124. Lucca: Maria Pacini Fazzi Editore, 2010.

Luzzini, Francesco. "Matrices, not Seeds. Vallisneri's Research on Mines: Between Empiricism and Philosophy." In *History of Research in Mineral Resources*, edited by José Eugenio Ortiz, Octavio Puche, Isabel Rábano, and Luis F. Mazadiego, 105–12. Madrid: Instituto Geológico y Minero de España, 2011.

Luzzini, Francesco. "Multa curiosa: Vallisneri's early studies on Earth sciences." *Nuncius* XXVI, no. 2 (2011): 334–54.

Luzzini, Francesco. *Il miracolo inutile. Antonio Vallisneri e le scienze della Terra in Europa tra XVII e XVIII secolo*. Florence: Olschki, 2013.

Luzzini, Francesco. "An Uncomfortable, yet Wonderful Journey. Antonio Vallisneri and his Exploration of the Northern Apennines." In *Nel nome di Lazzaro. Saggi di storia della scienza e delle istituzioni scientifiche tra il XVII e il XVIII secolo*, edited by Centro Studi Lazzaro Spallanzani, 207–20. Bologna: Edizioni Pendragon, 2014.

Luzzini, Francesco. "Through Dark and Mysterious Paths. Early Modern Science and the Search for the Origin of Springs from the 16th to the 18th Centuries." *Earth Sciences History* 34, no. 2 (2015): 169–89.

Luzzini, Francesco. "Sounding the Depths of Providence: Mineral (Re)Generation and Human-Environment Interaction in the Early Modern Period," *Earth Sciences History* 39, no. 2 (2020): 389–408.

Oldroyd, David R. *Thinking About the Earth: A History of Ideas in Geology*. London: Athlone, 1996.

Rappaport, Rhoda. *When Geologists were Historians, 1665–1760*. Ithaca: Cornell University Press, 1997.

Rudwick, Martin J. S. *The Meaning of Fossils: Episodes in the History of Paleontology*. Chicago, London: The University of Chicago Press, 1972.

Young, Davis A. *The Biblical Flood: A Case Study of the Church's Response to Extrabiblical Evidence*. Grand Rapids: William B. Eerdmans Publishing Company; Carlisle: The Paternoster Press, 1995.

CHAPTER 4

Flora Near and Far: Accumulating Knowledge on Plants in Eighteenth-Century Zurich

Meike Knittel

In 1740, while preparing the publication of his *Enumeratio methodica stirpium Helvetiae indigenarum*, which became the first comprehensive flora of Switzerland, Albrecht von Haller (1708–1777) wrote from Göttingen to his friend Johannes Gessner (1709–1790) in Zurich asking a favour. Knowing that Gessner had catalogued the "loca natalia" of many rare Swiss plants based on his in-depth first-hand knowledge, Haller wished to receive an excerpt of this record.[1] At the same time, Gessner himself was interested in plants from all over the world, as he was working on a universal history of plants ("historia plantarum"), intending to cover all aspects including localities, cultivation and uses, especially medical and economic ones.[2] Most importantly, Gessner planned to include illustrations of the characteristics needed to classify a plant according to the Linnaean classification system, which was becoming increasingly popular in those years.[3] To Gessner, this classification system seemed a good way to engage a large number of people in the creation of plant knowledge. Illustrations, he hoped, would facilitate access to Linnaean taxonomy

1 Letter, A. von Haller to J. Gessner, 10 June 1740. Henry E. Sigerist, *Albrecht von Haller's Briefe an Johannes Gesner (1728–1777)*, Abhandlungen der Gesellschaft der Wissenschaften 11 (Berlin: Weidmann, 1923), 146.

2 Autobiography J. Gessner, Zentralbibliothek Zürich, ZBZ, Hss., Ms. M 18.10, Urs Boschung, *Johannes Gessner (1709–1790): Der Gründer der Naturforschenden Gesellschaft in Zürich; Seine Autobiographie – aus seinem Briefwechsel mit Albrecht von Haller; Ein Beitrag zur Geschichte der Naturwissenschaften in Zürich im 18. Jahrhundert* (Alpnach-Dorf: KOPRINT AG, 1995), 44.

3 On the increasing popularity of the Linnaean system, or the resistance to it respectively, in different locations, see Miguel A. Puig-Samper, "Difusión e institucionalización del sistema linneano en España y América," in *Mundializacion de la ciencia y cultura nacional: Actas del congreso international "Ciencia, Descubrimento y Mundo Colonial,"* ed. Antonio Lafuente, Alberto Elena and María Luísa Ortega (Aranjuez: Doce Calles, 1993); Marco Beretta and Alessandro Tosi, ed., *Linnaeus in Italy: The Spread of a Revolution in Science*, (Sagamore Beach: Science History Publications, 2007); Andreas Önnerfors, "Auswärtige Saamen und Gewächse ingleichen zur Correspondence: Die Verbreitung Linnéscher Naturalhistorie in Schwedisch-Pommern im 18. Jahrhundert," in *Wissen im Netz: Botanik und Pflanzentransfer in europäischen Korrespondenznetzen des 18. Jahrhunderts*, ed. Regina Dauser et al. (Berlin: Akademie-Verlag, 2008).

and thereby recruit more and more people to help compile a complete record of the world's flora.

A few years later, Gessner commissioned the miniaturist Christian Gottlieb Geissler (1729–1814) to draw and engrave more than a thousand plants growing in Europe, Asia, Africa, and the Americas, for the purpose of illustrating fructification organs. The resulting plates were published posthumously, at the turn of the century, by Gessner's grandnephew Christoph Salomon Schinz (1764–1847) under the title *Tabulae phytographicae*.[4] The *Tabulae phytographicae* provided information on the habitat of many of the plants depicted.[5] In other words, they linked the particular species to a geographical place or an environmental setting. However, far from being standardised, these indications referred differently to different territories. Localities range from references to entire continents – such as "Africa" (often meaning South Africa) or "Asia," differentiating between "northern" and "southern" only for Europe and the Americas – over larger regions, to countries. Countries in Europe, for instance, included France ("Gallia"), Italy ("Italia"), Spain ("Hispania"), and of course Switzerland ("Helvetia"); Asian territories included Malabar, Ceylon, and India.[6]

The great variability of this habitat information given by the Zurich naturalist raises the question of perception of the plant world. More precisely: How did an 18th-century botanist understand and map the abundance of plant species he came to know? What did he know about the territories mentioned as habitats? This particular case is of interest because European naturalists like Gessner crucially contributed to the universalisation of plant knowledge

4 *Johannis Gessneri Tabulae phytographicae, analysin generum plantarum exhibentes / cum commentatione edidit Christ. Sal. Schinz, Zurich, 1795–1804*. Although the *Tabulae phytographicae* were published only posthumously, they circulated in advance. Gessner showed them to visitors in Zurich and sent tableaus IV, XIII, XLIV and XLIX to some of his correspondents. Albrecht von Haller, Carl Linnaeus, Jean-François Séguier and Abraham Gagnebin verifiably received printings during the 1750s and 1760s and were asked to evaluate them. On Gagnebin's use of the illustrations in his botanical study, see Rossella Baldi, "La circulation du savoir botanique par le texte et par l'image: Le Species plantarum d'Abraham Gagnebin," in *Rosseau Botaniste: Je vais devenir plante moi-même*, ed. Claire Jaquier and Timothée Léchot (Fleurier: Editions du Belvédère, 2012).

5 Information on the origins and growing conditions was of no importance to the classification of plants according to the Linnaean system, which organised species based on artificial criteria, according to external features. The *Tabulae phytographicae* contain only annotations of localities for the first thirty tableaus.

6 Tableau XIII, depicting representatives of the fifth Linnaean class, for example, shows species occurring "in Europe, Asia and Northern America" ("Habitat in Europae, Asiae, Americae septentrionalibus") as well as in Southern Europe ("in Europa australi"). *Tabulae phytographicae*, 102 and 107.

FIGURE 4.1 Tab. XIII of Gessner's *Tabulae phytographicae* depicting representatives of the fifth Linnaean class. Zentralbibliothek Zürich, Alte Drucke, NF 170

and the perception of distant political territories. Therefore, it is important to understand the factors influencing what he knew about the flora of the different regions of the world. Starting from the localities printed alongside the plant engravings within *Tabulae phytographicae*, this paper examines the Zurich botanist's knowledge of plants from far-away territories using the auction catalogue of Gessner's library and the herbarium he put together for the Zurich Physical Society, of which he was co-founder and an active member from 1746 until his death.[7]

The eighteenth century saw numerous expeditions supported by European trading companies and colonial administrations, aimed at gaining access to potentially useful plants from Asia, the Americas, and the islands in the South Seas. The journeys not only generated many records on these distant worlds, but also brought a plethora of plant seeds and dried specimens into the botanical gardens and herbariums of places like Madrid, Paris, and Kew.[8] For a long time, research focused on transfers of botanical knowledge within the individual European colonial empires.[9] While, in recent years, studies have shown that botanists collecting information and plant species regularly overcame imperial borders, little is currently known about what kind of knowledge of the flora from far-away regions reached the European hinterlands.[10] The perception of plant knowledge outside Atlantic port cities, globally influential trading

7 Gessner's library is documented in manuscript and print, both surviving in the ZBZ. 1) Manuscript Department, Ms. Z II 620 (manuscript of the catalogue handed down in the estate of the first director of the Swiss National Museum, Heinrich Angst (1847–1922); 2) Rare Book Department, Dr O 456.4 (Johann Füssli & Son, *Catalogus Librorum Bibliothecae Joannis Gessneri* […], Zurich 1798), referred to as *Catalogus* in the following. The herbarium is preserved in the Department of Systematic and Evolutionary Botany of the University of Zurich under the title *Hortus siccus Societatis Physicae Tigurinae, collectus et Linnaeana methodo dispositus a Joanne Gesnero, 1751*. On the members and activities of this society, see Sarah Baumgartner, "Das nützliche Wissen: Akteure, Tätigkeiten, Kommunikationspraxis und Themen der Naturforschenden Gesellschaft in Zürich, 1746 bis ca. 1830" (PhD diss., University of Bern, 2019).

8 David Miller and Peter H. Reill, ed., *Visions of Empire: Voyages, Botany and Representations of Nature* (Cambridge: Cambridge University Press, 1996); Mauricio Nieto, "Presentación gráfica, desplazamiento y aprobación de la naturaleza en las expediciones botánicas del siglo XVIII," *Asclepio* 47, no. 2 (1995); Catherine Gaziello, *L'expédition de Lapérouse 1785–1788: réplique française aux voyages de Cook* (Paris: CTHS, 1984).

9 For a study beyond European empires, see Yota Batsaki, Anatole Tchikine and Sarah Burke Cahalan, ed., *The Botany of Empire in the Long Eighteenth Century*, Dumbarton Oaks Symposia and Colloquia (Baltimore: Dumbarton Oaks Publishing, 2017).

10 For examples of 19th-century actors from Switzerland in an imperial context, see Bernhard C. Schär, *Tropenliebe: Schweizer Naturforscher und niederländischer Imperialismus in Südostasien um 1900* (Frankfurt am Main: Campus Verlag, 2015); Andreas Zangger, *Koloniale Schweiz: Ein Stück Globalgeschichte zwischen Europa und Südostasien (1860–1930)* (Bielefeld: transcript, 2011).

centres and university towns remains insufficiently understood.[11] Therefore, this contribution investigates how different plant worlds were perceived in the context of a Swiss city that neither was directly involved in overseas trade nor had a university, but was the centre for the training of Protestant clergy. Looking through the lens of an 18th-century botanist from Zurich, namely Johannes Gessner, this paper asks how a researcher working a long way from the flora's origin perceived and organised the ever-expanding plant world.

To do so, this contribution first seeks to understand how the Zurich naturalist's expert knowledge on Swiss plants was formed. Then, using the example of American plants, we argue that Gessner's perception of extra-European floras was determined by the joint presence of publications in languages he understood and specimens from those regions. Finally, the case of a collection of dried plants from South Africa sheds light on what happened in the absence of factors that were compelling in previous cases.

1 Knowing What Is Near

Like many others of his generation, Gessner spent most of his life where he was born. He left Zurich, only, at the age of seventeen, to study medicine at the Universities of Basel and Leiden, as well as in Paris. These stays had a long-term influence on Gessner's botanical activities and thus affected which plant worlds he became acquainted with. As we shall see in more detail below, his decision to study in Leiden was particularly momentous as Leiden had been a centre of botanical study and cultivation of plants from all over the world since the intensified presence of Europeans in non-Europeans areas.[12] The involvement of the Leiden townspeople in global trade meant that seeds were imported from Asia, South Africa, and the Americas in abundance to the Dutch town. As a result, botanists were able to gather detailed information on these "new" plants. During his studies there, Gessner visited the university garden as well as several private gardens, which cultivated and sold plants from every

11 An exception examining this question for an outstanding Swiss naturalist of the 16th century: Urs B. Leu, "Konrad Gessner und die Neue Welt," *Gesnerus* 49, no. 3–4 (1992).

12 On 16th-century botany in Leiden, see José Pardo-Tomás, "Two Glimpses of America from a Distance: Carolus Clusius and Nicolás Monardes," in *Carolus Clusius: Towards a Cultural History of a Renaissance Naturalist*, ed. Florike Egmond, G. Hoftijzer and Robert W. Visser (Amsterdam: Koninklijke Nederlandse Akademie van Wetenschappen, 2007); Florike Egmond, Esther van Gelder and Nicolas Robin, eds., *Flowers of Passion and Distinction: Practice, Expertise and Identity in Clusius' World* (Stuttgarrt: Franz Steiner Verlag, 2012). For an overview, see William T. Stearn, "The Influence of Leyden on Botany in the Seventeenth and Eighteenth Century," *The British Journal for the History of Science* 1, no. 2 (1962).

known corner of the world.[13] His choice of Leiden was very typical of the time, since Herman Boerhaave (1668–1738) attracted botanically interested medical students from all over Europe. Yet, by studying in Leiden, Gessner did no more than follow a local tradition; indeed, young people from Zurich had been students at the Protestant university since the 17th century.[14]

Having graduated as a medical doctor in Basel, Gessner returned to his hometown, where his family had had citizen status for more than two hundred years, and subsequently left the city only for short journeys. In Zurich, Gessner produced botanical illustrations of, and published dissertations on, the uses of local plants as well as plants growing in distant territories.[15] In addition, together with his close friend Albrecht von Haller from Bern (then professor at the University of Göttingen), Gessner pursued the objective of compiling a comprehensive flora of Switzerland, a work eventually completed and published by Haller as *Enumeratio* in 1742.[16] His expert knowledge of Swiss plants also emerges in the indications of localities within the *Tabulae phytographicae*, especially in cases where Gessner informs the reader that a depicted specimen grew, for example, "on meadows" ("pratis"), "on crags" ("ad rupes") or "on fields" ("segetes") in Switzerland, thereby coming much closer to indications of habitat in a modern sense than with the statements given for non-European plants (see below).[17] While Gessner largely followed the indications of habitat given by Linnaeus in his *Species plantarum*, for plants growing in Switzerland, his work provides further information (e.g. in the case of *Campanula thyrsoides* L. and *Jasminum officinale* L.) where he could add Switzerland or the Swiss Alps respectively as habitats.[18]

13 On the Zurich botanists' long-term relations to the merchant gardener Willem van Hazen, see Meike Knittel, "Netzwerke der Botanik: Johannes Gessner (1709–1790) und die botanische Forschung im 18. Jahrhundert" (PhD diss., University of Bern, 2018).

14 Sabine Heller, *Boerhaaves Schweizer Studenten: Ein Beitrag zur Geschichte des Medizinstudiums* (Zurich: Juris-Verlag, 1984). Hanspeter Marti, "Leiden," *Historisches Lexikon der Schweiz Online*, last modified November 20, 2008, https://hls-dhs-dss.ch/de/articles/006615/2008-11-20/ (accessed on 30 April 2020).

15 For an analysis of Gessner's dissertation on officinal plants see Meike Knittel, "'Dominus creavit ex terra medicamenta': Heilpflanzenwissen in Johannes Gessners Phytographia sacra," in *Wer das Gras wachsen hört: Wissensgeschichte(n) der pflanzlichen Ressourcen vom Mittelalter bis ins 20. Jahrhundert*, ed. Simona Boscani Leoni and Martin Stuber, Jahrbuch für Geschichte des ländlichen Raumes 14 (Innsbruck: Studien-Verlag, 2017).

16 For details on the collaboration, see Boschung, *Johannes Gessner*.

17 *Tabulae phytographicae*, 22; 44; 92.

18 Carl Linnaeus, *Species plantarum: exhibentes plantas rite cognitas, ad genera relatas, cum differentiis specificis, nominibus trivialibus, synonymis selectis, locis natalibus, secundum systema sexuale digestas*, (Holmiae: Impensis Laurentii Salvii, 1753), 167; *Tabulae phytographicae*, 103; *Species plantarum*, 7; *Tabulae phytographicae*, 7.

The Zurich naturalist had studied the Swiss flora for many decades. Travelling around the Swiss cantons, their subject and neighbouring territories, Gessner acquired extensive knowledge about the plants of what is today Switzerland. During his tour with Haller in 1728 alone, he studied the flora growing in the territory of Basel, around the cities of Biel, Neuchâtel, Lausanne, and Geneva. In the course of this study trip, he also explored the Alpine flora while crossing the passes in the Pennine, Bernese and Uri Alps.[19] For years before and after, Gessner collected plants in the mountains of Appenzell, and again in Central Switzerland as well as in the territory of the Three Leagues.[20] In the following decades as he was no longer able or in need to travel himself, he provided students from Zurich with funds to travel, so that they would bring him fresh plant material from these Alpine regions.[21] Accumulating, in Zurich, plants from the Swiss mountains and sharing them with his correspondents in other places made Gessner a reputable specialist in Alpine botany.

With his focus on Alpine plants, the Zurich naturalist follows a local research tradition whose most prominent representative was his teacher Johann Jakob Scheuchzer (1672–1733).[22] Besides being inspired by Scheuchzer's work, Gessner could also draw on plant material he had received from him as a gift, as the various comments in his herbarium indicate.[23] Early on, Gessner could

---

19 Christoph Salomon Schinz, Preface to the *Tabulae phytographicae*, VII.

20 Autobiography J. Gessner, ZBZ, Hss., Ms. M 18.10, Boschung, *Johannes Gessner*, 24–45.

21 On the students' tours on Gessner's commission, see Meike Knittel and Reto Nyffeler, "Flora Alpina," in *Montan-Welten: Alpengeschichte abseits des Pfades*, ed. Tina Asmussen, Æther 03 (Zürich: intercom Verlag, 2019), last modified May 2019, https://aether.ethz.ch/ausgabe/montan-welten/flora-alpina/, accessed on 26 April 2021; Meike Knittel, "Devenir botaniste au XVIIIe siècle: les premiers voyages de Gessner et des jeunes Zurichois dans les Alpes," *Bulletin de l'Association culturelle pour le voyage en Suisse* 20 (2019). On the relevance of exploration trips for scholarly careers, see Hanna Hodacs, "Linnaean Scholars Out of Doors: So Much to Name, Learn and Profit From," in *Naturalists in the Field: Collecting, Recording and Preserving the Natural World from the Fifteenth to the Twenty-First Century*, ed. Arthur MacGregor (Leiden: Brill, 2018).

22 The same holds true for Gessner's specialisation in grasses, becoming apparent in the indications for the plants depicted in tableau V, which locates species in marshy areas or stagnant waters, for example. His interest presumably was inspired by the research of Johann Scheuchzer (1684–1738), another Zurich naturalist of the previous generation. On Johann Jakob Scheuchzer, see Simona Boscani Leoni, "Men of Exchange: Creation and Circulation of Knowledge in the Swiss Republics of the Eighteenth Century," in *Scholars in Action: The Practice of Knowledge and the Figure of the Savant in the 18th Century*, ed. André Holenstein, Hubert Steinke, and Martin Stuber (Leiden: Brill, 2013).

23 "HB Scheuchzer" is among others noted next to a Veronica species in vol. 1 of the *Hortus siccus*. Gessner also adopted Scheuchzer's practices, especially the annotation of a specimen's origin. For Scheuchzer's practices see Urs B. Leu, ed., *Natura Sacra: Der Frühaufklärer Johann Jakob Scheuchzer (1672–1733)* (Zug: Achius Verlag, 2012).

study the collection of dried plants his teacher had put together in Zurich and continued to do so through Scheuchzer's heirs after his death.[24]

However, Gessner's interest was in no way limited to plants growing in the Swiss mountains. He was also well-informed about species growing in the French and Italian Alps ("in alpibus Galliae, Helvetiae, Italiae, etc.") and the Austrian ones ("Austriacis"), the Pyrenees ("in Alpibus Helveticis, Pyrenaeis, Austriacis") as well as the mountains of Lapland ("in Alpibus Lapponicis"), as the comments in the *Tabulae phytographicae* make clear.[25] Partly, this was the result of publications available (such as Linnaeus's); but Gessner also corresponded with other botanists interested in Alpine flora. His correspondents Jean-François Séguier (1703–1784) and Carlo Allioni (1728–1804), for example, were willing to share the specimens they had collected on Monte Baldo or in Piedmont respectively, in return for species from the Swiss Alps.[26] Thereby, they helped each other to become experts on plants growing on mountains in different regions.

It comes as no great surprise that a decades-long study of plants in the immediate neighbourhood of his hometown and his constant access to fresh plant material from the Swiss mountains were a sound foundation for Gessner's detailed knowledge of the flora of Zurich, the Old Swiss Confederacy and the Alps. However, it also became clear that the question of what a natural scientist was particularly interested and knowledgeable in was shaped by his contacts – local ones and those he made during his stays abroad. Does this hold true for plants from other world regions, for which the process of gathering knowledge was much less straightforward?

2 Understanding What Is Far

Looking at American plants, it becomes obvious that Gessner's perspective was influenced by him being a German-speaking protestant, as demonstrated by the publications he owned on Central and Southern America. These included, for example, the account of the Dominican Bartolomé de las Casas

24 On Gessner's access to Scheuchzer's collection, see also: Luc Lienhard, "'La Machine Botanique': Zur Entstehung von Hallers Flora der Schweiz," in *Hallers Netz: Ein europäischer Gelehrtenbriefwechsel zur Zeit der Aufklärung*, ed. Martin Stuber, Stefan Hächler, and Luc Lienhard (Basel: Schwabe, 2005).

25 *Tabulae phytographicae*, 17, 31, 55.

26 For Haller via Gessner: Letter, J.-F. Séguier to J. Gessner, 21 March 1751. ZBZ, Hss., Ms. M 18.25.3. For Carlo Allioni's collecting activities: Letter, J.-F. Séguier to J. Gessner, 24 January 1751. ZBZ, Hss., Ms. M 18.25.2.

(1484/85–1566) on the conquest of the "Indian countries" (*Beschreibung der Indianischen Länder*, 1665) in a German version, which evinced strong criticism of the Catholic Spanish colonial power. Besides, he owned the travelogue of New Spain by another cleric, Thomas Gage (c. 1603–1656), who had converted to Protestantism.[27] The language of the publications available was key for the perception of plant knowledge. Gessner actively used German and Latin for his writing and, in addition, was able to understand French texts like Charles Plumier's (1646–1704) illustrated *Description des plantes de l'Amérique* and Louis Feuillée's (1660–1732) work on medical plants of Peru and Chile.[28] However, within the library, the French publications are clearly outnumbered by Latin and German ones. Of the texts in Gessner's native language, many had originally been published in English, like the above-mentioned account of New Spain. Beyond that, German translations of Philip Miller's *Gardener's Dictionary* and Humphry Marshall's *The American Grove* enabled the Zurich naturalist to acquire knowledge on the cultivation of foreign species as well as on trees and shrubs native to North America.[29] However, it was his command of Latin that enabled Gessner to explore new plant worlds, allowing him to access botanical knowledge published in London and Saint Petersburg as well as in various Swiss, Italian, German, French, and Dutch cities. In particular, the multitude of Latin books published in Leiden acquainted the Zurich naturalist with the variety of species growing not only in the Dutch city itself, but in botanical gardens elsewhere, for example in Montpellier.[30] Consulting all the indices and treatises written in Latin, Gessner learned about the flora of Sweden as well as that of the East and West Indies.

Judging from the indications of habitats given for American plants in the *Tabulae phytographicae* Gessner's knowledge of the continent's flora seems considerably nuanced. Among others, they refer to political territories such as Brazil (then a Portuguese colony), Jamaica (under British rule), Mexico (an administrative unit in the Spanish Viceroyalty of New Spain) as well as the British colonies of Virginia and of Carolina (a collective term for North and

27 Catalogus, ZBZ, Dr O 456.4, 123.

28 Catalogus, ZBZ, Dr O 456.4, 38 and 42.

29 The German translation of Humphry Marshall's (1722–1801) work by Christian Friedrich Hoffmann (1762–1820) was published in 1788 under the title: *Beschreibung der wildwachsenden Bäume und Staudengewächse in den vereinigten Staaten von Nordamerika*. Catalogus, ZBZ, Dr O 456.4, 53.

30 Gessner owned several descriptions of Leiden's botanical garden: Paul Hermann, *Horti academici Lugduno-Batavi Catalogus* (Lugduni Batavorum: Apud Cornelium Boutesteyn, 1687) and his *Paradisus batavus* in two editions. As well as: Herman Boerhaave's records from 1710 and 1720 and Adriaan van Royen, *Flora Leidensis* (Lugduni Batavorum: Apud Samuelem Luchtmans,1740).

South Carolina).[31] Some indications are even more specific, as in the case of the so-called "fever bark" (*Cinchona officinalis*), also shown in tableau XIII, known to be growing in Loja in [the Viceroyalty of] Peru ("Loxa Peruviae").[32]

Such exact indications of habitats were facilitated by the strong interest of Gessner's contemporaries in the matter.[33] Not only had he been able to read up about this species in a variety of publications, but he had also discussed the use of the plant against various diseases with other members of the Zurich Physical Society. The use of the plant as a remedy against fever and other afflictions was a subject of conversation in the Society's meetings, as its minutes and its published *Abhandlungen* (Transactions) show.[34] Experiences with previously unknown plant species from the Americas, e.g. the Cinchona bark, fuelled expectations among Gessner's contemporaries that more and more plant riches could be expected from that part of the world.[35] Therefore, European botanical amateurs and professionals eagerly collected species and recorded information about plants from the Americas. The same holds true for passionflowers, for example: a coveted species among European botanists who tried to obtain dried specimens and seeds to grow. In the 1770s, the gardeners of the Physical Society succeeded in cultivating at least three different species of *Passiflora* in the Society's garden in Zurich: blue passionflower (*Passiflora caerulea*), *Passiflora foetida* with white flowers, and the yellow-flowering *Passiflora lutea*.[36] Gessner had sixteen species of passionflowers engraved for his *Tabulae phytographicae* (fig. 4.2) and assembled a collection of passionflowers which

31 *Tabulae phytographicae*, for example for species on Tab. XXIX.

32 *Tabulae phytographicae*, 107.

33 For a study of the beliefs and practices around the "fever bark": Stefanie Gänger, *A Singular Remedy: Cinchona Across the Atlantic World, 1751–1820* (Cambridge, New York: Cambridge University Press, 2020).

34 The bark was discussed in two sessions, in August and September 1760. Tagbuch der Naturforschenden Gesellschaft, Staatsarchiv Zürich (StAZ), B IX 181, 50–60. Many thanks to my colleague Sarah Baumgartner (Bern) for pointing me towards this source. The discussions within the meetings resulted in a publication in the Society's review: Johann Heinrich Rahn et al., "Bemerkungen von der Würkung der Fieberrinde in verschiedenen Krankheiten," *Abhandlungen der Naturforschenden Gesellschaft in Zürich* 1 (Zurich: Heidegger und Compagnie, 1760).

35 This is suggested by Johan Gustaf Hallman (1726–1795) in his account on passionflowers. Johan Gustaf Hallman, "Passiflora", in Carl von Linné, *Amoenitates Academicae* 1 (Holmiae et Lipsiae: Apud Godofredum Kiesewette, 1749).

36 Anonymus [Salomon Schinz], *Catalogus Horti Botanici Societatis Physicae Turicensis*, Zurich 1776.

FIGURE 4.2 Tab. LIV of Gessner's *Tabulae phytographicae* showing passion flowers. Zentralbibliothek Zürich, Alte Drucke, NF 170

the German pharmacist Johann Gerhard Reinhard Andreae (1724–1793), visiting it in the early 1760s, considered remarkably extensive.[37]

Both plants, the Peruvian bark and the passionflowers, were the subject of dissertations published in the *Amoenitates Academicae*, a collective publication of dissertations by students of Linnaeus father and son. In this context, the Swedish botanist Hallman discussed the medical and economic uses of, as well as the religious and poetic interest in, passionflowers. In his account on the *Passiflora*, Hallman distinguished 22 species of the genus growing in North and South America. Gessner managed to acquire ten representatives of the genus until 1763 and added half a dozen in the following years, all of which he classified in volume 29 of the herbarium.[38] Because many others besides him were fascinated by passionflowers, Gessner could learn many details about the different species. Through Hallman's publication, he found out where exactly which species grew, whether in Virginia, Mexico, the Caribbean, Guyana, Peru or Brazil.[39] Thus, knowledge on American plants landed in Zurich by a detour via Sweden.

The same can be said for knowledge on the flora of other territories. As Alix Cooper has pointed out, in the course of the eighteenth century, local flora went global.[40] While the genre of the local flora originated in German university towns with botanical gardens in the seventeenth century, in the following decades, inventories were made for numerous territories all around the world. Gessner owned both – the older accounts published within the German Baroque context and those appearing throughout the eighteenth century – especially those within in the *Amoenitates Academicae*. The dissertations of students at the University of Uppsala inventoried and discussed the flora of England, Palestine, Montpellier, and Denmark, of Kamchatka and the Cape of Good Hope. Moreover, they also recorded the plants growing in much smaller territories, i.e. private gardens like the garden of the Swedish merchant family Alströmer and the manor of Linnaeus's patron Carl Gustaf Tessin (1695–1770).[41]

37 Johann Gerhart Reinhart Andreae, *Briefe aus der Schweiz nach Hannover geschrieben in dem Jare 1763*, 2. Abdruck (Zurich, Winterthur: Heinrich Steiner und Comp. Buchhändlern, 1776), 64.

38 *Hortus siccus*, vol. 29 "Gynandria", 79–95.

39 Hallman, *Passiflora*, 31. The *Tabulae phytographicae* do not give indications of localities for all plants depicted, although Gessner knew about them. This is the case for the passionflowers shown in plate LIV.

40 Alix Cooper, *Inventing the Indigenous: Local Knowledge and Natural History in Early Modern Europe* (Cambridge, New York: Cambridge University Press, 2007), 52.

41 Carl von Linné, *Amoenitates Academicae*. Catalogus, ZBZ, Dr O 456.4, 34. The different volumes of the *Amoenitates Academicae* appeared in Stockholm, and Erlangen and Leipzig respectively, from 1749 onwards. Gessner cited them regularly in all his publications.

In addition, Gessner also received practical plant knowledge filtered through the Swedish lens. The German translation of the transactions of the Royal Swedish Academy of Sciences published in Leipzig gave him access to plant knowledge originally published in Swedish. From 1749 onwards, Abraham Gotthelf Kästner (1719–1800) made the *Vetenskapsakademiens Handlingar* (*Transactions of the Royal Swedish Academy of Sciences*) available for a German audience, because he was convinced that information was absorbed more easily when given in the reader's native language. At the same time, Kästner was in no doubt that knowledge produced in Sweden was useful beyond that country.[42] Thanks to the German-language version, Gessner was able to learn about experiments with American trees and herbs in Sweden as well as the cultivation of exotic plants in Stockholm's botanical garden.[43] The publication allowed him to gain an impression of growing conditions of foreign plants in Sweden and thus also enabled him to compare them to the situation in Zurich. Knowledge on plants like liquorice and melons thus became accessible for the Zurich naturalist through a combination of efforts by botanists and publishers in Sweden and Leipzig.

The local floras that figured within the *Amoenitates Academicae* were based on material that the students had themselves collected or others had sent to Linnaeus, for example the dissertation on the Surinamese plants based on the collection by Daniel Rolander (1723–1793) and Frédéric-Louis Allamand (1736–post 1803) – both assembled during their time in Dutch service.[44] More

42 Abraham Gotthelf Kästner, *Der Königl. Schwedischen Akademie der Wissenschaften neue Abhandlungen aus der Naturlehre, Haushaltungskunst und Mechanik; auf das Jahr 1741*. Vol. 3 (Leipzig, Hamburg: Heinsius, 1750). On the transfers of the *Handlingar* see Ingemar Oscarsson, "'… who has had the courage and ambition to learn Swedish': The Handlingar of the Swedish Academy of Sciences in 18th Century European Translations, Adaptations, and Reviews," *La Révolution française* 13 (2018), accessed 20 April 2021, doi:10.4000/lrf.1910.

43 Gessner took notice of: Mårten Triewald, "Anmerkungen über die Pflanzung ausländischer Frucht- und anderer Bäume in Schweden aus eigener Prüfung und Versuch vorgestellt von Martin Friedwald [sic!], Königl. Mechanico und Fortificationscapitain", in *Der Königl. Schwedischen Akademie der Wissenschaften neue Abhandlungen aus der Naturlehre, Haushaltungskunst und Mechanik, auf das Jahr 1739* (Leipzig, Hamburg: Heinsius, 1749); Triewald, "Ein glücklich abgelaufener Versuch, ob die *Glycyrrhiza* oder das spanische Süßholz in Schweden wachse, und unsern Winter aushalten kann", in *Der Königl. Schwedischen Akademie der Wissenschaften neue Abhandlungen aus der Naturlehre, Haushaltungskunst und Mechanik, auf das Jahr 1744*, vol. 6 (Leipzig, Hamburg: Heinsius, 1751) as his notes preserved in manuscript prove: Repertorium. ZBZ, Hss., Ms. Z VIII 12.

44 David G. Frodin, *Guide to Standard Floras of the World* (Cambridge, New York: Cambridge University Press, 2001), 323. On Frédéric-Louis Allamand (1736–1809), a ship-surgeon from

generally, when looking at the geographical coverage of these publications, it becomes obvious that most of the accounts on extra-European plants were based on collections carried out within the sphere of influence of the Dutch colonial trading companies.[45]

Such Dutch networks were effective in many ways. The numerous references to "Virginia" throughout the *Tabulae phytographicae* suggest that Gessner was intimately acquainted with the flora of this territory. On the one hand, he could refer to and rely on the *Flora Virginica* published in 1743 and 1762, which in its second edition contained even a map of the area where John Clayton (1694/5–1773), pioneer of the English colony's flora, had botanised.[46] Showing geographical features like mountains and watercourses, houses and trees, the map divides Virginia into smaller units. Numbers written alongside the rivers and bays provide information on the names of the significant geographical features and political entities.[47] However, Gessner's familiarity with plants from the English colony was not based solely on his reading of *Flora Virginica*.[48] Instead, it was inspired by the plant material he had acquired from that region. The Zurich naturalist had a considerable number of dried specimens from Virginia in the herbarium, labelled "Claytono lecta in Virginia" and "Claytono ad Gronovium ex Virginia missa" referring to the Virginian specimens' journey from the Netherlands.[49] This proves that Gessner stored plant

Vaud (then subject-territory of Bern), see Ernst Schlunegger, "Allamand, Frédéric-Louis," in *Historisches Lexikon der Schweiz Online*, last modified April 26, 2021, https://hls-dhs-dss.ch/de/articles/043932/2021-04-26/ (accessed on 30 April 2021). For a study of foreign surgeons in Dutch service, see Iris Bruijn, *Ship's Surgeons of the Dutch East India Company: Commerce and the Progress of Medicine in the Eighteenth Century*, (Leiden: Leiden University Press, 2009).

45 On the use of the Linnaean binomial nomenclature in colonial contexts: Staffan Müller-Wille, "Walnuts at Hudson Bay, Coral Reefs in Gotland: The Colonialism of Linnean Botany," in *Colonial Botany: Science, Commerce, and Politics in the Early Modern World*, ed. Londa Schiebinger and Claudia Swan (Philadelphia: University of Pennsylvania Press, 2005).

46 Catalogus, ZBZ, Dr O 456.4. On Clayton: Edmund Berkeley and Dorothy Smith Berkeley, *John Clayton: Pioneer of American Botany* (Chapel Hill: The University of North Carolina, 1963).

47 However, it was only in the first decades of the 19th century that botanical distribution maps were developed within a community of amateurs interested in applied botanical knowledge. Even then, they rather mapped cultivated than wild plants, as the former were easier to depict and of interest to a wider public. Nils Güttler, "Drawing the Line: Mapping Cultivated Plants and Seeing Nature in Nineteenth-Century Plant Geography," in *New Perspectives on the History of Life Sciences and Agriculture*, ed. Denise Phillips and Sharon Kingsland (Cham: Springer International Publishing, 2015), 29.

48 Gessner owned both editions (1738 and 1762). Catalogus, ZBZ, Dr O 456.4, 42 and 47.

49 *Hortus siccus*, vol. 1 for example, comprises species from the genera *Circaea* and *Monarda*.

specimens collected by a specialist on Virginian flora. The Dutch botanist Jan Frederik Gronovius (1686–1762), whom he had met during his studies in Leiden, provided Gessner with plants that had been collected in Virginia by Clayton himself.

Gessner received many plants for the herbarium from Gronovius with notes connecting them to Virginia but to other places, too. Alongside a representative of the genus *Ziziphora*, for example, is noted "HB Gronovium" and next to the exemplar of *Sclarea Mexicana* L., a *Salvia* species, it is written "ex Herbario Gronov.," indicating that the specimens had been sent by Gessner's Dutch correspondent from his collection of dried plants. Gronovius also sent specimens from his garden, for example a representative of the genus *Verbena*.[50] The Dutch botanist had a plethora of plants from different corners of the world; and, obviously, he was eager to share them with his Swiss correspondent hoping to receive Alpine plants in return.[51] In his letters, Gronovius also shared information about the plants at his disposal, telling Gessner about his attempts to classify Clayton's specimens and his plans to publish *Flora Virginica* based on Clayton's findings.[52] Hence, this direct contact to the publisher and the fact that he received dried specimens from this region encouraged Gessner's interest in Virginia's plant world and led to his familiarity with it.

The same applies to other regions recorded by Gessner as a particular "habitat". The case of Vera Cruz confirms the finding that familiarity with the flora of a certain territory was significantly enhanced by personal contacts with an explorer and the reception of plant material from there.[53] Thus, for example, *Boerhavia erecta* in tableau VIII of the *Tabulae phytographicae* was labelled "Habitat in Vera Cruce."[54] At the same time, Gessner obtained an exemplar of this species for his herbarium which he marked as "Collecta Vera Cruce a D. Houston," referring to the British physician William Houstoun (1695?–1733)

50 *Hortus siccus*, vol. 1.

51 For a more detailed study of the relations between Zurich and Dutch botanists, see Knittel, "Netzwerke der Botanik."

52 Letter, J. F. Gronovius to J. Gessner, 12 April 1738. ZBZ, Ms. M 18.18.4.

53 Direct contacts with plant material also inspired Gessner's intensive involvement with the Siberian flora: Gessner made the acquaintance of Johann Georg Gmelin (1709–1755), who had explored that region during the Second Kamchatka Expedition between 1733 and 1743, when Gmelin visited Zurich in 1748. Subsequently, Gmelin became a supporter of the Physical Society's botanical garden and provided the botanists in Zurich with seeds and dried specimens for the herbarium, for example species belonging to the genera *Salicornia* and *Blitum*. *Hortus siccus*, vol. 1. For Gmelin's visit, see Boschung, *Johannes Gessner*, 96–98.

54 *Tabulae phytographicae*, 66 (*Boerhavia erecta*), 210–11 (*Yucca aloifolia*).

who collected in Central America.[55] In this case, too, the Dutch botanist Gronovius figured as a broker of specimens, inspiring Gessner's curiosity in the plant world of this far-away territory.[56] Subsequently, the study of the dried plants combined with the publications on their plant world, i.e. *Flora Virginica* and *Reliquiae Houstonianae* published by Joseph Banks (1743–1820) equipped the Zurich naturalist with specific knowledge of these regions he lacked for other ones.[57] Through the specimens sent by his Dutch correspondent, the Zurich naturalist became familiar first-hand with plants from American territories like Virginia and Vera Cruz, which sparked his interest to learn more about them.

The case of the American plants proves that the Zurich naturalist's knowledge of plants from far-away territories was crucially influenced by the botanical collecting and studies of his Dutch and Swedish contemporaries. Plant material and publications from Leiden and Amsterdam, Stockholm, and Uppsala shaped Gessner's perception of the world's flora. Together, these examples show that it was the result of two factors: first, access to published literature pertaining to a given (and often not clearly defined) region; second, the ability to interact with collectors who had more privileged or first-hand access to plant materials from these regions than the botanist residing in the European hinterland.

3 Puzzling over the Inaccessible

The importance of the two factors identified in the previous section cannot be understood simply by studying far-away regions for which Gessner eventually succeeded in developing a nuanced understanding. In fact, it is even better illustrated by regions for which he failed to do so, where one of the factors was clearly lacking. Namely, the herbarium Gessner put together for the Zurich Physical Society, included four hundred plants declared to be "African" and perceived as something special by his contemporaries as well as later generations.[58] Gessner had bought them from the estate of another collector,

55 *Hortus siccus*, vol. 1.

56 Letter, J. F. Gronovius to J. Gessner, 20 February 1734. ZBZ, Hss., Autogr Ott, Gronovius; Letter, J. Gessner to J.-F. Séguier, Zürich, 31 July 1752. Bibliothèque municipale de Nîmes, ms 498.14.

57 Catalogus, ZBZ, Dr O 456.4, 43.

58 This number is given by Ferdinand Rudio in his 1896 publication. Ferdinand Rudio, "Festschrift der Naturforschenden Gesellschaft in Zürich; 1746–1896," *Vierteljahrsschrift der Naturforschenden Gesellschaft in Zürich* 41 (1896).

who had picked them up around the Cape of Good Hope. While posing an enormous challenge to the Zurich naturalist, this collection of African plants offers an informative insight on the difficulties in his learning about the flora of far-away regions.

In 1753, with the help of his correspondents, Gessner managed to buy the collection of the recently deceased physician Laurent Garcin (1683–1752) from (then Prussian) Neuchâtel. As a young man, Garcin had travelled via the Cape of Good Hope to India, Java, Sumatra and Ceylon as a surgeon in the service of the Dutch East India Company (VOC).[59] The collection of dried plants he assembled during this journey gained some reputation in the following decades and by the time of his death had become a valuable good. Having been told about it by his correspondent Abraham Gagnebin (1707–1800), Gessner had cast an eye on the collection for some time, and wanted to obtain some specimens.[60] Additionally, the Zurich naturalist had learned about Garcin's findings from the memoirs published in the *Philosophical Transactions* of the Royal Society and Johannes Burmann's *Thesaurus Zeylanicus*, which he held in his library.[61] Thus, he knew that Garcin's collection consisted of hundreds of species from the Cape of Good Hope and the Indies, including some rare pieces.

After Garcin's death, Gessner rushed to buy the collection without first ascertaining its condition. As his correspondents acted fast, he was successful in buying the material he wished. Unfortunately, it proved difficult to make use of this acquisition. Not only had the dried specimens been shaken during the transport, but they also arrived in Zurich without labels.[62] Subsequently, Gessner tried to obtain written information on the newly acquired specimens For weeks, he pushed his correspondent Samuel Engel (1702–1784) to put

59 The plants he had collected during this journey were described by Nicolaas Laurens Burman (1734–1793) in his *Flora Indica* (Amsterdam 1768). For the Indian plants of Garcin, see Alexandra Cook, "Laurent Garcin, M.D. F.R.S: A Forgotten Source for N. L. Burman's Flora Indica (1768)," *Harvard Papers in Botany* 21, no. 1 (2016).

60 Letter, A. Gagnebin to J. Gessner, 27 August 1749. ZBZ, Autogr Ott, Gagnebin. On Gagnebin: Baldi, "Circulation."

61 Laurent Garcin, II. Memoirs communicated by Mons. Garcin to Mons. St. Hyacinthe, F.R.S. […], in *Philosophical Transactions* 36 (1730); Garcin, "The Settling of a new Genus of Plants, called after the Malayans, MANGOSTANS," *Philosophical Transactions* 38 (1733). Johannes Burman, *Thesaurus Zeylanicus* (Amstelaedami: Apud Janssonio-Waesbergios, & Salomonem Schouten, 1737). The posthumously published auction catalogue of Gessner's library proves ownership: ZBZ, Dr O456.4, 129; 41.

62 Letter, S. Engel to J. Gessner, 21 April 1753. ZBZ, Hss., Ms. M 18.28.3.

pressure on his contacts close to Garcin's heirs to send more information.[63] Gessner hoped that there were notes in the estate that would enable him to learn what kind of plants exactly he had bought and where they came from. However, over time it became clear that hope was in vain and Garcin had never named the plants in his collection.

Consequently, identifying the plants proved extremely difficult: not only because the Cape of Good Hope, where Garcin had collected the specimens, remained inaccessible to Gessner, but also because printed material on those regions was limited and difficult to obtain. A botanist striving to identify material like this had to have contacts that would supply him with the publications and be able to afford the often-expensive prints.[64] In the following decades, Gessner took on that task and never stopped purchasing new publications on the region where Garcin had collected, for example the *Descriptiones plantarum ex capite Bonae Spei* by Peter Jonas Bergius (1730–1790) comprising engravings of the South African flora.[65] Despite all his efforts to expand his knowledge of the plant world of these far-away regions, Gessner did not succeed. All he could do was to classify parts of the dried specimens from Garcin's collection, while many have remained unidentified. For those plants he could not sort according to the Linnaean taxonomy, Gessner created a separate herbarium volume entitled *Plantae vagae*, relinquishing his ambitions to classify them. Hence, although many European travellers botanised around the Cape of Good Hope, the information he could accumulate in Zurich was not enough for Gessner to classify the specimens purchased from Garcin's estate. Some of the specimens were in an extremely poor state; for others the Zurich naturalist probably found no description in the publications he had access to. As Gessner could neither travel to the region nor query the collector or anybody

63 Correspondence with Samuel Engel in July 1753: Letter, S. Engel to J. Gessner, 2 July 1753. ZBZ, Hss., Ms. M 18.28.8; Letter, S. Engel to J. Gessner, 12 July 1753. ZBZ, Hss., Ms. M 18.28.9. For details of the discussion see Knittel, "Netzwerke der Botanik."

64 This is also expressed by the young Jakob Christoph Ramspeck (1722–1797) from Basel, who only saw the learned Albrecht von Haller capable of identifying the plants collected by Garcin. While the young Ramspeck obviously wanted to flatter the reputable naturalist Haller with this statement, it reveals that knowledge on these still little-known African plants remained reserved to botanists well equipped with books and correspondents. Letter, J. Ch. Ramspeck to A. von Haller, 8 January 1754. Boschung, *Johannes Gessner*, 105.

65 Nicolaas Laurens Burman, *Flora Indica: cui accedit series zoophytorum Indicorum, nec non prodromus florae Capensis* (Amsterdam, 1768); Peter Jonas Bergius, *Descriptiones plantarum ex capite Bonae Spei: cum differentiis specificis, nominibus trivialibus, et synonymis auctorum justis: secundum systema sexuale / ex autopsia concinnavit atque solicite digessit Petrus Jonas Bergius* (Stockholm, 1767). *Catalogus*, ZBZ, Dr O 456, 41 and 46.

else familiar with the plants of the Cape of Good Hope about the localities, they remained "vague" to him.

The difficulties Gessner encountered with the specimens acquired were not a unique, individual problem. In fact, European naturalists had for centuries struggled to properly annotate collected specimens. Written instructions existed on how to gather and preserve plant specimens, requiring the specification of each object's provenance.[66] In his *Instructions,* the English naturalist John Woodward (1665–1728) asked collectors to write on the outside of the quires of brown paper designed to contain the specimens: "*in what Country the inclosed Collection of Plants were gathered.*" Additionally, he recommended a spatial separation of the specimens of different countries so that, should the markings get lost, the grouping of samples would reference their common provenance.[67] While still on site, the collectors would map the plants in a way that was comprehensible to their readers and visitors, predominantly unfamiliar with the collecting sites. In his book *Collecting the World*, James Delbourgo has shown that, in many cases, names of plantations or of their owners, ports and trading posts were used as reference points, *en passant* employing a colonial order that subsequently affected the perception of the island's plant world.[68]

Together, these examples show that European botanists' knowledge about the flora of distant territories they could not visit in person depended on their ability to request details from those who assembled the herbarium collection. The mere availability of dried specimens as well as access to the growing body of literature on their regions of origin were not sufficient for that. Rather, the possibility of bringing specimens like "Garcin's African plants" together with book knowledge available in Europe strongly depended on each collector's diligence in recording at the scene.

66 For a history of these instructions in different political contexts, see the contribution of Simona Boscani Leoni in this volume. Especially collectors in Europe that invested money into the gathering of *naturalia* sought to enhance their chances of receiving well-preserved specimens enriched with as much information as possible, For the instructions of naturalists from London, see Mark Laird and Karen Bridgman, "American Roots: Techniques of Plant Transportation and Cultivation in the Early Atlantic World," in *Ways of Making and Knowing: The Material Culture of Empirical Knowledge*, ed. Pamela H. Smith, Amy Meyers, and Harold Cook (Ann Arbor: University of Michigan Press, 2014); Charles E. Jarvis, "'Take with You a Small Spudd or Trowell': James Petiver's Directions for Collecting Natural Curiosities," in *Naturalists in the Field*, ed. Arthur MacGregor, (Leiden: Brill, 2018).

67 John Woodward, *Brief Instructions for Making Observations in All Parts of the World: As also for Collecting, Preserving, and Sending over Natural Things* (London: Printed for Richard Wilkin, 1696), 12.

68 James Delbourgo, *Collecting the World: The Life and Curiosity of Hans Sloane* (London: Allen Lane, 2017), 101.

4 Conclusion

Based on the example of Johannes Gessner in Zurich, this contribution has focused on the factors influencing the plant knowledge accumulated by 18th-century naturalists working away from botanical centres such as university towns and port cities. As shown, the Zurich botanist not only was an expert in local Swiss plants, especially those growing in mountainous areas of the eastern cantons, but he also had a more profound knowledge of plants from some far-away territories. The analysis revealed that three factors in particular were crucial in determining which territories he succeeded in developing a detailed understanding of.

First, the large number of books and journals he accumulated in his library shows that Gessner was a wealthy botanist with a working correspondence network. Hence, although he conducted his work away from a large botanical centre, he managed to keep pace with the emerging plant knowledge outside his home ground. However, this applies mainly to publications available in languages he easily understood, more specifically treatises in Latin as well as German translations of accounts originally published in vernaculars such as Swedish and English. Furthermore, Gessner's knowledge of the habitats of specific plants such as *Cinchona* or passionflowers was facilitated by his contemporaries' intensive concern with them. Their therapeutic benefit or their visually appealing flowers made these plants relatively well-studied objects. Consequently, botanists in Zurich and elsewhere could learn about their distribution, though they had only limited access to them.

Second, Gessner's more detailed study of plants from certain regions was prompted by the supply of dried specimens gathered by experts on the spot. This helped him gain a first-hand understanding of a territory`s otherwise unknown flora. However, the impact of these dried specimens on Gessner`s knowledge was not limited to these same species. Rather, the study of the dried specimens encouraged him to purchase any new publications on these areas, which in turn would help him further consolidate his knowledge of these regions' flora more generally.

Third and most importantly, we have shown that it is precisely the interplay of these two factors that enabled Gessner to develop knowledge of a specific region. Knowledge of plants from distant territories remained rudimentary as long as written information was missing and available plant exemplars provided inadequate comparison with published species. The case of the herbarium specimens purchased from Garcin demonstrated that only scattered exemplars were identified and matched to flora of the Cape. So, the bulk of the plants remained only very vaguely associated with Africa. And yet Gessner had access to a growing literature on the Cape region.

Because he depended on both – namely accessible literature and access to dried specimens from a given region – Gessner's 'botanical world map' was far from universal. Rather, he had an extremely specific but in no way unusual perspective on the plants of far-away regions. The Zurich botanist saw them through Dutch-Swedish glasses, or more accurately binoculars. For knowledge in print on the flora of distant regions he drew mainly on publications authored in Dutch and Swedish university towns, whether in their original Latin version or in German translation. More exactly, the bulk of the publications Gessner used originated in a Linnaean context, i.e. they were published by early adopters and supporters of the system, or were written by students of Linnaeus and as such followed a particular pattern. At the same time, much of this plant knowledge on far-away territories was generated within the sphere of influence of the Dutch colonial trading companies. In conclusion, by using this knowledge, the Zurich botanist contributed to the universalisation of a specific perception of territories and their flora.

Acknowledgements

This contribution is based on research funded by the Swiss National Science Foundation and carried out in the context of the project "Kulturen der Naturforschung" (Cultures of Natural History), Project-Nr. 144731, led by Simona Boscani Leoni at the University of Bern (2014–2018).

Bibliography

Manuscripts

University of Zurich (UZH), Department of Systematic and Evolutionary Botany

Hortus siccus Societatis Physicae Tigurinae, collectus et Linnaeana methodo dispositus a Joanne Gesnero, 1751.

Zentralbibliothek Zürich, Handschriftensammlung

Autogr Ott, Gagnebin. Letter, A. Gagnebin to J. Gessner, 27 August 1749.

Autogr Ott, Gronovius. Letter, J. F. Gronovius to J. Gessner, 20 February 1734.

Ms. M 18.10. Autobiography J. Gessner.

Ms. M 18.18. Correspondence of J. F. Gronovius with J. Gessner.

Ms. M 18.25. Correspondence of J.-F. Séguier with J. Gessner.
Ms. M 18.28. Correspondence of S. Engel with J. Gessner.
Ms. Z VIII 12. Repertorium J. Gessner.

Bibliothèque municipale de Nîmes

Ms 498.14 Letter, J. Gessner to J.-F. Séguier, Zürich, 31.07.1752.

Staatsarchiv Zürich (StAZ)

B IX 181 Tagbuch der Naturforschenden Gesellschaft Zürich, 3.

Printed Sources

Andreae, Johann Gerhart Reinhart. *Briefe aus der Schweiz nach Hannover geschrieben in dem Jare 1763*, 2. Abdruck. Zurich, Winterthur: Heinrich Steiner und Comp. Buchhändlern, 1776.

Anonymus [Salomon. Schinz], *Catalogus Horti Botanici Societatis Physicae Turicensis*, Zurich 1776/72.

Burman, Johannes. *Thesaurus Zeylanicus, exhibens Plantas in Insula Zeylana nascentes* [...], Amstelaedami: Apud Janssonio-Waesbergios, & Salomonem Schouten, 1737.

Catalogus. ZBZ, Rare Book Department, Dr O 456.4 (Johann Füssli & Son, Catalogus Librorum Bibliothecae Joannis Gessneri [...], Zurich 1798).

Garcin, Laurent. "II. Memoirs communicated by Mons. Garcin to Mons. St. Hyacinthe, F.R.S." *Philosophical Transactions* 36 (1730), 377–94.

Garcin, Laurent. "The Settling of a new Genus of Plants, called after the Malayans, MANGOSTANS." *Philosophical Transactions* 38 (1733): 232–242.

Hallman, Johan Gustaf. "Passiflora." In *Amoenitates Academicae* 1, Holmiae et Lipsiae: Apud Godofredum Kiesewette, 1749.

Kästner, Abraham Gotthelf. *Der Königl. Schwedischen Akademie der Wissenschaften neue Abhandlungen aus der Naturlehre, Haushaltungskunst und Mechanik; auf das Jahr 1741*. Leipzig, Hamburg: Heinsius,1750.

Linnaeus, Carl. *Species plantarum: exhibentes plantas rite cognitas, ad genera relatas, cum differentiis specificis, nominibus trivialibus, synonymis selectis, locis natalibus, secundum systema sexuale digestas*, Holmiae: Impensis Laurentii Salvii, 1753.

Rahn, Johann Heinrich et al. "Bemerkungen von der Würkung der Fieberrinde in verschiedenen Krankheiten." *Abhandlungen der Naturforschenden Gesellschaft in Zürich* 1 (1760): 189–210.

Schinz, Christoph Salomon. *Johannis Gessneri Tabulae phytographicae, analysin generum plantarum exhibentes cum commentatione edidit Christ. Sal. Schinz*. Zurich: Fuessli, 1795–1804.

Triewald, Mårten. "Anmerkungen über die Pflanzung ausländischer Frucht- und anderer Bäume in Schweden aus eigener Prüfung und Versuch vorgestellt von

Martin Friedwald, Königl. Mechanico und Fortificationscapitain." *Der Königl. Schwedischen Akademie der Wissenschaften neue Abhandlungen aus der Naturlehre, Haushaltungskunst und Mechanik, auf das Jahr 1739*, Leipzig, Hamburg: Heinsius, 1749: 250–54.

Triewald, Mårten. "Ein glücklich abgelaufener Versuch, ob die Glycyrrhiza oder das spanische Süßholz in Schweden wachse, und unsern Winter aushalten kann." *Der Königl. Schwedischen Akademie der Wissenschaften neue Abhandlungen aus der Naturlehre, Haushaltungskunst und Mechanik, auf das Jahr 1744*, Hamburg, Leipzig: Heinsius, 1751: 226–30.

Woodward, John. *Brief Instructions for Making Observations in All Parts of the World: As also for Collecting, Preserving, and Sending over Natural Things*. London: Printed for Richard Wilkin, 1696.

Literature

Baldi, Rossella. "La circulation du savoir botanique par le texte et par l'image: le 'Species plantarum' d'Abraham Gagnebin." In *Rousseau Botaniste: Je vais devenir plante moi-même*, edited by Claire Jaquier and Timothée Léchot, 15–24. Fleurier: Editions du Belvédère, 2012.

Batsaki, Yota, Anatole Tchikine, and Sarah Burke Cahalan, eds. *The Botany of Empire in the Long Eighteenth Century*. Baltimore: Dumbarton Oaks Publishing, 2017.

Baumgartner, Sarah. "Das nützliche Wissen: Akteure, Tätigkeiten, Kommunikationspraxis und Themen der Naturforschenden Gesellschaft in Zürich, 1746 bis ca. 1830" (PhD diss., University of Bern, 2019).

Beretta, Marco, and Alessandro Tosi, eds. *Linnaeus in Italy: The Spread of a Revolution in Science*. Sagamore Beach: Science History Publications, 2007.

Berkeley, Edmund, and Dorothy S. Berkeley. *John Clayton: Pioneer of American Botany*. Chapel Hill: The University of North Carolina, 1963.

Boscani Leoni, Simona. "Men of Exchange: Creation and Circulation of Knowledge in the Swiss Republics of the Eighteenth Century." In *Scholars in Action: The Practice of Knowledge and the Figure of the Savant in the 18th Century*, edited by André Holenstein, Hubert Steinke, and Martin Stuber, 507–33. Leiden: Brill, 2013.

Boschung, Urs. *Johannes Gessner (1709–1790): Der Gründer der Naturforschenden Gesellschaft in Zürich; Seine Autobiographie – aus seinem Briefwechsel mit Albrecht von Haller; Ein Beitrag zur Geschichte der Naturwissenschaften in Zürich im 18. Jahrhundert*. Alpnach-Dorf: KOPRINT AG, 1995.

Bruijn, Iris. *Ship's Surgeons of the Dutch East India Company: Commerce and the Progress of Medicine in the Eighteenth Century*. Leiden: Leiden University Press, 2009.

Cook, Alexandra. "Laurent Garcin, M.D. F.R.S: A Forgotten Source for N. L. Burman's Flora Indica (1768)." *Harvard Papers in Botany* 21, no. 1 (2016): 31–53.

Cooper, Alix. *Inventing the Indigenous: Local Knowledge and Natural History in Early Modern Europe*. Cambridge, New York: Cambridge University Press, 2007.

Delbourgo, James. *Collecting the World: The Life and Curiosity of Hans Sloane*. London: Allen Lane, 2017.

Egmond, Florike, Esther van Gelder, and Nicolas Robin, eds. *Flowers of Passion and Distinction: Practice, Expertise and Identity in Clusius' World*. Stuttgart: Franz Steiner Verlag, 2012.

Frodin, David G., *Guide to Standard Floras of the World*. Cambridge, New York: Cambridge University Press 2001.

Gänger, Stefanie. *A Singular Remedy: Cinchona Across the Atlantic World, 1751–1820*. Cambridge, New York: Cambridge University Press, 2020.

Gaziello, Catherine. *L'expédition de Lapérouse 1785–1788: réplique française aux voyages de Cook*. Paris: CTHS, 1984.

Güttler, Nils. "Drawing the Line: Mapping Cultivated Plants and Seeing Nature in Nineteenth-Century Plant Geography." In *New Perspectives on the History of Life Sciences and Agriculture*, edited by Denise Phillips and Sharon Kingsland, 27–52. Cham: Springer International Publishing, 2015.

Heller, Sabine. *Boerhaaves Schweizer Studenten: Ein Beitrag zur Geschichte des Medizinstudiums*. Zurich: Juris-Verlag, 1984.

Hodacs, Hanna. "Linnaean Scholars Out of Doors: So Much to Name, Learn and Profit From." In *Naturalists in the Field: Collecting, Recording and Preserving the Natural World from the Fifteenth to the Twenty-First Century*, edited by Arthur MacGregor, 240–57. Leiden: Brill, 2018.

Jarvis, Charles E. "'Take with you a small Spudd or Trowell': James Petiver's Directions for Collecting Natural Curiosities." In *Naturalists in the Field: Collecting, Recording and Preserving the Natural World from the Fifteenth to the Twenty-First Century*, edited by Arthur MacGregor, 212–39. Leiden: Brill, 2018.

Knittel, Meike. "Netzwerke der Botanik: Johannes Gessner (1709–1790) und die botanische Forschung im 18. Jahrhundert" (PhD diss., University of Bern, 2018).

Knittel, Meike and Reto Nyffeler. "Flora Alpina." In *Montan-Welten: Alpengeschichte abseits des Pfades*, edited by Tina Asmussen. Æther 03. Zürich: intercom Verlag, 2019. Last modified: May 2019. https://aether.ethz.ch/ausgabe/montan-welten/flora-alpina/, accessed on 26 April 2021.

Knittel, Meike. "'Dominus Creavit Ex Terra Medicamenta': Heilpflanzenwissen in Johannes Gessners Phytographia sacra." In *Wer das Gras wachsen hört: Wissensgeschichte(n) der pflanzlichen Ressourcen vom Mittelalter bis ins 20. Jahrhundert*, edited by Simona Boscani Leoni and Martin Stuber. Jahrbuch für Geschichte des ländlichen Raumes 14. Innsbruck: Studien-Verlag, 2017.

Knittel, Meike. "Devenir botaniste au XVIIIe siècle: les premiers voyages de Gessner et des jeunes Zurichois dans les Alpes." *Bulletin de l'Association culturelle pour le voyage en Suisse* 20, 2019: 5–9.

Laird, Mark and Karen Bridgman. "American Roots: Techniques of Plant Transportation and Cultivation in the Early Atlantic World." In *Ways of Making and Knowing: The Material Culture of Empirical Knowledge*, edited by Pamela H. Smith, Amy Meyers and Harold Cook, 164–93. Ann Arbor: University of Michigan Press, 2014.

Leu, Urs, ed. *Natura Sacra: Der Frühaufklärer Johann Jakob Scheuchzer (1672–1733)*. Zug: Achius Verlag, 2012.

Leu, Urs. "Konrad Gessner und die Neue Welt." *Gesnerus* 49, 3–4 (1992): 279–309.

Lienhard, Luc. "'La machine botanique': Zur Entstehung von Hallers Flora der Schweiz," in *Hallers Netz: Ein europäischer Gelehrtenbriefwechsel zur Zeit der Aufklärung*, edited by Martin Stuber, Stefan Hächler and Luc Lienhard, 371–410. Basel: Schwabe, 2005.

Marti, Hanspeter. "Leiden." *Historisches Lexikon der Schweiz Online*. Last modified November 20, 2008. https://hls-dhs-dss.ch/de/articles/006615/2008-11-20/, accessed on 30 April 2021.

Miller, David P. and Peter H. Reill, eds. *Visions of Empire: Voyages, Botany and Representations of Nature*. Cambridge, New York: Cambridge University Press, 1996.

Müller-Wille, Staffan. "Walnuts at Hudson Bay, Coral Reefs in Gotland: The Colonialism of Linnean Botany." In *Colonial Botany: Science, Commerce, and Politics in the Early Modern World*, edited by Londa Schiebinger and Claudia Swan, 34–48. Philadelphia: University of Pennsylvania Press, 2005.

Nieto, Mauricio. "Presentación gráfica, desplazamiento y aprobación de la naturaleza en las expediciones botánicas del siglo XVIII," *Asclepio* 47, no. 2 (1995): 91–107.

Önnerfors, Andreas. "Auswärtige Saamen und Gewächse ingleichen zur Correspondence: Die Verbreitung Linnéscher Naturalhistorie in Schwedisch-Pommern im 18. Jahrhundert," in *Wissen im Netz: Botanik und Pflanzentransfer in europäischen Korrespondenznetzen des 18. Jahrhunderts*, edited by Regina Dauser, Stefan Hächler, Michael Kempe, Franz Mauelshagen and Martin Stuber, 91–111. Berlin: Akademie-Verlag, 2008.

Oscarsson, Ingemar. "'… who has had the courage and ambition to learn Swedish': The Handlingar of the Swedish Academy of Sciences in 18th Century European Translations, Adaptations, and Reviews." *La Révolution française* 13 (2018): 1–23. doi:10.4000/lrf.1910, accessed 20 April 2021.

Pardo-Tomás, José. "Two Glimpses of America from a Distance: Carolus Clusius and Nicolás Monardes." In *Carolus Clusius: Towards a Cultural History of a Renaissance Naturalist*, edited by Florike Egmond, P. G. Hoftijzer and Robert P. W. Visser, 173–94. Amsterdam: Koninklijke Nederlandse Akademie van Wetenschappen, 2007.

Puig-Samper, Miguel A. "Difusión e institucionalización del sistema linneano en España y América." in *Mundializacion de la ciencia y cultura nacional: Actas del congreso international "Ciencia, Descubrimento y Mundo Colonial,"* edited by Antonio Lafuente, Alberto Elena and María Luísa Ortega, 349–59. Aranjuez: Doce Calles, 1993.

Rudio, Ferdinand. "Festschrift der Naturforschenden Gesellschaft in Zürich; 1746–1896." *Vierteljahrsschrift der Naturforschenden Gesellschaft in Zürich* 41 (1896).

Schär, Bernhard C. *Tropenliebe: Schweizer Naturforscher und niederländischer Imperialismus in Südostasien um 1900*. Frankfurt am Main: Campus Verlag, 2015.

Schlunegger, Ernst. "Allamand, Frédéric-Louis." Historisches Lexikon der Schweiz Online. Last modified April 26, 2021. https://hls-dhs-dss.ch/de/articles/043932/2021-04-26/, accessed on 30 April 2021.

Sigerist, Henry E. *Albrecht von Haller's Briefe an Johannes Gesner (1728–1777)*. Berlin: Weidmann, 1923.

Stearn, William T. "The Influence of Leyden on Botany in the Seventeenth and Eighteenth Century." *The British Journal for the History of Science* 1, no. 2 (1962): 137–58.

Zangger, Andreas. *Koloniale Schweiz: Ein Stück Globalgeschichte zwischen Europa und Südostasien (1860–1930)*. Bielefeld: transcript, 2011.

CHAPTER 5

The Secrets of Indians: Native Knowers in Enlightenment Natural Histories of the Southern Americas

Stefanie Gänger

The "primitive inhabitants of Peru," as José Hipólito Unánue (1755–1833) put it in his 1791 essay on the nascent discipline of botany in the Viceroyalty of Peru, had long been "excellent herbalists (*excelentes herbolarios*)." "Devoted to empirical agriculture and medicine," they had come to discover the virtues of many plants and most of the country's historians were agreed that the Indians' plant knowledge, passed on "from parents to children by the ministry of the word (*de padres a hijos por el ministerio de la palabra*)," was superior to that even "of the medical profession (*médicos de profesión*)." For Unánue and his intellectual community of naturalists that knowledge was virtually lost, however. For, in his view, "in the wake of the upheaval of the conquest," the Indians' temper had become "mysterious, stubborn and mistrustful" and led them "to conceal their knowledge from the Europeans." Neither flattery nor bribery nor threats of violence would induce them to betray their secrets and it was only sometimes, through the imprudence or treason of one of them, that their knowledge about the virtues of medicinal plants was ever unveiled to outsiders.[1]

The topos of the reluctant Indian informer, unwilling to share his or her excellent, empirical knowledge of botany, medicine, agronomy, and other areas of natural historical inquiry, was one of the earliest and most enduring elements of conventional wisdom about the inhabitants of the New World – an obsession "fixed in a pattern," as the art historian Natalia Majluf put it, traceable to the very beginning of the colonial period, and "repeated and recreated obsessively" thereafter.[2] Although countless studies have in recent years examined

1 José Hipólito Unanue, "Botánica. Introducción a la descripción científica de las plantas del Perú, vol. II, núms. 43 y 44, 68–86," in *El Mercurio Peruano: 1790–1795. Antología*, ed. Jean-Pierre Clément (Frankfurt am Main: Vervuert, 1998 (1791)), 96–97.

2 In the context of her study on the theme of the Indian in modern Peruvian painting, Natalia Majluf examined representations of "the widespread stereotype" "of the silent Indian" "who refuses to communicate with the Creole." She argues that it "can be traced back to the colonial period" but "gained new relevance within the late 18th century interest in archaeology

DOI:10.1163/9789004412477_006

one or another aspect of native "skill in nature" in and beyond Spanish and Portuguese America and the Caribbean – its stereotypical nature,[3] profitable commodification,[4] or prejudiced erasure and dismissal[5] – the roots and history of the "inscrutable" native knower as an epistemic category has yet to be the focus of historians' attention. This contribution is concerned with the intellectual ancestry and genealogy of the topos of the secretive Indian knower of New World territories and nature, which flourished in enlightened natural histories and travelogues concerned with the Southern Americas. It argues that the widespread and long-lived trope of the "mistrustful" native knower was, like other archetypes of indigeneity – the "man-eating" Indian,[6] the non-acquisitive, "idle" Indian, who needed to be coerced into labour,[7] or "the ecological Indian," living "in harmony," and sustainably, with his natural environment[8] – a historicizeable cultural discourse with identifiable roots in Christian and humanist legacies. In a first part, the paper sketches the contours of the topos; in a second part it traces its resemblance to, and intellectual kinship with, other inscrutable, "illiterate knowers" in late-medieval and early

and the description of the country's geography." Natalia Majluf, *The* "Creation of the Image of the Indian in 19th-Century Peru: The Paintings of Francisco Laso (1823–1869)" (PhD diss., University of Texas, 1996), 350–51.

3 On the attribution of 'skill in nature' to the "ecological Indian," see Shepard Krech, *The Ecological Indian: Myth and History* (New York: W. W. Norton, 1999), 19. Exotic man's "impenetrable inscrutability" is generally acknowledged to be an enlightened stereotype. George S. Rousseau and Roy Porter, "Introduction: Approaching Enlightenment Exoticism," in *Exoticism in the Enlightenment*, ed. George S. Rousseau and Roy Porter (Manchester, New York: Manchester University Press, 1990), 4–5. Natalia Majluf first pointed to the stereotypical nature of the "inscrutable Indian" in Andean South America. Majluf, "The Creation of the Image of the Indian in 19th-Century Peru." For North America, see Susan Scott Parrish, *American Curiosity: Cultures of Natural History in the Colonial British Atlantic World* (Chapel Hill: University of North Carolina Press, 2006), 217.

4 On the history of colonial "bioprospecting," see Londa Schiebinger, *Plants and Empire* (Cambridge, MA, London: Harvard University Press, 2004); Schiebinger, "Prospecting for Drugs. European Naturalists in the West Indies," in *The Postcolonial Science and Technology Studies Reader*, ed. Sandra Harding (Durham, London: Duke University Press, 2011).

5 See, for instance, Kathleen S. Murphy, "Translating the Vernacular: Indigenous and African Knowledge in the Eighteenth-Century British Atlantic," *Atlantic Studies* 8, no. 1 (2011).

6 The cannibal epithet has been applied to different human groups over the centuries – more prominently, America's Aztec and Tupinambá societies in the wake of the Iberian conquest. See, for instance, the classic work by William Arens, *The Man-Eating Myth: Anthropology and Anthropophagy* (Oxford: Oxford University Press, 1979), 10–13.

7 "Idleness," long attributed in humanist thought to the "poor" and to beggars, was transferred to an entire society – the Indians – in the wake of the conquest. See Nicolas Sanchez-Albornoz, "El trabajo indigena en los Andes: Teorías del siglo XVI," *Revista ecuatoriana de historia económica* 2 (1987): 173.

8 Krech, *The Ecological Indian: Myth and History*.

modern Europe – "rustics" and peasants, "simpletons," and women –; in a last part, it studies the grounds of the stereotype's worldwide propagation and dissemination during the Enlightenment. Though the essay traces the association between indigeneity, secrecy, and skill in nature across the Americas and into our present, its analytical focus rests on the liminal space of post-conquest Spanish America. As shall be shown, it was in this first period of "upheaval," decades before Portugal, England, and France came to the exploration of New World nature,[9] that early modern observers first sought to make sense of, and classify, the knowledge of America's new-found inhabitants, to suit their epistemic paragons and to make it familiar.[10]

1 Unalienable Truths

The figure of the Indian informer, anxious to "conceal" his or her "excellent," "empirical" knowledge of the land, was an iconic form in natural historical writings concerned with the Southern Americas in the late 1700s and early 1800s–a field of enquiry that, in the Iberian tradition at least, encompassed the collecting and study of mined silver, medicinal bezoars, "curious" antiquities, and valuable timber alike in the period.[11] The trope was prevalent in creole writings and some of the earliest foreign travelogues, which contained various anecdotes about how "the Indians," out of "hatred," had concealed from officials the "situation of a mine,"[12] or the site of "ancient monuments" dating back to the time before the arrival of the Spanish.[13] It flourished particularly, however, in natural historical writings devoted to the healing and poisonous properties

9 Scott Parrish, *American Curiosity*, 28.

10 For a discussion of the vast literature on this matter, see Karen Ordahl Kupperman, "Introduction: The Changing Definition of America," in *America in European Consciousness, 1493–1750*, ed. Karen Ordahl Kupperman (Chapel Hill, London: University of North Carolina Press, 1995). "News out of the *newfound world*" is a reference to the English translation of Monardes's history: Nicolás Monardes, *Joyfull Newes out of the Newe Founde Worlde* (London: Willyam Norton, 1577).

11 Paula De Vos, "The Rare, the Singular, and the Extraordinary: Natural History and the Collection of Curiosities in the Spanish Empire," in *Science in the Spanish and Portuguese Empires, 1500–1800*, ed. Daniela Bleichmar, Paula de Vos, Kristin Huffine, and Kevin Sheehan (Stanford: Stanford University Press, 2009).

12 Johann Jakob von Tschudi, *Peru: Reiseskizzen aus den Jahren 1838–1842*, 2 vols., vol. 2 (St. Gallen: Scheitlin und Zollikofer, 1846), 138–139.

13 D.F.D.P.D.L.M.L., "Carta remitida a la Sociedad, que publica con algunas notas," *Mercurio Peruano*, no. 344 (1794): 257. Also cited in Majluf, "The Creation of the Image of the Indian in 19th-Century Peru," 351.

of plants. From the Spanish and Portuguese American Viceroyalties of Peru, New Spain and Brazil to the colonial Caribbean, countless writers eulogized the "marvellous cures" of "wild Indians", who possessed greater "skill in simples, and the virtue of plants," "than the whites," but who "guarded their secrets" and could almost never be "persuaded" – that is, flattered, bribed, or coerced – to instruct naturalists "coming from Europe."[14] Enlightened Jesuit chroniclers,[15] French expeditionaries[16] and Mexican physicians[17] alike were all familiar with one or another anecdote about how the "secrets" of antidotes to arrow poison, of the "salutary virtue" of cinchona – a febrifuge – or of the healing properties of lizard meat, had been extorted from "the Indians" by way of deceit, furtive observation, or "the arts of intimacy." As Alexander von Humboldt (1769–1859), a paragon for later travellers in many respects, put it following his stay with a "master of curare (*amo del curare*)" in the upper Orinoco valley, "a veil of secrecy" lay everywhere over the "wild Indians" (*Wilden*) knowledge of "poisons and antidotes."[18] "The Indians" only rarely "betrayed" of their own volition the "secrets" they had, by all accounts, "sworn to keep among themselves, because of their hatred of their enemies."[19] Those who did were mostly "wom[e]n" who,

14 Eighteenth-century Brazilian surgeons like Luís Gomes Ferreira (1686–1764), for instance, confided in the plant knowledge and experience of the inhabitants of the Sertão, in particular the Carijós. Júnia Ferreira Furtado, ed. *Luís Gomes Ferreira, Erário mineral* (Rio de Janeiro: Editora FIOCRUZ, 2002 [1735]), 28–29. For examples from the Caribbean, see Schiebinger, "Prospecting for Drugs," 113–20.

15 Friar José Guevara (1719–1806) suspected that the Spanish, desirous to learn the antidote, injured an Indian prisoner with a poisoned arrow and freed him so he would look for the medicinal plant and thus betray his knowledge. José Guevara, *Historia de la conquista del Paraguay, Rio de la Plata y Tucuman* (Buenos Aires: E. Ostwald, 1882 [around 1766]), 208.

16 Charles-Marie de la Condamine, "Sur l'arbre du quinquina," *Mémoires de l'Academie Royale* MDCCXL (1738 (1737)): 233–34.

17 On the "secret uses" of lizard meat, see José Vicente Garcia de la Vega, Discurso critico que sobre el uso de las lagartijas, como especifico contra muchas enfermedades, produjo D. Joseph Vicente García de la Vega, Profesor de Medicina en la Imperial Corte de Mexico (Mexico: Imprenta de D. Felipe de Zúñiga y Ontiveros, 1782), 18. Cited in Miruna Achim, "Los remedios de los indios 'nuevamente descubiertos'," in Lagartijas medicinales: Remedios americanos y debates científicos en la Ilustración, ed. Miruna Achim (Mexico: Conaculta/UAM-C, 2008), 75.

18 Alexander von Humboldt, *Reise in die Äquinoktial-Gegenden des Neuen Kontinents*, 2 vols., vol. 2 (Frankfurt am Main, Leipzig: Insel-Verlag, 1991 [1814]), 1190.

19 François Citte, "De l'usage du quinquina et des règles de son application dans les fièvres intermittentes," in *Collection des thèses soutenues a l'école de médicine de Montpellier pendant l'an XII*, ed. L'École de Médicine de Montpellier (Montpellier: G. Izar et A. Richard, 1804), 4. For a similar remark see Francisco José de Caldas, *Memoria sobre el estado de las quinas en general y en particular sobra la de Loja*, Anales de la Universidad Central de Quito (Quito: Tipografía y Encuadernación Saletiana, 1907 [1809]), 242.

"enamoured with a Spaniard, and afraid to see [their] lover dying," "revealed to him (*lui découvrit*) the means of finding relief."[20] The theme of the Peruvian woman betraying her medical knowledge on account of love, friendship, or compassion even found its way into the realm of literature and the theatre. In Félicité de Genlis' (1746–1830) 1818 novel "Zuma, or the Discovery of Cinchona" (*Zuma, ou la découverte du quinquina*), which was later adapted for the stage and the opera in England, France, and the Netherlands, Zuma, "the most beautiful of Indian women" and the faithful protégé of a dying Spanish Countess, revealed the secret of the medicinal properties of cinchona to save her benefactress' life.[21]

The trope of the mistrustful Indian knower of nature certainly predated the late 1700s and early 1800s, the decades in which Unánue authored his essay on the discipline of botany in Peru. "Indians" who knew "many herbs to cure" illnesses but continuously put off "curious gentlemen" who sought to learn their secrets – even though they threatened the Indians, exerted violence, or promised them "such good pay, with flattery and kindness" – had already made their appearance in various prominent seventeenth-century naturalist treatises, such as the Dutch physician Willem Piso's (1611–1678) 1648 *Historia naturalis Brasiliae* (Natural History of Brazil),[22] the Jesuit friar Bernabé Cobo's (1580–1657) 1653 *Historia del Nuevo Mundo* (History of the New World)',[23] and Sebastianus Badus's 1663 *Anastasis Corticis Peruviae* (Redemption of

20 Citte, "De l'usage du quinquina et des règles de son application dans les fièvres intermittentes," 4. For an example of a male traitor, see George Motherby, *A New Medical Dictionary* (London: J. Johnson, 1775), Entry CORTEX PERUVIANUS.

21 Stéphanie Félicité du Crest de Saint-Aubin, Countess de Genlis, *Zuma, ou la découverte du quinquina* (Paris: Maradan, Libraire, 1818), 10–11; 21. For examples of theatrical and operatic adaptations, see, for instance, the comic opera "Zuma, or the Tree of Health," first performed in 1818 in London, Covent Garden. Margaret Ross Griffel, "ZUMA, or, the Tree of Health," in *Operas in English: A Dictionary*, ed. Margaret Ross Griffel (Plymouth: Scarecrow Press, 2013), 556. For a Dutch adaptation, see Jan de Quack, *Zuma, of De ontdekking van den kina-bast: tooneelspel in vier bedrijven* (Amsterdam: J. C. van Kesteren, 1819).

22 On Willem Piso's difficulties in acquiring information on poisonous plants from the natives in Dutch Brazil, see Júnia Ferreira Furtado, "Tropical Empiricism: Making Medical Knowledge in Colonial Brazil," in *Science and Empire in the Atlantic World*, ed. James Delbourgo and Nicholas Dew (New York: Routledge, 2008), 136.

23 Bernabé Cobo, certain that the Indians knew "many herbs to cure" their illnesses, had heard from his correspondents about instances of secrecy in which curious gentlemen had "promised the Indian[s] such good pay, with flattery and kindness" to learn their medical secrets but were put off by them. Bernabé Cobo, *Inca Religion and Customs*, ed. John Howland Rowe, trans. Roland Hamilton (Austin: University of Texas Press, 1990 (1653)), 222.

the Peruvian Bark).[24] Scholars, novelists and playwrights from Britain, the Netherlands, or France – countries where the "black legend" popularized the myth of Spain's unique brutality in the conquest of the New World[25] – were particularly eloquent on instances of Spanish coercion, threat, and violence. Hans Sloane (1660–1753), for instance, in his 1707 *Voyage to the Islands Madera, Barbadoes, Nieves, St Christophers, and Jamaica* related how "the Indians about Guiana had first discovered [the] Vertue [of Contra Yerva] to the Spaniards," when these threatened an Indian prisoner "to wound him with one of their own venemous [sic] Arrows, if immediately he did not declare their Cure for that Disease, upon which the Indian immediately chaw'd some of this *contra yerva*, and put it into the wound, and it healed."[26]

By all appearances, the skilful yet secretive Indian knower of nature had held his own in the New World right from the arrival of the first Spanish and Portuguese conquerors. Already the earliest natural history of the New World, Gonzalo Fernández de Oviedo y Valdés's (1478–1557) 1535 *Historia general y natural de las Indias* (General, and Natural History of the Indies) made reference to how "the Indians' were 'covetous (*avara*)" about their knowledge of medicinal "herbs and plants," and "the secrets of nature," "especially about whatever could be useful to the Christians," because "that way this science is [...] their dominion (*su señorío*)."[27] The evidence Unánue cited for his assertion that the Indians concealed their knowledge from the Europeans dated likewise from the sixteenth century. Unánue referred to a 1568 letter from a Lima-based soldier – Pedro de Osma – transcribed in part two of the 1574 "Medicinal History" of the New World (*Primera y Segunda y Tercera Partes de la Historia Medicinal*

24 Sebastianus Badus' treatise contained the tale of how "the natives" were believed to have hidden their knowledge of the medicinal virtues of the Peruvian bark from the Spanish', and kept it a secret, for a long time. Sebastianus Badus, *Anastasis corticis Peruviae; seu chinae chinae defensio* (Genoa: Typis Petri Joannis Calenzani, 1663), 21–24.

25 Walter D. Mignolo, Margaret R. Greer, and Maureen Quilligan, "Introduction," in *Rereading the Black Legend: Discourses of Religious and Racial Difference in the Renaissance Empires*, ed. Walter D. Mignolo, Margaret R. Greer, and Maureen Quilligan (Chicago, London: The University of Chicago Press, 2007).

26 Hans Sloane, *A voyage to the islands Madera, Barbados, Nieves, S. Christophers and Jamaica, with the Natural History of the Herbs and Trees, Four-footed Beasts, Fishes, Birds, Insects, Reptiles, &c. of the last of those islands*, 2 vols., vol. 1 (London: B. M., 1707), LV. See also Schiebinger, "Prospecting for Drugs."

27 Gonzalo Fernández de Oviedo y Valdés, *Primera parte de la historia natural y general de las indias, yslas e tierra firme del mar oceano* (Sevilla: Iuam Cromberger, 1535), book XI, chapter V. On the original manuscript's way into print, see Henry Lowood, "The New World and the European Catalog of Nature," in *America in European Consciousness, 1493–1750*, ed. Karen Ordahl Kupperman (Chapel Hill, London: University of North Carolina Press, 1995), 308–09.

de las Cosas que se traen de nuestras Indias Occidentales que sirven en Medicina) authored by Nicolás Monardes (1493–1588), a Seville physician and trader who had never crossed the Atlantic.[28] De Osma, supposedly a reader of the first part of Monardes' medicinal history who had followed the physician's exhortation to investigate native medicinal knowledge in the New World, allegedly encountered much reluctance among the Indians to divulge their "admirable" understanding of the virtues of medicinal plants and animal substances. When he was seeking to "discover the part in which the bezoar stone is engendered in the vicuñas," de Osma lamented that he could obtain no intelligence from the Indians, "because they are bitter towards us and do not want us to know their secrets." It was an Indian boy, "ten or twelve years of age," who "revealed to" de Osma and his companions the method to extract the bezoar stones and "that instant, his countrymen threatened to slay him."[29] "Regardless of kind words and gifts, and of fierce words and threats," de Osma wrote, the Indians, "being wicked and our enemies, will not disclose one secret, or one virtue of a herb, even if they were to see us dying." Other than from children, de Osma told Monardes in the 1568 letter, the conquerors only learned about the medicinal virtues of America's plants from those "Indian women, who get involved with the Spaniards and reveal and tell them everything they know."[30]

2 Ancestors and Kins[wo]men

The veracity and accuracy of Fernández de Oviedo y Valdés's account is doubtful and so is that of de Osma's, or rather Monardes's, assertions. As a matter of fact, historians of Iberian science have cited the 1568 letter with some reserve, since there is no evidence to prove it was genuine: de Osma may well have been a creature of Monardes's imagination.[31] The early and ready acceptance

---

28 The letter was transcribed in part II of Nicolás Monardes's "Medicinal History (*Historia Medicinal*)." See Nicolás Monardes, *Primera y segunda y tercera partes de la historia medicinal, de las cosas que se traen de nuestras Indias Occidentales, que siruen en Medicina* (Sevilla: Alonso Escriuano, 1574), 73–77. For biographical notes on Monardes, see Marcy Norton, *Sacred Gifts, Profane Pleasures: A History of Tobacco and Chocolate in the Atlantic World* (Ithaca, London: Cornell University Press, 2008), 107–12.

29 Monardes, *Primera y segunda y tercera partes de la historia medicinal*, 74.

30 Ibid., 77.

31 Historians of Iberian science have cited de Osma's letter with some reserve. See, for instance, Daniela Bleichmar's remark on "the letter's laudatory tone and the way it dramatizes the importance and utility" of Monardes's work. Daniela Bleichmar, "Books, Bodies, and Fields: Sixteenth-Century Transatlantic Encounters with New World *Materia Medica*," in *Colonial Botany. Science, Commerce, and Politics in the Early Modern World*,

of the idea that the Indians possessed and concealed an "admirable" knowledge of the nature of the New World would appear inconsistent, given its foundation on rather unstable ground, were it not for the possibility that in this, as in so many other instances, familiar, ancient Old World paragons exercised a strong hold over early modern observers' minds and their representations of the New World.[32] Indeed, the resemblance between the reticent Indian informer and other "illiterate knowers" in late-medieval and early modern thought is conspicuous, and likely eloquent. From the earliest days of colonial rule in the Americas, the Indians "from the fields" – not any Indian, for the category encompassed, into the early nineteenth century, Hispanized, noble or urban persons[33] – had been likened to Europe's peasantry. Both were "natural men," society-less, poor, uncultured creatures "close in condition, if not in kind, to the animals among which they worked."[34] A university-trained physician versed in scholastic medicine like Monardes, or a Romance humanist like Fernández de Oviedo y Valdés, would have been familiar with the notion

ed. Londa Schiebinger and Claudia Swan (Philadelphia: University of Pennsylvania Press, 2005), 92. Other historians have, however, stressed the relevance of Monardes's epistolary networks, in ways that suggest the letter may well have been real. Norton, *Sacred Gifts, Profane Pleasures*, 117. On elements of imaginativeness in Fernández de Oviedo y Valdés's chronicle, see, for instance, Carlo Klauth, *Geschichtskonstruktion bei der Eroberung Mexikos: Am Beispiel der Chronisten Bernal Diáz del Castillo, Bartolomé de las Casas und Gonzalo Fernández de Oviedo* (Hildesheim: Georg Olms Verlag, 2012), 184.

32 On the method of "working from affinities with recognizable," Old World "forms" in natural history, see Lowood, "The New World and the European Catalog of Nature," 298. See also Roland Greene, "Petrarchism among the Discourses of Imperialism," in *America in European Consciousness, 1493–1750*, ed. Karen Ordahl Kupperman (Chapel Hill: University of North Carolina Press, 1995), 135. On the same predicament, see Anthony Grafton, *New Worlds, Ancient Texts. The Power of Tradition and the Shock of Discovery* (Cambridge, MA: Harvard University Press, 1992), 6; Anthony Pagden, *The Fall of Natural Man: The American Indian and the Origins of Comparative Ethnology* (Cambridge, New York: Cambridge University Press, 1982), 1–6, 11–12.

33 On elite, or noble, Indians in South and Mesoamerica, see David Cahill, "Colour by Numbers: Racial and Ethnic Categories in the Viceroyalty of Peru, 1532–1824," *Journal of Latin American Studies* 26, no. 2 (1994); Scarlett O'Phelan Godoy, *Kurakas sin sucesiones. Del cacique al alcalde de indios. Perú y Bolivia 1750–1835* (Cuzco: CBC, 1997). In the viceroyalties of Peru and New Spain alike, herbalists usually were described as "simple folks" – men and women who lived "far-off," away from the city, in "Indian villages." See, for instance, Miruna Achim, "From Rustics to Savants: Indigenous *Materia medica* in Eighteenth-Century Mexico," *Studies in History and Philosophy of Biological and Biomedical Sciences* 42 (2011): 282.

34 The meaning of "natural man" changed over the centuries, from that of a society-less creature, "something less than human," in the sixteenth and seventeenth centuries, to Enlightenment's "natural man," someone whose mind is unfettered by the moral and intellectual constraints of civil society. Pagden, *The Fall of Natural Man*, 8–9; 97–98; 21–22.

that many useful remedies and practices could be learned from "ordinary" laypersons – "rustics" and peasants, "simpletons," or women – whose knowledge was, unlike that of learned scholars, who inquired into "causes or first principles" in the physical operation of natural substances, based on the "accurate but un-theorized observation of phenomena accessible to the senses."[35] Indeed, accounts of Indian poverty, illiteracy, and simplicity – an image fashioned particularly by the Franciscan order[36] – would have resonated with early modern thought which, in its reliance on the truths spoken by "illiterate knowers," drew on a genre of Christian piety that placed the best hope of salvation for mankind in the poor and ordinary people, who were closer not only to nature but also to God than the mighty.[37] From the earliest days of Christianity, the "divine fool" – simple, illiterate, and humble – had told truths and offered hope of redemption.[38] Writers like Monardes and Fernández de Oviedo y Valdés would have been aware, too, that many "illiterate knowers" – women in particular – communicated their knowledge only to their close associates – to other women – orally and in the vernacular, and that they would conceal it from men, since it represented one important way of exerting power over them. They, as well as their readership, would have known also that one of the most important tasks of literate physicians and naturalists consisted of retrieving the "secrets" of women and other artisanal practitioners and committing them to Latin script to ensure their permanence and relative publicity within

35 Katherine Park, *Secrets of Women: Gender, Generation, and the Origins of Human Dissection* (New York: Zone Books, 2006), 84–85. On how sixteenth-century physicians took seriously what they learnt from laypersons – empirics', artisans, or women –, see Michael Stolberg, "Learning from the Common Folks. Academic Physicians and Medical Lay Culture in the Sixteenth Century," *Social History of Medicine* 27, no. 4 (2014). On the "mechanic classes" as bearers of valuable empirical knowledge in the seventeenth century, which, however, needed to be "drawn out, freely and disinterestedly reflected upon, and codified for its value to be realized," see also Steven Shapin, *A Social History of Truth: Civility and Science in Seventeenth-Century England* (Chicago, London: The University of Chicago Press, 1994), 396.

36 Julia McClure, *The Franciscan Invention of the New World* (Basingstoke: Palgrave, 2016), 178.

37 Harold J. Cook, *Matters of Exchange: Commerce, Medicine and Science in the Age of Empire* (Hyderabad: Orient Longman, 2008), 34. See also Laurence Brockliss and Colin Jones, *The Medical World of Early Modern France* (Oxford: Clarendon Press, 1997), 276.

38 The discursive tradition of "divine idiocy," of "holy folly," grew out of the crossings of numerous traditions – early Christian thought, the Enlightenment, and Romanticism, among others. Dana Heller, "Holy Fools, Secular Saints, and Illiterate Saviors in American Literature and Popular Culture," *CLCWeb: Comparative Literature and Culture* 5, no. 3 (2003): 4, 7.

an elite scientific community.[39] And while there may have been a grain of truth in assertions about the Indians' skill in nature and secrecy – during the early years of the Spanish conquest, secrecy was a conduct it would have been perfectly prudent to adopt[40] – they were likely something other, if not rather, than an observation. The stereotype of the "inscrutable Indian knower" may well have been the continuance and transfer of a European topos of long standing to the Americas: one of a plethora of Christian and humanist legacies that the Old World bequeathed to the New. Monardes's *Medicinal History* was translated into Latin by Carolus Clusius (1526–1609) in 1582 and soon after became available in French, German, Italian and English translations, which were re-edited several times during the sixteenth and seventeenth centuries. As historians of Iberian medicine have often noted, it was through Monardes that many of early modern Europe's physicians, apothecaries and naturalists were first introduced to the natural world of the Americas.[41] And, it was probably along this same vein, among others, that the "secretive Indian," the reluctant informer about the New World, put down roots in natural historical discourse.

The image of the reticent Indian informer hibernated in natural histories of the seventeenth century, but re-emerged with new vigour in the Enlightenment when it was reprised by writers like Unánue. In fact, the latter was most likely looking, once more, to the European tradition of 'inscrutable rustics' Monardes had tapped into back in the sixteenth century, when he was trying to come to terms with and classify the knowledge of America's new-found inhabitants. Historians have sometimes noted the striking similarity between colonial encounters in the West Indies and those described by enlightened European naturalists with "wise women," "rustics," or the peasantry.[42] And indeed, many enlightened writers from Sweden, England or Prussia alike – Carl Linnaeus (1707–1778) among them – were "keen ethnobotanist[s]," as Lisbet Koerner put it, who studied "the plant lore" and "therapeutic secrets" of "simple folks," "anxious to protect their skills," in ways very much akin to those

39 Parks, *Secrets of Women*, 84.

40 See, for instance, Suzanne Austin Alchon's reflections on how Indian healing would often actually have occurred in secrecy because of its association with idolatry and consequent repression. Suzanne Austin Alchon, "Tradiciones médicas nativas y resistencia en el Ecuador," in *Saberes andinos: Ciencia y tecnología en Bolivia, Ecuador y Perú*, ed. Marcos Cueto (Lima: IEP, 1995), 27; 36. On the role of the inquisition, see Achim, "Los remedios de los indios 'nuevamente descubiertos'," 85.

41 Achim, "From Rustics to Savants," 278.

42 Schiebinger, "Prospecting for Drugs," 121. See also Daniel Rood, "Toward a Global Labor History of Science," in *Global Scientific Practice in an Age of Revolutions, 1750–1850*, ed. Patrick Manning and Daniel Rood (Pittsburgh: University of Pittsburgh Press, 2016), 265.

of their contemporaries in Latin America.[43] The "empirical," "mysterious" Indians' likeness to reticent Swedish peasants was likely anything but accidental. Not only were Spanish and Portuguese American scholars under the influence of European writings – Unánue, for instance, was an avid reader and promoter of Linnaeus' treatises in Spanish America[44] – by the time of the Enlightenment, naturalists and physicians from countries like Sweden or France fashioned their own peasantry, Sami "savage," or "wise woman" after the Iroquois, Barbary, or Amazonian indigenes they encountered in travelogues and early "ethnographic" observations.[45] What is more, cosmopolitan naturalists like Linnaeus, Sloane, or Unánue partook of the same epistemic principles and discourses: the same Christian and humanist legacies, to be sure, but also the same empiricist dictates – the idea that all knowledge ought to be based on experience derived from the senses – , which made early modern physicians and naturalists from Habsburg Prague to viceregal Brazil open to, and even dependent upon, those simple, and "rustic" men and women who were seen to possess the most experiential and sensuous, the keenest and most immediate, apprehension of the natural world.[46] Most importantly, perhaps, they also shared the same enlightened principles and discourses – a nostalgic, Ovidian ruralism and critique of civilization[47] as well as the demand that science be applied, economic and vernacular, which accorded renewed significance to

43 Lisbet Koerner, "Women and Utility in Enlightenment Science," *Configurations* 3, no. 2 (1995): 238; 50–51; Schiebinger, "Prospecting for Drugs," 121–22.

44 Unánue's 1791 essay introduced Peruvian readers to Linnaeus's system of plant classification and binomial nomenclature. Unanue, "Botánica," 68–86.

45 For examples from Sweden, see Lisbet Koerner, *Linnaeus. Nature and Nation* (Cambridge MA, London: Harvard University Press, 1999), 62. For examples from France, see Brockliss and Jones, *The Medical World of Early Modern France*, 277. On the co-constitution of ideas about "indigeneity" in the early modern period more broadly, and on how European naturalists re-enacted practices from overseas, see Alix Cooper, *Inventing the Indigenous: Local Knowledge and Natural History in Early Modern Europe* (Cambridge, New York: Cambridge University Press, 2007); Cooper, "The Indigenous versus the Exotic: Debating Natural Origins in Early Modern Europe," *Landscape Research* 28, no. 1 (2003).

46 Scott Parrish, *American Curiosity*, 16. On Habsburg Prague, see Stolberg, "Learning from the Common Folks," 652. See also Andreas-Holger Maehle, *Drugs on Trial: Experimental Pharmacology and Therapeutic Innovation in the Eighteenth-Century* (Amsterdam: Editions Rodopi, 1999), 251–52; Martha Baldwin, "Expanding the Therapeutic Canon: Learned Medicine Listens to Folk Medicine," in *Cultures of Communication from Reformation to Enlightenment*, ed. James Van Horn Melton (Ashgate: Aldershot, 2002). On Mexico, see Achim, "From Rustics to Savants." On Iberian empiricism, see also Richard H. Grove, "Indigenous Knowledge and the Significance of South-West India for Portuguese and Dutch Constructions of Tropical Nature," *Modern Asian Studies* 30, no. 1 (1996): 130. On "tropical empiricism" in Brazil, see Ferreira Furtado, "Tropical Empiricism."

47 Koerner, *Linnaeus: Nature and Nation*, 75–77.

unlettered "local knowers of nature" the world over.[48] Enlightened engineers, physicians and antiquaries from Shropshire to Lima were persuaded there was something vital to be learned from "old women," Uzbek herders and South Asian non-Brahminic low-castes, "wild Indians," as well as enslaved men and women of African descent, about the care of animals, the meaning of antiquities, and, time and again, the uses of plants.[49] Indeed, "Negro Doctors" and slave "wenches" in particular were generally believed to possess and "jealously" guard, much like "the Indians," many "marvellous" "Secrets", particularly "in the Art of Physick," which naturalists "coming from Europe" could only "obtain" through "stratagem" or "menace".[50]

Global historians of science have often noted and denounced colonial naturalists' efforts to "extract knowledge from reticent informants" – recounting how "protestations of friendship, shameless flattery, monetary reward, and

48 Koerner, "Women and Utility in Enlightenment Science," 251.

49 In 1775, the physician William Withering encountered a "receipt for the cure of the dropsy," which "had long been kept a secret by an old woman in Shropshire." William Withering, *An Account of the Foxglove, and Some of its Medical Uses* (Birmingham: M. Swinney, 1785), 2. For a contemporary voice from Lima, encouraging the "zoologist," the "botanist," and the "physician" to seek the advice of "humble indigenes," and "mistreated slaves," see Mariano Eduardo de Rivero y Ustariz, "Prospecto," *Memorial de ciencias naturales, y de industria nacional y extranjera* 1, no. 1 (1827). On the Indians' privileged knowledge of antiquities, see José Hipólito Unanue, "Idea general de los monumentos del Antiguo Perú, e introducción a su estudio," *El Mercurio Peruano* 1 (1791): 202. Historians have argued that medico-botanical knowledge in British India also tended to privilege non-Brahminic, low-caste epistemologies. Grove, "Indigenous Knowledge and the Significance of South-West India for Portuguese and Dutch Constructions of Tropical Nature," 128–33. For Uzbek herders, see Koerner, "Women and Utility in Enlightenment Science," 251; Koerner, *Linnaeus. Nature and Nation*, 68.

50 Thomas Heney, MD of St David's, in the Island of Jamaica, reported in 1791 how he discovered "the secret" of the cure effected by a "negro wench" by "strategem." Thomas Heney, "On the Efficacy of the Zanthoxylon," *The Medical and Physical Journal* 2 (1799): 34. Also cited in Londa Schiebinger, *Secret Cures of Slaves. People, Plants, and Medicine in the Eighteenth-Century Atlantic World* (Stanford: Stanford University Press, 2017), 47–48. For more examples, see Schiebinger, "Prospecting for Drugs," 119. See also the various narratives in circulation about "quassia," generally thought to have been "a secret remedy" before its disclosure to Swedish naturalists, employed by "a negro named Quassi," "with uncommon success" in the Dutch plantation colony of Surinam. John Mason Good, Olinthus Gregory, and Newton Bosworth, *Pantologia: A new Cyclopaedia, Comprehending a Complete Series of Essays, Treatises, and Systems*, vol. x (London: G. Kearsley et al., 1813), QUA. On narratives about secretive "negro herbalists," see also Francisco Guerra, "Medical Almanacs of the American Colonial Period," *Journal of the History of Medicine* XVI, no. 3 (1961): 247–48; Richard B. Sheridan, *Doctors and Slaves: A medical and Demographic History of Slavery in the British West Indies, 1680–1834* (Cambridge, New York: Cambridge University Press, 1985), 80–81.

eavesdropping, as well as imprisonment and torture were all tried" – but they have generally taken that reticence at face value, even accounting for it with the "suspicion" and "resentment" common to a colonial context.[51] Indeed, practitioners in the field have recently suggested "recast[ing] the global history of science as a coercive form of labour management" on those grounds.[52] There is certainly no denying the fact that naturalists' dealings with their informers were uneven and at times coercive; and, as during the early years of the Spanish conquest, there may well have been truth to assertions about "reticent informants" – secrecy was still a conduct it would have been prudent to adopt around 1800. As before, however, and perhaps even more so, narrations of encounters between "rustics" and naturalists were also fraught with literary topoi, fragments of discourse that reveal "networks of production,"[53] threaded together by men and women from various world regions who had long engaged with and relied upon the same humanist, empiricist, and enlightened, paradigms and narratives.[54] All of them consequently interpreted and made sense of the world in expected ways, imbibed from their readings, which determined their narrative choice and content. A renewed penchant for "mistrustful," "empirical" Indian knowers in natural histories of the Americas over the later 1700s and early 1800s was likely one result of that close epistemic connection.

3 Coda

The secretive Indian herbalist came into being in the late sixteenth century, and, like other images and ideas that travelled along the veins of colonial dominion and through time, it recurred like a wave thereafter, with ebbs of

---

51 Rood, "Toward a Global Labor History of Science," 265. See also Schiebinger, "Prospecting for Drugs," 123. See also Austin Alchon, "Tradiciones médicas nativas y resistencia en el Ecuador," 36.

52 Rood, "Toward a Global Labor History of Science," 266.

53 On practices of equation in the history of medicine, see Carla Nappi, "Winter Worm, Summer Grass: *Cordyceps*, Colonial Chinese Medicine, and the Formation of Historical Objects," in *Crossing Colonial Historiographies: Histories of Colonial and Indigenous Medicines in Transnational Perspective*, ed. Anne Digby, Projit B. Muhkarji, and Waltraud Ernst (Newcastle upon Tyne: Cambridge Scholars, 2010), 29–30.

54 Karen Stolley, *Domesticating Empire: Enlightenment in Spanish America* (Nashville: Vanderbilt University Press, 2013), 3. For the realm of naturalism, see also Juan Pimentel, "The Iberian Vision: Science and Empire in the Framework of a Universal Monarchy, 1500–1800," *Osiris: A Research Journal Devoted to the History of Science and its Cultural Influences* 15 (2000); Jorge Cañizares Esguerra, "Iberian Colonial Science," *Isis* 96 (2005).

latency and dispersal even beyond recognition in some times and places, and flows of incidence and salience in others.

Spanish America's vibrant Creole scientific community of the late-eighteenth century, the context in which Unánue authored his essay, for instance, saw a heyday of the association between secrecy, skill in nature, and indigeneity, partly because the "mistrustful" Indian knower spoke to a creole anxiety that the Indians were the only lawful owners and heirs of American nature. Creoles like Unánue were fashioning an American identity for themselves – in contrast to peninsular-born Spaniards – as the sons and daughters of the land in the late-colonial period.[55] They were well aware, however, that their Americanness was called into question by the presence of those other, "original natives," the Indians – the men and women who had, as Unánue put it, inherited their knowledge about nature and history straight "from their ancestors."[56] Indeed, the trope of the secretive Indian would gain a forceful life of its own in nineteenth-century antiquarian and archaeological writings,[57] related decreasingly to the Indians' intimacy with nature and increasingly with that creole anxiety, an implicit recognition that the Indians were "the only rightful heirs of the pre-Columbian past and its archaeological remains."[58] Secretive Indians were familiar and popular figures in early modern northern European – Dutch, English, or French – natural historical treatises about the virtues of Spanish American medicinal plants too, partly because the Black Legend ensured the popularity and willing acceptance of the idea that America's "natives" had guarded their knowledge out of their "revulsion" against the "barbarism of their conquerors (*révoltés de la barbarie de leurs vainqueurs*)."[59] Colourful accounts of "mysterious" Indian "discoverers" of plant medicines also thrived with the

55 On creole identity, see David Brading, *The First America: The Spanish Monarchy, Creole Patriots, and the Liberal State 1492–1867* (Cambridge, New York: Cambridge University Press, 1991). See also Bernard Lavallé, *Las promesas ambiguas: Ensayos sobre el criollismo en los Andes* (Lima: IRA, 1993).

56 Majluf, "The Creation of the Image of the Indian in 19th-Century Peru", 24. See also Stefanie Gänger, *Relics of the Past: The Collecting and Study of Pre-Columbian Antiquities in Peru and Chile, 1837–1911*, Oxford Studies in the History of Archaeology (Oxford: Oxford University Press, 2014), 136–37.

57 For narratives about how "the Indians" "guarded" the secrets of "antiquities," "inherited from their ancestors," see Mariano Eduardo de Rivero y Ustariz and Johann Jacob von Tschudi, *Peruvian Antiquities*, trans. Francis L. Hawks, 2nd ed. (New York: Putnam, 1857), 104–07; 29; Antonio Raimondi, "Itinerario de los viajes de Raimondi: De Lima á las montañas de Huancayo, Tarma, Pampa de Junín y Cerro de Pasco," *Boletín de la Sociedad Geográfica de Lima* 5, no. 4–6 (1896).

58 Majluf, "The Creation of the Image of the Indian in 19th-Century Peru", Chapter 8.

59 Thomas Collingwood, "Observations on the Peruvian Bark," *Medical Commentaries* x (1785): 265–66; Julien Dufau, *Essai sur l'application du quinquina dans le traitement des*

increasing commercialisation and pluralism of medical and pharmacological practice in the wake of the 1700s: from London to Lima, apothecaries, drug merchants, and physician-writers included stories about "wild Indians" – of whom, in their unlettered simplicity and closeness to nature mankind could reasonably expect the revelation of nature's most coveted and "excellent" remedies – in advertisements, almanacs, and medical treatises. It seemingly became an effective way of popularizing and marketing remedies and standing out within a bustling "medical marketplace."[60] Cosmopolitan creoles like Unánue – in the Spanish and British American colonies alike – might also have chosen to stress the Indians' secrecy and closeness to nature at a time when they were seeking to carve out a role and a space for themselves in the period's transatlantic Republic of Letters. Whereas in the European context, early modern naturalists in the wake of the seventeenth century fashioned themselves very much as discoverers of the secrets of nature that lay hidden,[61] naturalists in the Americas frequently substituted that discourse for a "rhetoric of approximation" – the "arts of intimacy" – and brokerage with "the Indians."[62] As various historians have argued, creole experts in the natural saw their peculiar privilege and province in mediation – not between the enlightened sciences and nature, but, much like scholastic physicians and Romance humanists with "rustics," peasants, or women, between the enlightened sciences and the inscrutable, withdrawn Indians; as intermediaries between the Indians' "empirical medicine" and "metropolitan abstraction" and ratification,[63] as "go-betweens," and

---

fièvres intermittentes, Collection des theses soutenues a l'ecole de médicine de Paris (Paris: Didot Jeune, 1805), 10.

60 On the fetishisation of indigenous and African medical knowledge, particularly by mestizo and Creole sectors of society in Spanish America, see Steven Palmer, *Doctors, Healers, and Public Power in Costa Rica, 1800–1940* (Durham, London: Duke University Press, 2003), 31. On North America, see Guerra, "Medical Almanacs of the American Colonial Period," 247–48. Irina Podgorny has found that charlatans sometimes claimed to have lived among indigenous groups who had introduced them to the secret healing powers of America's nature. Irina Podgorny, "From Lake Titicaca to Guatemala: The Travels of Joseph Charles Manó and his Wife of Unknown Name," in *Nature and Antiquities: The Making of Archaeology in the Americas*, ed. Philip L. Kohl, Irina Podgorny, and Stefanie Gänger (Tucson: University of Arizona Press, 2014).

61 Daniela Bleichmar, *Visible Empire: Botanical Expeditions and Visual Culture in the Hispanic Enlightenment* (Chicago, London: The University of Chicago Press, 2012), 48. See also Carolyn Merchant, *Reinventing Eden: The Fate of Nature in Western Culture* (New York, London: Routledge, 2013), xiii–xiv.

62 Achim, "From Rustics to Savants," 280. For the term and concept of the "rhetoric of approximation" see Majluf, *The Creation of the Image of the Indian in 19th-Century Peru*, 308.

63 Scott Parrish, *American Curiosity*, 217; 38.

"filters of testimony," whose position was enhanced and rendered all the more necessary by the Indians' "permanent alterity," "epistemological distance," and inscrutability.[64] Indeed, the dictates of empiricism in some measure gave colonial naturalists the means of making up for their limited access to the newest publications, prestigious institutions, or their lower rung in a – political, environmental, or economic – hierarchy.[65]

The trope of the skilful yet secretive Indian knower of nature did not come to an end with the colonial period. Rather, the association between indigeneity, secrecy, and skill in nature thrived in natural histories, artworks, and travelogues of the 19th and 20th centuries. Across the Americas, secretive Indians testified, forever reluctantly, about various aspects of the territory: the topography of the land, the migrational patterns of birds, the way to gold mines and ruins, and, time and again, the healing and poisonous properties of plants.[66] Indeed, the belief in a closely-kept, intuitive skill in nature peculiar to the Indians became in time a fundamental element of the repertoire of "romanticized ecological discourses" and "the received global wisdom of what constitutes Indianness," common to environmental politics and much anthropological literature on indigenous knowledge at present.[67] The trope's enduring popularity was presumably both a creature and a vehicle of the hardening of an increasingly racialized dichotomy between indigenous and western knowledge that took shape in the wake of the late 1700s and early 1800s–an age of "Global Imperialism" and the "Great Divergence," in which northern European

64 For the argument that "creole attempts at approximation presuppose[d] the stereotype of the inscrutable Indian," see Majluf, "The Creation of the Image of the Indian in 19th-Century Peru", 356. Miruna Achim, *Lagartijas medicinales: Remedios americanos y debates científicos en la Ilustración* (Mexico: Conaculta/UAM-C, 2008), 75. On the settlers as mediators, in the face of native "permanent alterity" and "epistemological distance," who may have liked to stress the "Indian" herbalists' "secrets" to prove their "hard-won and rare value," see Scott Parrish, *American Curiosity*, 16; 216 17; 22; 29 30; 53–56. On creole scholars as trans-Atlantic "brokers," and intermediaries, see also Stefanie Gänger, "Disjunctive Circles: Modern Intellectual Culture in Cuzco and the Journeys of Incan Antiquities, c. 1877–1921," *Modern Intellectual History* 10, no. 2 (2013).

65 See Scott Parrish, *American Curiosity*, 16. On Peru, see Gänger, "Disjunctive Circles," 409.

66 See, for instance, Joshua David Bellin, "Taking the Indian Cure: Thoreau, Indian Medicine, and the Performance of American Culture," *The New England Quarterly* LXXIX, no. 1 (2006): 23; Scott Parrish, *American Curiosity*; Majluf, *The Creation of the Image of the Indian in 19th-Century Peru*, 351.

67 Arun Agrawal, "Dismantling the Divide between Indigenous and Scientific Knowledge," *Development and Change* 26 (1995): 416–17; Michael R. Dove, "Indigenous People and Environmental Politics," *Annual Review of Anthropology* 35 (2006): 194.

empires first gained worldwide military, political and economic superiority.[68] For while the 19th and 20th centuries allocated western knowledge the prerogative of universality, abstraction, impersonality, and mobility, they dealt the indigenous the reverse:[69] knowledge that was reliant on "intuition," "instinct," and evidence "directly available to the senses," for one thing, and, for another, "closed," passed down but "by word of mouth," "from generation to generation," that is, "bound" to the lives of the people who generated it.[70] Indeed, secrecy presumably became an ever more deeply entrenched and predictable element of conventional wisdom about the Indians, as the natural secondary effect of attempts by outsiders to remove knowledge inherently destined to be immovable – meant, by that logic, to be stationary, bound, and unalienable. The Indians' closely kept skill in nature was likely a topos, fashioned after a plethora of Christian and humanist legacies, in the wake of the Spanish conquest. It survived, and indeed thrived, however, because it precisely fit the logic of the epistemic hierarchies and global topography of knowledge that took shape thereafter.

Bibliography

Printed Sources

Badus, Sebastianus. *Anastasis corticis Peruviae; seu chinae chinae defensio, Sebastiani Badi Genvensis patrij vtriusque Nosochomij olim medici, et Publicae Sanitatis in Ciuitate Consultoris. Contrà ventilationes Joannis Jacobi Chifletii, gemitusque Vopisci Fortunati Plempii, Illustrium Medicorum*. Genoa: Typis Petri Joannis Calenzani, 1663.

Caldas, Francisco José de. *Memoria sobre el estado de las quinas en general y en particular sobra la de Loja*, Anales de la Universidad Central de Quito. Quito: Tipografía y Encuadernación Saletiana, 1907 (1809).

Citte, François. "De l'usage du quinquina et des règles de son application dans les fièvres intermittentes." In *Collection des thèses soutenues a l'école de médicine de Montpellier pendant l'an XII*, edited by L'École de Médicine de Montpellier, 3–23. Montpellier: G. Izar et A. Richard, 1804.

68 On "The First Age of Global Imperialism," Christopher A. Bayly, "The First Age of Global Imperialism, c. 1760–1830," *The Journal of Imperial and Commonwealth History* XXVI, no. 2 (1998). On the "Great Divergence," see Kenneth Pomeranz, *The Great Divergence: China, Europe, and the Making of the Modern World* (Princeton: Princeton University Press, 2000).

69 Marwa Elshakry, "When Science Became Western: Historiographical Reflections," *Isis* 101. Focus. Global Histories of Science (2010).

70 Agrawal, "Dismantling the Divide between Indigenous and Scientific Knowledge," 416–17.

Collingwood, Thomas. "Observations on the Peruvian Bark." *Medical Commentaries* x (1785): 265–75.

Condamine, Charles-Marie de la. "Sur l'arbre du quinquina." *Mémoires de l'Academie Royale* MDCCXL (1738 (1737)): 226–44.

Crest de Saint-Aubin, Stéphanie Félicité du, Countess de Genlis. *Zuma, ou la découverte du quinquina*. Paris: Maradan, Libraire, 1818.

D.F.D.P.D.L.M.L. "Carta remitida a la Sociedad, que publica con algunas notas." *Mercurio Peruano*, no. 344 (1794): 255–62.

Dufau, Julien. *Essai sur l'application du quinquina dans le traitement des fièvres intermittentes*, Collection des theses soutenues a l'ecole de médicine de Paris. Paris: Didot Jeune, 1805.

Fernández de Oviedo y Valdés, Gonzalo. *Primera parte de la historia natural y general de las indias, yslas e tierra firme del mar oceano; escripta por el capitan Gonçalo Hernández de Oviedo y Valdés, alcayde de la fortaleza de la ciudad de Sancto Domingo de la ysla Española, y cronista de la sacra, cesárea y catholicas majestades del emperador don Carlos quinto de tal nombre, rey de España, e de la sereníssima e muy poderosa reyna doña Juana su madre nuestros señores. Por cuyo mandado el auctor escrivió las cosas maravillosas que hay en diversas yslas e partes destas indias e imperio de la corona real de castilla, según lo vido e supo en veynte e dos años e más que ha que vive e reside en aquellas partes. La qual historia comiença en el primero descubrimiento destas indias, y se contiene en veynte libros este primero volumen*. Sevilla: Iuam Cromberger, 1535.

Garcia de la Vega, José Vicente. *Discurso critico que sobre el uso de las lagartijas, como especifico contra muchas enfermedades, produjo D. Joseph Vicente García de la Vega, Profesor de Medicina en la Imperial Corte de Mexico*. Mexico: Imprenta de D. Felipe de Zúñiga y Ontiveros, 1782.

Guevara, José. *Historia de la conquista del Paraguay, Rio de la Plata y Tucuman*. Buenos Aires: E. Ostwald, 1882 (around 1766).

Heney, Thomas. "On the Efficacy of the Zanthoxylon." *The Medical and Physical Journal* 2 (1799): 31–34.

Humboldt, Alexander von. *Reise in die Äquinoktial-Gegenden des Neuen Kontinents*, 2 vols., vol. 2. Frankfurt, Leipzig: Insel-Verlag, 1991 (1814).

Mason Good, John, Olinthus Gregory, and Newton Bosworth. *Pantologia. A new Cyclopaedia, Comprehending a Complete Series of Essays, Treatises, and Systems, Alphabetically Arranged; with a General Dictionary of Arts, Sciences, and Words: the Whole Presenting a Distinct Survey of Human Genius, Learning, and Industry*, vol. x. London: G. Kearsley et al., 1813.

Monardes, Nicolás. *Joyfull Newes out of the Newe Founde Worlde*. London: Willyam Norton, 1577.

Monardes, Nicolás. *Primera y segunda y tercera partes de la historia medicinal, de las cosas que se traen de nuestras Indias Occidentales, que siruen en Medicina*. Sevilla: Alonso Escriuano, 1574.

Motherby, George. *A New Medical Dictionary; Or General Repository of Physic Containing an Explanation of the Terms, and a Description of the Various Particulars Relating to Anatomy, Physiology, Physic, Surgery, Materia Medica, Pharmacy &c. &c. &c.* London: J. Johnson, 1775.

Quack, Jan de. *Zuma, of De ontdekking van den kina-bast: tooneelspel in vier bedrijven*. Amsterdam: J. C. van Kesteren, 1819.

Raimondi, Antonio. "Itinerario de los viajes de Raimondi. De Lima á las montañas de Huancayo, Tarma, Pampa de Junín y Cerro de Pasco." *Boletín de la Sociedad Geográfica de Lima* 5, no. 4–6 (1896): 122–74.

Rivero y Ustariz, Mariano Eduardo de. "Prospecto." *Memorial de ciencias naturales, y de industria nacional y extranjera; redactado por M. de Rivero y de N. de Piérola* 1, no. 1 (1827): 1–3.

Rivero y Ustariz, Mariano Eduardo de, and Johann Jacob von Tschudi. *Peruvian Antiquities*, translated by Francis L. Hawks, 2 ed. New York: Putnam, 1857.

Sloane, Hans. *A voyage to the islands Madera, Barbados, Nieves, S. Christophers and Jamaica, with the Natural History of the Herbs and Trees, Four-footed Beasts, Fishes, Birds, Insects, Reptiles, &c. of the last of those islands; to which is prefix'd, an Introduction, Wherein is an Account of the Inhabitants, Air, Waters, Diseases, Trade, &c. of that Place, with some Relations concerning the Neighbouring Continent, and Islands of America*, 2 vols., vol. 1. London: B.M., 1707.

Tschudi, Johann Jakob von. *Peru: Reiseskizzen aus den Jahren 1838–1842*, 2 vols., vol. 2. St. Gallen: Scheitlin und Zollikofer, 1846.

Unanue, José Hipólito. "Idea general de los monumentos del Antiguo Perú, e introducción a su estudio." *El Mercurio Peruano* 1 (1791/03/17 1791): 201–08.

Unanue, José Hipólito. "Botánica: Introducción a la descripción científica de las plantas del Perú, vol. II, núms. 43 y 44 (29 de mayo y 2 de junio de 1791), 68–86." In *El Mercurio Peruano: 1790–1795. Antología*, edited by Jean-Pierre Clément, 93–114. Frankfurt am Main: Vervuert, 1998 (1791).

Withering, William. *An Account of the Foxglove, and Some of its Medical Uses: with Practical Remarks on Dropsy, and Other Diseases*. Birmingham: M. Swinney, 1785.

Literature

Achim, Miruna. "From Rustics to Savants: Indigenous Materia medica in Eighteenth-Century Mexico." *Studies in History and Philosophy of Biological and Biomedical Sciences* 42 (2011): 275–84.

Achim, Miruna. *Lagartijas medicinales: Remedios americanos y debates científicos en la Ilustración*. Mexico: Cohjnaculta/UAM-C, 2008.

Achim, Miruna. "Los remedios de los indios 'nuevamente descubiertos'." In *Lagartijas medicinales: Remedios americanos y debates científicos en la Ilustración*, edited by Miruna Achim, 73–119. Mexico: Conaculta/UAM-C, 2008.

Agrawal, Arun. "Dismantling the Divide between Indigenous and Scientific Knowledge." *Development and Change* 26 (1995): 413–39.

Arens, William. *The Man-Eating Myth: Anthropology and Anthropophagy*. Oxford, New York: Oxford University Press, 1979.

Austin Alchon, Suzanne. "Tradiciones médicas nativas y resistencia en el Ecuador." In *Saberes andinos: Ciencia y tecnología en Bolivia, Ecuador y Perú*, edited by Marcos Cueto, 15–36. Lima: IEP, 1995.

Baldwin, Martha. "Expanding the Therapeutic Canon: Learned Medicine Listens to Folk Medicine." In *Cultures of Communication from Reformation to Enlightenment*, edited by James Van Horn Melton, 239–56. Ashgate: Aldershot, 2002.

Bayly, Christopher A. "The First Age of Global Imperialism, c. 1760–1830." *The Journal of Imperial and Commonwealth History* XXVI, no. 2 (1998): 28–47.

Bellin, Joshua David. "Taking the Indian Cure: Thoreau, Indian Medicine, and the Performance of American Culture." *The New England Quarterly* LXXIX, no. 1 (2006): 3–36.

Bleichmar, Daniela. "Books, Bodies, and Fields: Sixteenth-Century Transatlantic Encounters with New World *Materia Medica*." In *Colonial Botany: Science, Commerce, and Politics in the Early Modern World*, edited by Londa Schiebinger and Claudia Swan, 83–99. Philadelphia: University of Pennsylvania Press, 2005.

Bleichmar, Daniela. *Visible Empire: Botanical Expeditions and Visual Culture in the Hispanic Enlightenment*. Chicago, London: The University of Chicago Press, 2012.

Brading, David. *The First America: The Spanish Monarchy, Creole Patriots, and the Liberal State 1492–1867*. Cambridge, New York: Cambridge University Press, 1991.

Brockliss, Laurence and Colin Jones. *The Medical World of Early Modern France*. Oxford: Clarendon Press, 1997.

Cahill, David. "Colour by Numbers: Racial and Ethnic Categories in the Viceroyalty of Peru, 1532–1824." *Journal of Latin American Studies* 26, no. 2 (1994): 325–46.

Cañizares Esguerra, Jorge. "Iberian Colonial Science." *Isis* 96 (2005): 64–70.

Cook, Harold J. *Matters of Exchange: Commerce, Medicine and Science in the Age of Empire*. Hyderabad: Orient Longman, 2008.

Cooper, Alix. "The Indigenous versus the Exotic: debating natural origins in early modern Europe." *Landscape Research* 28, no. 1 (2003): 51–60.

Cooper, Alix. *Inventing the Indigenous: Local Knowledge and Natural History in Early Modern Europe*. Cambridge, New York: Cambridge University Press, 2007.

De Vos, Paula. "The Rare, the Singular, and the Extraordinary: Natural History and the Collection of Curiosities in the Spanish Empire." In *Science in the Spanish and*

Portuguese Empires, 1500–1800, edited by Daniela Bleichmar, Paula de Vos, Kristin Huffine, and Kevin Sheehan, 271–89. Stanford: Stanford University Press, 2009.

Dove, Michael R. "Indigenous People and Environmental Politics." *Annual Review of Anthropology* 35 (2006): 191–208.

Elshakry, Marwa. "When Science Became Western: Historiographical Reflections." *Isis* 101. Focus. Global Histories of Science (2010): 98–109.

Ferreira Furtado, Júnia, ed. *Luís Gomes Ferreira, Erário mineral*. Rio de Janeiro: Editora FIOCRUZ, 2002 (1735).

Ferreira Furtado, Júnia. "Tropical Empiricism: Making Medical Knowledge in Colonial Brazil." In *Science and Empire in the Atlantic World*, edited by James Delbourgo and Nicholas Dew, 127–51. New York. London: Routledge, 2008.

Gänger, Stefanie. "Disjunctive Circles: Modern Intellectual Culture in Cuzco and the Journeys of Incan Antiquities, c. 1877–1921." *Modern Intellectual History* 10, no. 2 (2013): 399–414.

Gänger, Stefanie. *Relics of the Past: The Collecting and Study of Pre-Columbian Antiquities in Peru and Chile, 1837–1911*, Oxford Studies in the History of Archaeology. Oxford: Oxford University Press, 2014.

Grafton, Anthony. *New Worlds, Ancient Texts: The Power of Tradition and the Shock of Discovery*. Cambridge MA, London: Harvard University Press, 1992.

Greene, Roland. "Petrarchism among the Discourses of Imperialism." In *America in European Consciousness, 1493–1750*, edited by Karen Ordahl Kupperman, 130–65. Chapel Hill, London: University of North Carolina Press, 1995.

Grove, Richard H. "Indigenous Knowledge and the Significance of South-West India for Portuguese and Dutch Constructions of Tropical Nature." *Modern Asian Studies* 30, no. 1 (1996): 121–43.

Guerra, Francisco. "Medical Almanacs of the American Colonial Period." *Journal of the History of Medicine* XVI, no. 3 (1961): 234–55.

Heller, Dana. "Holy Fools, Secular Saints, and Illiterate Saviors in American Literature and Popular Culture." *CLCWeb: Comparative Literature and Culture* 5, no. 3 (2003): 1–15.

Klauth, Carlo. *Geschichtskonstruktion bei der Eroberung Mexikos: Am Beispiel der Chronisten Bernal Diáz del Castillo, Bartolomé de las Casas und Gonzalo Fernández de Oviedo*. Hildesheim, Zurich, New York: Georg Olms Verlag, 2012.

Koerner, Lisbet. *Linnaeus: Nature and Nation*. Cambridge MA, London: Harvard University Press, 1999.

Koerner, Lisbet. "Women and Utility in Enlightenment Science." *Configurations* 3, no. 2 (1995): 233–55.

Krech, Shepard. *The Ecological Indian: Myth and History*. New York: W. W. Norton, 1999.

Lavallé, Bernard. *Las promesas ambiguas: Ensayos sobre el criollismo en los Andes*. Lima: IRA, 1993.

Lowood, Henry. "The New World and the European Catalog of Nature." In *America in European Consciousness, 1493–1750*, edited Karen Ordahl Kupperman, 295–323. Chapel Hill, London: University of North Carolina Press, 1995.

Maehle, Andreas-Holger. *Drugs on Trial: Experimental Pharmacology and Therapeutic Innovation in the Eighteenth-Century*, Clio Medica/The Wellcome Institute Series in the History of Medicine. Amsterdam: Editions Rodopi, 1999.

Majluf, Natalia. "The Creation of the Image of the Indian in 19th-Century Peru: The Paintings of Francisco Laso (1823–1869)." PhD diss., University of Texas, 1996).

McClure, Julia. *The Franciscan Invention of the New World*. Basingstoke: Palgrave, 2016.

Merchant, Carolyn. *Reinventing Eden: The Fate of Nature in Western Culture*. New York, London: Routledge, 2013.

Mignolo, Walter D., Margaret R. Greer, and Maureen Quilligan. "Introduction." In *Rereading the Black Legend: Discourses of Religious and Racial Difference in the Renaissance Empires*, edited by Walter D. Mignolo, Margaret R. Greer, and Maureen Quilligan, 1–27. Chicago, London: The University of Chicago Press, 2007.

Murphy, Kathleen S. "Translating the Vernacular: Indigenous and African Knowledge in the Eighteenth-Century British Atlantic." *Atlantic Studies* 8, no. 1 (2011): 29–48.

Nappi, Carla. "Winter Worm, Summer Grass: *Cordyceps*, Colonial Chinese Medicine, and the Formation of Historical Objects." In *Crossing Colonial Historiographies: Histories of Colonial and Indigenous Medicines in Transnational Perspective*, edited by Anne Digby, Projit B. Muhkarji, and Waltraud Ernst, 21–35. Newcastle upon Tyne: Cambridge Scholars, 2010.

Norton, Marcy. *Sacred Gifts, Profane Pleasures: A History of Tobacco and Chocolate in the Atlantic World*. Ithaca, London: Cornell University Press, 2008.

O'Phelan Godoy, Scarlett. *Kurakas sin sucesiones: Del cacique al alcalde de indios. Perú y Bolivia 1750–1835*. Cuzco: CBC, 1997.

Ordahl Kupperman, Karen. "Introduction: The Changing Definition of America." In *America in European Consciousness, 1493–1750*, edited by Karen Ordahl Kupperman, 1–32. Chapel Hill, London: University of North Carolina Press, 1995.

Pagden, Anthony. *The Fall of Natural Man: The American Indian and the Origins of Comparative Ethnology*. Cambridge, New York: Cambridge University Press, 1982.

Palmer, Steven. *Doctors, Healers, and Public Power in Costa Rica, 1800–1940*. Durham, London: Duke University Press, 2003.

Park, Katherine. *Secrets of Women: Gender, Generation, and the Origins of Human Dissection*. New York: Zone Books, 2006.

Pimentel, Juan. "The Iberian Vision: Science and Empire in the Framework of a Universal Monarchy, 1500–1800." *Osiris: A Research Journal Devoted to the History of Science and its Cultural Influences* 15 (2000): 17–30.

Podgorny, Irina. "From Lake Titicaca to Guatemala: The Travels of Joseph Charles Manó and his Wife of Unknown Name." In *Nature and Antiquities: The Making of*

Archaeology in the Americas, edited Philip L. Kohl, Irina Podgorny, and Stefanie Gänger, 125–44. Tucson: University of Arizona Press, 2014.

Pomeranz, Kenneth. *The Great Divergence: China, Europe, and the Making of the Modern World*. Princeton: Princeton University Press, 2000.

Rood, Daniel. "Toward a Global Labor History of Science." In *Global Scientific Practice in an Age of Revolutions, 1750–1850*, edited by Patrick Manning and Daniel Rood, 255–74. Pittsburgh: University of Pittsburgh Press, 2016.

Ross Griffel, Margaret. "ZUMA, or, the Tree of Health." In *Operas in English: A Dictionary*, edited by Margaret Ross Griffel, 556. Plymouth: Scarecrow Press, 2013.

Rousseau, George S. and Roy Porter. "Introduction: Approaching Enlightenment Exoticism." In *Exoticism in the Enlightenment*, edited by George S. Rousseau and Roy Porter, 1–22. Manchester, New York: Manchester University Press, 1990.

Rowe, John Howland, ed. *Inca Religion and Customs by Father Bernabe Cobo*. Austin: University of Texas Press, 1990 (1653).

Sanchez-Albornoz, Nicolas. "El trabajo indigena en los Andes: teorías del siglo XVI." *Revista ecuatoriana de historia económica* 2 (1987): 153–81.

Schiebinger, Londa. *Plants and Empire*. Cambridge MA, London: Harvard University Press, 2004.

Schiebinger, Londa. "Prospecting for Drugs: European Naturalists in the West Indies." In *The Postcolonial Science and Technology Studies Reader*, edited by Sandra Harding, 110–26. Durham, London: Duke University Press, 2011.

Schiebinger, Londa. *Secret Cures of Slaves: People, Plants, and Medicine in the Eighteenth-Century Atlantic World*. Stanford: Stanford University Press, 2017.

Scott Parrish, Susan. *American Curiosity: Cultures of Natural History in the Colonial British Atlantic World*. Chapel Hill: University of North Carolina Press, 2006.

Shapin, Steven. *A Social History of Truth: Civility and Science in Seventeenth-Century England*. Chicago, London: The University of Chicago Press, 1994.

Sheridan, Richard B. *Doctors and Slaves: A medical and Demographic History of Slavery in the British West Indies, 1680–1834*. Cambridge, New York: Cambridge University Press, 1985.

Stolberg, Michael. "Learning from the Common Folks: Academic Physicians and Medical Lay Culture in the Sixteenth Century." *Social History of Medicine* 27, no. 4 (2014): 649–67.

Stolley, Karen. *Domesticating Empire: Enlightenment in Spanish America*. Nashville: Vanderbilt University Press, 2013.

PART 2

Authorities' and Societies' Strategies

CHAPTER 6

Change and Continuity: The Bureaucracy of Knowledge in South America

Irina Podgorny

Much of nineteenth-century historiography affirmed the idea, spread during the Spanish American Revolution and Independence period that the metropolis had abandoned its former territories to a state of darkness in terms of science.[1] Paradoxically, this understanding was consolidated thanks to the very nature of the most important innovations implemented by the Spanish monarchy: the technology of long-distance government, bureaucracy, and the archive, and, within this framework, the secretive nature of the administration of the Indies.[2]

Over the past thirty years, Spanish and Mexican scientific historians have reversed this vision, studying the dynamics of the expeditions sponsored by the Crown and the enormous quantity of materials about the three kingdoms of nature that were collected in the Royal Cabinet, the Royal Pharmacy of Madrid, and the cabinets created in major cities of the viceroyalties during the last third of the eighteenth century. An analysis of the products from the Indies in medicine, commerce, and mining as well as the envoy of experts to the area reveals a history of knowledge as long and complex as those of the Dutch or British.[3]

1 This chapter was previously published in a similar form: "Bureaucracy, Instructions, and Paperwork: The Gathering of Data about the Three Kingdoms of Nature in the Americas, 1770–1815," *Nuevo Mundo Mundos Nuevos* [Online], Paper posted on February 19, 2019. https://doi.org/10.4000/nuevomundo.75454. The text has been adapted to the reflections developed in this volume.

2 During the period from 1810 to 1826, the political map of Spanish America was redrawn by the process of liberation from colonial rule triggered by the Napoleonic wars. In the Rio de la Plata Provinces, Independence was declared in 1816.

3 From studies of the work of Francisco Hernández (physician to Felipe II), the business of Nicolás Monardes in Seville, the royal expeditions of the eighteenth and early nineteenth centuries (Royal Botanical Expedition, Royal Antiquities Expedition, Ulloa's expedition), and even the border missions, contemporary bibliography acknowledges the mobilization of resources and individuals to study the Indies. These topics, developed first in the work of José María López Piñero, his students and collaborators, and then that of Juan Pimentel, Antonio Lafuente, and Nuria Valverde, have recently been rediscovered by US-American researchers.

That earlier historiographical notion ignored not only the series of dispositions from Madrid to gather data about the nature and commerce of the Americas, but also – and more importantly for our argument – the continuity between the "old" ways of the Spanish Empire and the new dispositions by American revolutionary governments. Indeed, the discourse on the rupture of the colonial order and the history of new scientific and academic institutions often forgets that the agents in charge of describing the natural and social American world changed flags, but not names or interests. The same priests, physicians, topographers, pilots, and engineers who botanized, made maps, and gathered insects and meteorological data before Independence then swore loyalty to the new Republics and used the same methodology to continue their work, trying with varying degrees of success to find a state who would sponsor them.[4] In doing so, intentionally or not, they brought the practices and knowledge learned in the academies, schools, and administrative offices of the late colonial Ibero-American world to the new political order.

By observing several episodes during the late colonial period and the early decades of the nineteenth century, this paper describes this system of production and circulation of knowledge linked to bureaucracy and the Atlantic trade. Based on primary sources from the General Archive of the Indies, the General Archive of the Nation (Argentina), the Archive of the Royal College of Surgeons (London), and on secondary bibliography, in particular the work of Javier I Sánchez Almazán on Pedro Franco Dávila's *Instructions for the Royal Cabinet of Madrid*, this chapter, rather than focusing on the objects collected, considers the documents that resulted from the "necessity" of collecting minerals, plants and animals. In this way, it reveals the true protagonists of this story: the pathways of bureaucracy and the flow of paperwork where data about nature and man in the Americas were generated and took shape. There, on official stationery, the instructions and networks established by the offices for remote governance would appear along with yet unknown animals, plants, rivers, ancient cities and islands. What expectations did this world of papers and bureaucracy create so that the agents of the royal service would go out to collect plants, minerals, and animals? This question becomes more significant after the overthrow of the colonial administration because these people continued with the same activities despite the fact that those expectations could no longer be met.

To consider these topics, this article will examine the instructions for the collection of samples from the natural world written before and after the process

4 Jorge Gelman, *Un funcionario en busca del Estado-Pedro Andrés García y la cuestión agraria bonaerense, 1810–1822* (Buenos Aires: Universidad Nacional de Quilmes, 1997).

of Independence. In particular, it will discuss those related to the provision of the Royal Cabinet of Madrid (1776), which illustrate how the history of natural history practices articulates the history of the sovereign's political curiosity (to know and control "everything") and the interests of those individuals who, as Arndt Brendecke recalls, appeal to this curiosity to combine the promise of new knowledge with the opportunity to promote their own projects. The history of these instructions also illuminates how the interests and expectations of those individuals are shaped by the practices in which they are immersed and how the instructions become independent of their "author" and continue to impact other institutions and subjectivities. In this sense, the instructions, as forms of bureaucracy, defy any attempt to assign them to a certain historical period based on the ruptures of political order, and instead demonstrate the continual recombination of knowledge and patterns from different domains and moments in time. Thus, this chapter refers to those contingent contexts beneficial to collecting activities that help to learn about a certain territory.

1 Instructions

In 1771, after various failed attempts, Pedro Franco Dávila (1711–1786) was able to persuade Carlos III to establish a Royal Cabinet of Natural History in Madrid with the objects he and Pedro Vicente Maldonado had collected in the Old and New Worlds.[5] Dávila, the son of a cacao producer from Guayaquil, had lived with his collections in Paris since 1745. Trained in the rules of Spanish commerce, he had visited ports and regions throughout the Americas, becoming familiar with the procedures that characterized overseas transport control. Once established in Europe, he maintained contact with family members overseas, other residents of Nueva Granada, and several European academics with whom he exchanged data, favours and objects.

The Royal Cabinet was inaugurated in November 1776, in the context of the so-called Bourbon reforms, which mobilized plants, antiquities, and animals

5 Juan Pimentel, *Testigos del Mundo: Ciencia, literatura y viajes en la ilustración* (Madrid: Marcial Pons, 2003); Javier I. Sánchez Almazán, *Pedro Franco Dávila (1711–1786): De Guayaquil a la Royal Society. La época y la obra de un ilustrado criollo* (Madrid: CSIC, 2012); Miguel Villena Sánchez-Valero et al., *El gabinete perdido: Pedro Franco Dávila y la historia natural del Siglo de las Luces* (Madrid: CSIC, 2009); Maria Ángeles Calatayud Arinero, *Catálogo crítico de los documentos del Real Gabinete de Historia Natural (1787–1815)* (Madrid: CSIC, 2000); Agustín Jesús Barreiro, *El Museo Nacional de Ciencias Naturales (1771–1935)* (Madrid: Doce Calles, 1992 [1944]). On Maldonado's collections, see Neil Safier, *Measuring the New World: Enlightenment Science and South America* (Chicago, London: The University of Chicago Press, 2008), 167–68.

to Madrid. These reforms also fuelled the flow of paper that characterized Spain's administration of their marine territories since the Habsburgs, when long distance government required them to incorporate written documents and a bureaucratic structure that compensated for king's to be physically present in the new territories. In the sixteenth century, the Council of Indies promoted the description of the New World through instructions and questionnaires and, starting in the 1570s, reports had to follow a series of questions to be answered on site.[6] After the reforms of Juan de Ovando during Felipe II's reign, every American bureaucrat was duty-bound to permanently describe the overseas Spanish territories. In this sense, the instructions, charts, and surveys were a means to construct a common space of knowledge.[7] These procedures aimed at educating the traveller and guiding them in the collection of data and objects or in the use of instruments. When the Bourbons in the eighteenth century reformulated these instructions, they tried to perfect those of Ovando, combining them with new interests and placing the collection of objects in a public or semi-public space (represented by the Royal Cabinet or Royal Pharmacy).

Thus, in May 1776, Joseph de Gálvez y Gallardo, newly designated Secretary of State of the Universal Office of the Indies, had the "Royal Order notice for the remission of Natural History curiosities to the Royal Cabinet" (1776–1777) printed and circulated, accompanied by the following note:

> The King has established in Madrid a Cabinet of Natural History in which the Animals, Vegetables, Minerals, rare Stones and whatever Nature

6 Jose María López Piñero, *El arte de navegar en la España del Renacimiento* (Madrid: Labor, 1979), 83–97.

7 Bernhard Siegert, *Passagiere und Papiere: Schreibakte auf der Schwelle zwischen Spanien und Amerika* (München: W. Fink, 2006); Wolfgang Schäffner, "Die Verwaltung der Endlichkeit: Zur Geburt des neuzeitlichen Romans in Spanien," in *Die Endlichkeit der Literatur*, ed. Eckart Goebel and Martin von Koppenfels (Berlin: Akademie Verlag, 2002), 1–12; Arndt Brendecke, *Imperium und Empirie: Funktionen des Wissens in der Spanischen Kolonialherrschaft* (Köln, 2009); also Guillaume Gaudin, *Penser et gouverner le nouveau monde au XVIIe siècle: l'empire de papier de Juan Díez de la Calle, commis du Conseil des Indes* (Paris: L'Harmattan, 2013). At the same time, the famous *Methodus apodemica* was developed in works such as those of Theodor Zwinger (Basel, 1577), see Irina Podgorny and Wolfgang Schäffner, "La intención de observar abre los ojos," *Prismas* 4 (2000): 218. In England, the first of these instructions, in the form of a traveller's survey, dates back to the second half of the seventeenth century. It was continued in the eighteenth century by various European academics and learned societies. See Marie-Noëlle Bourguet, "La collecte du monde: voyage et histoire naturelle (fin XVIIème siècle – début XIXème siècle)," in *Le Muséum au premier siècle de son histoire*, ed. Claude Blanckaert et al. (Paris: Muséum National d'Histoire Naturelle, 1997). See Simona Boscani Leoni's article in this volume.

produces in his Majesty's vast Domains will be gathered, as well as whatever is possible to acquire from strange lands. To complete and enrich the series and collections of the Royal Museum in each one of its classes, it is necessary that the subjects who govern in the Provinces and Towns of the Spanish Kingdoms, now and in the future take care to gather any curious pieces that they find in their districts and send them to the Cabinet to Natural History.

Hereby including you in His Majesty's order for your intelligence and completion in the part that involves you, persuading you that the King will look at your performance of this order as singular proof of the zeal of your service and love of the public good and so that you personally understand what curiosities are desired and the manner of their conservation, I include copies of the Instruction so that this effort that the King has ordered is extended, leaving it to you to distribute them to the subjects that are concerned with this effort, without having to limit this only to Justices of the Towns, it could also include Priests and other individual people who you choose and will be able to accurately carry this out.

Finally, I express to you that the King wants those who follow you in this position to continue to take care to collect and send the rare pieces that are discovered so that in this way, the different series of the Cabinet in the three Kingdoms of Nature will be completed and renovated, avoiding that such an important and useful establishment will fall apart.[8]

As Susan Socolow signaled several years ago, the "love of Royal Service," spread throughout the channels and networks created by the post and the body of civil servants, and included a series of characters whose biographies can barely be traced beyond the papers that they once signed or sent.[9] An instruction, like the one meant to enrich the Royal Cabinet, arrived for the viceroys or governors who were then supposed to retransmit it throughout their geographical space via judiciary officials, priests, and anyone else necessary to meet the order's objective, in a network already defined by Ovando's instructions. Moreover, the instruction was supposed to be maintained through time, left in place so subsequent groups of civil servants would also comply with it.

8 "Real Orden circular para la remisión de curiosidades de Historia Natural para el Real Gabinete, ejemplares de la instrucción (1776–7)", Indiferente General, folio 669 (Archivo General de Indias). My translation. See Ernesto Lemoine Villicaña, "Instrucciones para aumentar las colecciones del gabinete de historia natural de Madrid," *Boletín del Archivo General de la Nación* 2, no. 2 (1961).

9 Susan Socolow, *The Bureaucrats of Buenos Aires, 1769–1810: Amor al Real Servicio* (Durham: Duke University Press, 1987).

As long as the instruction was not cancelled, it continued to act and model practices, subjectivities, customs, objects, and papers. This explains why the governor of Buenos Aires, Juan José de Vértiz, responded from Montevideo on September 28, 1776, "offering to collect and direct when appropriate": the five copies of the instruction were in his power "with the care to proceed with the mere correspondence of complying with his Majesty's orders in all its parts," that is, to procure the animals, plants and minerals whose presence was required in Madrid and pack and prepare them in such a way that they arrive at their destination.[10]

The instructions that Gálvez circulated were not based as much on the King's personal desire as they were on a draft elaborated by Pedro Franco Dávila, who had been named in 1771 as lifetime director of the Royal Cabinet of Madrid.[11] In February 1776, with the cabinet already open to the public, Dávila wrote a "List so that the Viceroys, Governors, Chief Magistrates, Mayors and Provincial Superintendents in all of His Majesty's Domains can make, collect, prepare and send to Madrid all of Nature's productions that they find in the lands and towns of their districts." There, he detailed the minerals, plants, and animals of the Indies with which he was familiar through his own experience, literary references and/or observation of the collections held in Parisian cabinets, indicating the places where they should be sought and their equivalents in the Old World. The definitive instructions received by the viceroys added and omitted species and included the request for "petrified objects" and "curiosities of art" (dresses, weapons, instruments, furniture, machines, and idols).

Dávila is a good example of how an individual collector was able to move into the royal domain with that collection; later, as director of the cabinet, he utilized the Crown's communication and governmental structure to expand a collection that was linked to his own identity, an identity shaped by the practices of both bureaucracy and commerce. Dávila combined these practices with the catalogues that described the natural history cabinets proliferating in Paris and with the lists of merchandise transported by ships. Thus, the instructions that he wrote are similar to others produced in eighteenth-century Spain that were still valid at the time, such as those written in Aranjuez on June 6, 1752, and sent to the viceroys of Mexico, Lima, and the Kingdom of New

10 Bourguet, "La collecte"; María Eugenia Constantino and Antonio Lafuente, "The Hidden Logistics of Longinos's Novohispanic Cabinet," *Nuncius* 27, no. 2 (2012), and Marcelo F. Figueroa, "Packing Techniques and Political Obedience as Scientific Issues: 18th-Century Medicinal Balsams, Gums and Resins from the Indies to Madrid," *HOST* 5 (2012).

11 Sánchez Almazán, *Pedro Franco Dávila*, 23–145. This whole section is based on that work and on Lemoine Villicaña, "Instrucciones." My translation.

Granada for the collection of all types of natural products and the building of a "*Royal Cabinet of Natural History of the Mines that were found in the domains of His Majesty in America.*"

In this way, the instructions and the formalisms of commerce and traffic across the Atlantic produced a flow of objects while also reinforcing each other, thereby generating new forms and subjectivities that incorporated state bureaucracy and commercial registers into how the world and nature were ordered. As Jakob Tanner has underscored, this is about the "performing power" of bureaucracy and its "archive."[12] Thanks to these protocols, "unknown" things could be seen, such as the ancient houses of the ruins of Palenque in Chiapas or the "rare and corpulent animal" of the Luján River.[13] Moreover, they created a way for some priests, military engineers, governors, mayors, and viceroys in the overseas territories to take advantage of them so that through a skeleton, mineral or plant, they could draw attention to the ordinary and extraordinary productions of their lands as well as to their own loyalty and devotion to the Crown.[14]

2 The Skeleton of Luján

The instructions, as previously mentioned, indicated that every viceroy should send things revealed or discovered in the territories they governed back to Madrid. Thus, in 1788, the Viceroy of Buenos Aires, the Marquis of Loreto, accompanied a shipment of objects with two documents for Minister Porlier: one, dated in March, attached the design "in parts and according to how it should look after the skeleton is put together" of a very corpulent and unknown animal, and announced that the bones would follow in a future shipment. The other, from May, ratified the first and was a guide to the shipment of seven boxes on board the mail Frigate la Cantabria. In those letters, the Marquis reported:

12 Jakob Tanner, "Akteure, Akten und Archive," in *Was Akten bewirken können: Integrations- und Ausschlussprozesse eines Verwaltungsvorgangs*, ed. Claudia Kaufmann and Walter Leimgruber (Zurich: Seismo, 2008).

13 Irina Podgorny, "The Reliability of the Ruins," *Journal of Spanish Cultural Studies* 8, no. 2 (2007).

14 Until now, no one has studied whether these packages received any type of compensation that responded to the expectations of the person who sent the novelties. However, I understand it to be within the same system as the gifts, donations, and favours analysed by Viviana L. Grieco, *The Politics of Giving in the Viceroyalty of Rio de la Plata: Donors, Lenders, Subjects, and Citizens* (Albuquerque: The University of New Mexico Press, 2014).

Don Manuel Warnes first Alcalde of this city informs us that Brother Manuel de Torres of the Order of Predicadores has discovered the skeleton of a corpulent animal chiefly unknown in this part of America, where no species is to be found that it could be compared to, adding besides that the friar intended to acquaint me of it. I immediately decided to hear him, upon which, by his mere narrative I found the discovery to be a valuable one and that it might also help to prove erroneous the belief some had entertained in previous occasions when some loose bones happened to have been found that they belonged to the human species notwithstanding the enormous stature they were supposed and I determined therefore that this skeleton should be dug up from the place in which it was found and should be brought to my abode in the fort with the least possible damage, because the dampness of the soil caused some the parts to be very apt to mold of, especially the head and ribs.

All this was carefully entrusted to the said Friar […], I have determined all to be packed up in seven cases, that as they must necessarily be large on account of the size of some of the bones and the great amount of the straw necessary for their preservation if they were to be reduced to a smaller number would make it more difficult to be transported.

In each of the cases the contents are labeled in conformity with what is expressed and numbered in the first leaf of the annexed memorandum. The second leaf shrews the figure that it is supposed to have if joined together and in this leaf notations are made on the place in which the discovery took place and the inferences made to consider it as a novelty compared with other known animals.

Lately some of the Caciques or chiefs of the infidels of the pampa and the Sierra have happened to come to this city. I took care they should see these bones in the manner they had been placed in order to complete the shape of this animal, and they seemed to be astonished asserting afterwards they could not be of this country as they had no knowledge of them and they had always been under the belief that some bones that had been found belonged to their forefathers. But it is very natural to infer that the latter if the animal were mischievous and not numerous should have destroyed them when they were the sole possessors of this land.[15]

15 This is a translation of Spanish papers purchased in Seville c. 1833, when the remaining furniture from the Marquis of Loreto's house was sold after a fire on July 17, 1827, had destroyed most of the property's valuable items. The translation was sent by Manuel Williams, son of Julian Benjamin Williams, British Vice Consul in Seville, and an active dealer in Spanish works of art, to the Royal College of Surgeons Archives where it was kept in their file on *Megatherium*; see Vicente Lleó Cañal, "Julian Benjamin Williams y el

Friar Manuel Torres – one of the many agents who responded to the instructions – made his discovery of enormous bones on the Luján River, about five miles from the town of the same name and about sixty kilometres from Buenos Aires. Prior to removing the bones, the friar requested that the viceroy send a draughtsman to "extract them to paper," for all the work might be otherwise ruined. The Marquis of Loreto granted the request and further ordered that the articulations and parts in the picture also be numbered in order to identify the corresponding bone. The dimensions were to be taken and the skeleton described in detail, giving the name and distance from the nearest town and to the Río de la Plata. The skeleton was so immense that speculation began as to what the animal's body mass might have been with flesh and hide. Until that time, there had been no other reports in the Americas of a creature with similar characteristics, and it was not known whether this was an amphibious or aquatic animal, although it was assumed to be terrestrial based on the size of its nails. It bore no resemblance to the elephant, except in terms of size, nor to the rhinoceros or the anta, namely the South American tapir. However, animal bones had been found in the vicinity from smaller specimens of these same species. The gigantic bones were packed in hides and sent to Buenos Aires, where several experts reassembled them as a mounted skeleton. Finally, they were shipped off to Madrid in seven boxes, where the Royal Cabinet preparers submitted the skeleton for exhibition.[16] (Fig. 6.1 a, b and c)

The drawings, signed by the lieutenant Francisco Pizarro, draftsman of the Royal Body of Artillerymen, combined the discoverer's desires with the viceroy's instructions. The viceroyalty's bureaucracy, as seen in the official letters, sent off the bones and the drawings, leaving copies in the Buenos Aires archive. All of the people involved were instilled with the importance of the instructions and with their devotion to the Royal Service, which in addition to producing papers, also created and made visible new objects, giving potential to new searches. The Marquis sent the seven boxes to Galicia and then Madrid, where the Royal Cabinet preparers mounted it for exhibition. When the King was apprised of this, he reportedly requested a "live or stuffed" specimen of the rare, corpulent animal, a sketch of which eventually ended up in the hands of Georges Cuvier. Based on this, in the comparative anatomy laboratories of Paris, the Beast of Luján would be classified as belonging to a new genus:

comercio de arte en la Sevilla del XIX," *Boletín de la Real academia Sevillana de Buenas Letras: Minervae Baeticae* 36 (2008). The original document in Spanish is kept in the Archivo General de Indias, Indiferente General, Buenos Aires, 76, folio 31, Buenos Ayres, 2 de marzo de 1788, "Remesa de osamentas de un animal mui corpulento".

16 Manuel R. Trelles, "El Padre Manuel Torres," *Revista de la Biblioteca Pública de Buenos Aires* 4 (1882).

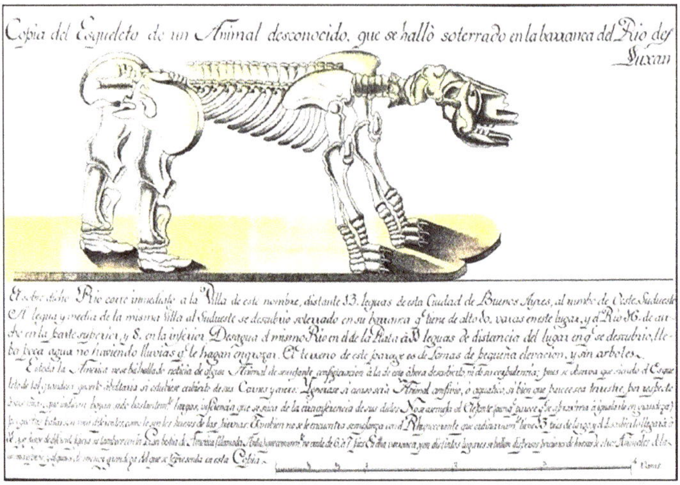

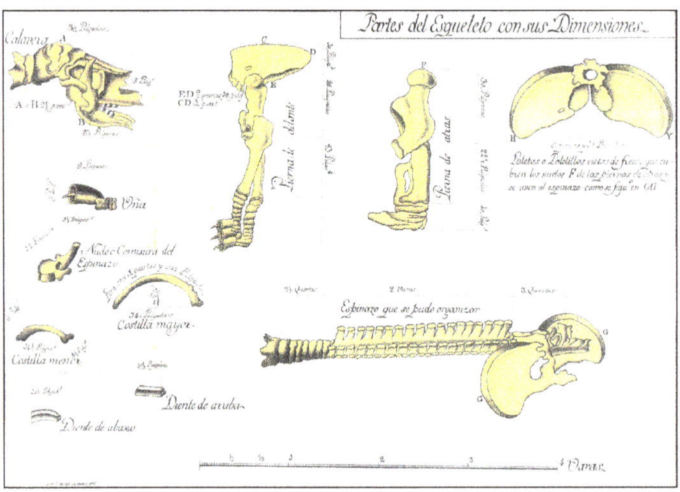

Numero, Peso, Dimensiones, y Contenido en cada uno delos siete Caxones que se Remiten

7. Caxones

NUMEROS.	PESO.	BULTO.	CONTENIDO
1.	12 arrobas 10 libras	Largo ... 1. vara [illegible] ancho ... $\frac{1}{2}$ vara 6 pulgs. alto ... $\frac{3}{4}$ ydem 4 pulgs.	Caveza, Vertebras, y huesos del Espinazo.
2.	9. arrobas	Largo ... 1. vara 6. pulgs. ancho ... $\frac{1}{2}$ ydem 6. pulgs. alto ... $\frac{1}{2}$ ydem $6\frac{1}{2}$ pulgs.	Dorsales, y Lombares.
3.	14 arrobas 3 libras	Largo ... 1. vara 4 pulgs. ancho ... $\frac{3}{4}$ ydem $3\frac{1}{2}$ pulgs. alto ... 1. vara 1. pulgs.	Hueso Sacro.
4.	7. arrobas 7. libras.	Largo ... 1. vara 7 pulgs. ancho ... $\frac{1}{2}$ vara $4\frac{1}{2}$ pulgs. alto ... $\frac{1}{2}$ vara 3 pulgs.	Brazo derecho.
5.	7. arrobas 7. libras.	Largo ... 1. vara $7\frac{1}{2}$ pulgs. ancho ... $\frac{1}{2}$ vara $4\frac{1}{2}$ pulgs. alto ... $\frac{1}{2}$ vara 3 pulgs.	Brazo yzquierdo.
6.	8. arrobas 23. libras.	Largo ... 1. vara 8 pulgs. ancho ... $\frac{1}{2}$ vara $4\frac{1}{2}$ pulgs. alto ... $\frac{1}{2}$ vara 4 pulgs.	Pierna derecha.
7.	9. arrobas 9. libras.	Larga ... 1. vara $7\frac{1}{2}$ pulgs. ancho ... $\frac{1}{2}$ vara $4\frac{1}{2}$ pulgs. alto ... $\frac{1}{2}$ vara $4\frac{1}{2}$ pulgs.	Pierna yzquierda.

FIGURE 6.1 Paperwork accompanying the shipment to Spain of the skeleton of the mighty animal, as printed in Manuel R. Trelles, "El Padre Manuel Torres," Revista de la Biblioteca Pública de Buenos Aires, 1882, vol. 4, 1882

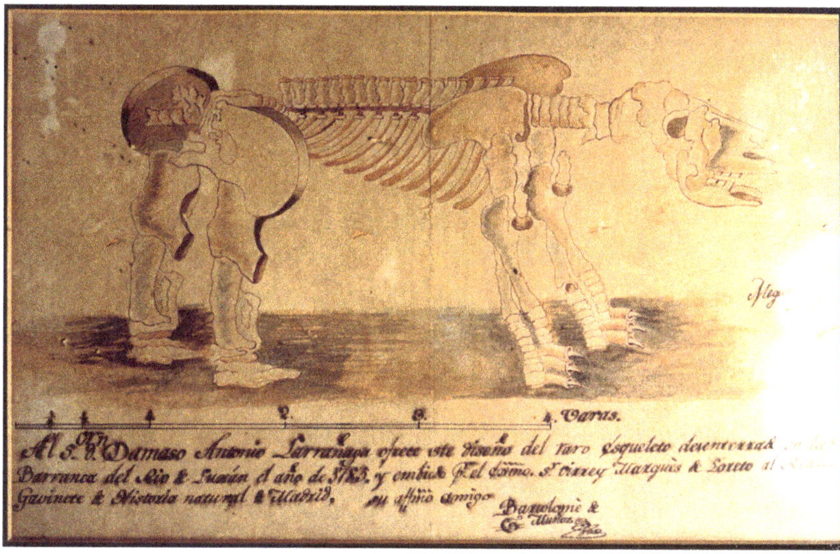

FIGURE 6.2 Copy made by Muñoz of the sketch of the skeleton kept in the archives of Buenos Aires (Archivos Museo de La Plata)

Megatherium. The name was invented specifically to describe its grandeur and to substantiate the idea of "vanished species," which was fundamental to the sciences in the coming century.

The papers that were produced were not only used in the metropolis, but also studied in the Río de la Plata. They were examined by other priests who were interested in the region's natural history, being published for the first time in Buenos Aires in 1835 by the topographer Juan María Gutiérrez (1809–1878) to illustrate an article in *Museo Americano*, a periodical directed by César Hipólito Bacle (1794–1838), fig. 6.2.

Gutiérrez gave several individuals credit for putting together the skeleton, including the collector, politician and journalist, Joaquín José de Araujo (1762–1835), whose papers included a copy of the drawing attributed to Brigadier José Custodio de Sá y Faría († 1792), a Portuguese engineer who was very active in marking the borders of the Río de la Plata. Not much is known about the discoverer, the artist, the Indians, and those in charge of mounting the skeleton in Buenos Aires; their names were barely registered in the administration's records. However, these few indicators reveal the type of agents and actions involved in such a discovery: After Dávila's death, the Royal Cabinet continued to gather petrifactions and other items from the three kingdoms of nature sent by individuals who were willing to respond to the requirements of the 1776 instructions. Not only that, as we will see below, the papers created by that bureaucracy would lead a parallel life and would end up appearing in the comparative anatomy debates of the 1830s.[17]

After the dissolution of the Spanish Empire, several colonial bureaucrats and public servants whose work involved producing and safeguarding papers and files, stayed in Buenos Aires and Montevideo, transformed into "civil servants in search of a State," according to the expression coined by Jorge Gelman.[18] The survival and use of colonial bureaucratic archives and the adaptation of the former technical bodies to the new political order and emerging scientific disciplines of the nineteenth century remains an issue deserving of further research. Thus, the instructions from 1812 together with the attempt to establish a museum in Buenos Aires and the circulation of the drawings and papers related to *Megatherium* are excellent opportunities to examine both their use and the type of understanding that was being established around the knowledge they generated in the final years of the colonial era. The following section studies a circle of erudite priests, such as Dámaso Larrañaga and Bartolomé Muñoz, involved with the founding of museums and libraries in the Río de la Plata region, and their interaction with French and English and the new publicists who arrived on their shores. Moreover, it studies how, through these priests and their access to old archives, colonial knowledge was introduced into the comparative anatomy debates of the Old World.

3 The Museum of Buenos Aires and the Collector-Priests

On June 27, 1812, the Buenos Aires Revolutionary government gave instructions to compile information on the flora and fauna of the various jurisdictions of the former viceroyalty. The common defence went hand in hand with the promotion of "establishments whose influence, together with the help of all Citizens with a love of good taste, will, when the moment of our sweet emancipation soon comes, also provide the means to attain the level of the Learned Peoples, which the destructive hand of the Peninsular Government

17 Irina Podgorny, "El camino de los fósiles: Las colecciones de mamíferos pampeanos en los museos franceses e ingleses," *Asclepio* 53, no. 2 (2001); Podgorny, "De ángeles, gigantes y megaterios: Saber, dinero y honor en el intercambio de fósiles en las provincias del Plata en la primera mitad del Siglo XIX," in *Los lugares del saber: Contextos locales y redes transnacionales en la formación del conocimiento moderno*, ed. Ricardo Salvatore (Rosario: Beatriz Viterbo, 2007); Fernando Ramírez Rozzi and Irina Podgorny, "La metamorfosis del megaterio," *Ciencia Hoy* 11, no. 61 (2001).

18 Gelman, *Un funcionario*.

had thus far deprived us of."[19] The aim was to encourage people to observe the continent's mineral, plant, and animal realms, "which today is without a doubt one of the most worthy occupations of Scholars all around the world, who, enjoying their knowledge and acquisition of the precious talents that are not easily found in our Mother Country, would surely be shocked to find that this, too, we have neglected." These instructions, sent to commanders of the various outposts, sought to establish a museum that would gather together all of the region's rare and unique products that were "worthy of being placed in such a repository, by encouraging all citizens who have them to give them as a gift." For example, in August 1812, an inventory was sent from Concepción in the Misiones province detailing the region's flora and fauna, and promising to send a list of medicinal herbs, "although not all are known." In his response, the commander at Concepción, Celedonio José del Castillo, wrote: "Upon receiving Your Excellency's High Order of 27 June last, I passed it on to the towns in this Department to let everyone know about it, offering the inhabitants to compensate anyone who brought us an animal or some other strange or peculiar thing."[20] This old colonial practice of hierarchically distributing instructions, from the centre of collection to the people who collected the objects, would live on informally in the traffic consolidated over the course of the nineteenth century of fossils, animals, and artifacts, and formally in museums' instructions to their providers and travelling naturalists. Celedonio del Castillo's response, in remarks concerning the initiatives of the previous administration, contradicted the wording of the instructions.

The instructions of 1812, in addition to seeking to establish a local centre to highlight regional products, set out a collective and integrating enterprise. The museum was to be created with the help of citizens, while at the same time encouraging private individuals to donate whatever oddities they held in their own estates. Thus, in 1813, as a matter of personal pride and patriotic honour, citizen Bartolomé de Muñoz (?–1831), a native of Spain, priest, and vicar general of the army of Montevideo, donated a variety of objects and acquisitions that he had procured for his own private use over a period of "twenty years of trying

19 To the Military Commander of Patagones, Buenos Aires, 27 June 1812, Archivo General de la Nación (Buenos Aires). My translation.

20 Del Castillo sent along a list of the "peculiarities" that he himself had come across in his thirty years of living in the area. This included birds, mammals, minerals, plants, medicinal herbs ("there is no single expert who knows them all"), yerba mate plantations ("the mines of this province"), and Araucaria trees ("whose trunks could be used for ships' masts").

diligence."[21] Muñoz's gift included a number of maps, dictionaries, Lavoisier's *Treatise on Chemistry*, "Wiedemann's Treatise on Oryctognosy," as well as natural history objects (a shell collection, natural zoophytes, mineral samples, and prints) and instruments (a microscope, prisms, and a thermometer).[22]

Muñoz's donations reflect the central role that priests, initially inspired by the colonial instructions, played in collecting culture from all over Spanish America. Priests, doctors, and military engineers responded to the government requests; however, motivated by personal interest, love of God or Royal Service, they invested their time and their own resources in purchasing books, scientific instruments, and anything else they needed to carry out the work and keep up to date with advances in the field of natural history. The instructions from 1812 are striking in terms of the clear confidence placed in the institution, and in its utilitarian conception. The museum, far from emerging as an institution governed by a council of the wise, issued the instructions to gather the objects. The same people taking part in the national government – people trained in the army or in law and colonial administration – would decide how to archive and classify the information coming in from the various jurisdictions and command headquarters of the districts of the River Plate region. This initiative did not prosper, and the individual collections continued to exist, albeit disconnected from both the viceroy's instructions and the failed revolutionary ones. A historiography that favours the notion of Spanish isolation and darkness has treated these collections as mere personal initiatives, forgetting that they also grew out of the attempts to coordinate them by first the metropolitan and, then, the revolutionary government.

In this context, the priests Dámaso A. Larrañaga (1771–1848) and again Bartolomé D. Muñoz stand out as avid readers and consumers of the books that, through different agents, arrived from Europe and from Rio de Janeiro.[23] The interest was such that European visitors filled their suitcases with the novelties published in Paris and London, sure to be able to sell them at a good

21 Muñoz, quoted in Irina Podgorny and Maria Margaret Lopes, *El desierto en una vitrina: Museos e historia natural en la Argentina, 1810–1890* (Mexico: Limusa, 2008), 36.

22 Muñoz sent his donation in a trunk in September 1813, and in October it was transferred to the library, where it was noticed that some items were missing. The donation was officially accepted in 1814.

23 Roberto Di Stéfano, *El púlpito y la plaza: Clero, sociedad y política de la monarquía católica a la república rosista* (Buenos Aires: Siglo XXI, 2004). Rio de Janeiro was the headquarters of the Portuguese court and as such, housed a natural history cabinet; see Maria Margaret Lopes, *O Brasil descobre a pesquisa científica: Os museus e as ciências naturais no século XIX* (São Paulo: Hucitec, 1997).

profit.[24] Collectors of manuscripts, devices and instruments, meteorological observations, plants, petrified objects and animals, the clerics exchanged data, papers, and drawings. After 1810 and the break with Spain, many of these religious men would divide their time between their commitment to revolution and their endeavours as naturalists. They were put in charge of the public libraries in Buenos Aires and Montevideo (1816) and/or museums or the initiatives to create them. While Muñoz, with patriotic zeal, donated his collections to establish a museum that never opened its doors, Larrañaga's private collections would attract all those interested in the past and future of these regions. They both compiled, read, translated and copied drawings and manuscripts, accumulating and organizing a vast corpus of documentary materials. They also extracted articles from the many books they purchased.[25]

Larrañaga, in particular, had a special predilection for plants: the world of the Bourbons had heightened awareness among clerics and pharmacists as to the benefits of studying texts on the medicinal properties of indigenous plants in the Americas and the potential for boosting the economy by capitalizing on the country's fruits and vegetables.[26] The *Flora Peruana*, or the plant collections of the botanists from the expedition of Alejandro Malaspina, arrived in Buenos Aires, not necessarily in book form, but rather through the accounts and manuscript copies that were gathered and transcribed from the collections of the region's curates. This copyist culture, forming as it did part of the clergy's education, shaped the study of natural history, in which the new methods of observation were combined with the practices of reading and extracting notes from manuscripts and printed materials. These priests drafted various treatises on natural history in dictionary form, bringing together different points of view organized in alphabetical order in booklets, embracing the many innovations that they added with each new reading. They also prepared *tableaux*,

24 So, when Aimé Bonpland (1773–1858) arrived in Buenos Aires in 1816, in addition to various commercial endeavours that he had in mind, he brought an enormous collection of books on natural history that he offered to collectors and to libraries in Buenos Aires, Santiago de Chile, and Montevideo. See Stephen Bell, *A Life in Shadow: Aimé Bonpland in Southern South America, 1817–1858* (Palo Alto: Stanford University Press, 2010).

25 Among others: the *Encyclopaedia Britannica, or A dictionary of arts, sciences, and miscellaneous literature; enlarged and improved* (Edinburgh, 1810), the *Dictionnaire d'Histoire Naturelle* (Paris, 1804), the *Florae peruvianae et chilensis* by Hipólito Ruiz and José Pavón (Madrid, 1794), the 13th edition of Carl Linnaeus's *Systema Naturae*, published between 1788 and 1793 by Johann Friedrich Gmelin, the natural histories of Félix de Azara and of Georges L. Leclerc, count of Buffon.

26 Rafael Algorta Camusso, *El Padre Dámaso Antonio Larrañaga: Apuntes para su Biografía* (Montevideo, 1922); Fernando Mañé Garzón, *El Glorioso montevideano: Vida y obra del Doctor José Manuel Pérez Castellano (1742–1815)* (Montevideo, 1998–2003).

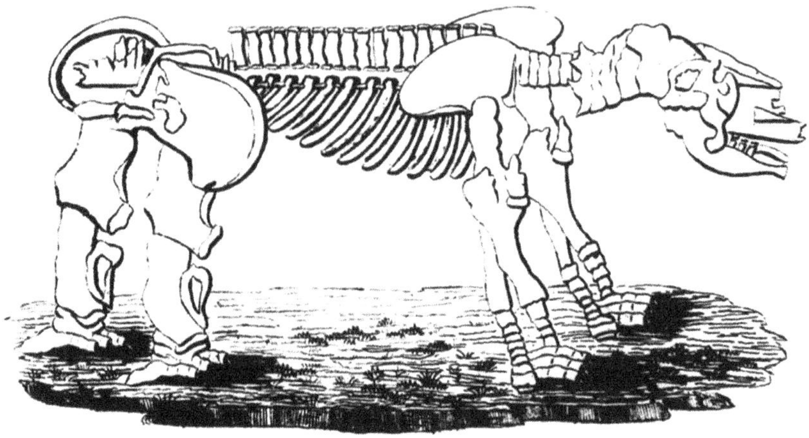

FIGURE 6.3 The colonial sketch as published in *El Museo Americano*

classification tables for the zoology, botany and mineralogy of the country, which were amended with the inclusion of new specimens. Furthermore, they illustrated and coloured their observations. Flowers, insects and birds sprang back to life through the clergy's ink.

One such drawing that they copied from the colonial archives in Buenos Aires, was the aforementioned colossal skeletal figure, when Larrañaga recorded in his journal in 1814 a new discovery of "bones of *Megatherium*." He subsequently translated a description published in the *Encyclopaedia Britannica* in 1810 that emphasised the paradoxical nature of this beast and an English article from 1806 that discussed the similarities of the animal to sloths and elephants. Muñoz, in copying the drawing from the Buenos Aires archive, omitted the description of the brigadier to instead note that the beast was now known as "megatherium," (fig. 6.3)

Hence, through the colonial archives and publications, they became aware of the debates about the controversial zoological affinity of this entity. From 1808 on, these priests devoted themselves to the study of comparative anatomy and to the classification of mammals, combining various systems with their observations and those of former colonial Spanish military engineers. In the early 1820s Larrañaga was convinced that the animal from Luján could have had a shell similar to that of the armadillos.

Larrañaga displayed his collections in the house-museum on his family's property, where he received several travelling captains and naturalists, to whom he would show his most precious objects and engage in debate. These visitors, who included Aimé Bonpland, the French botanist Auguste Saint-Hilaire (1779–1853), and Friedrich Sellow (1789–1831), commissioned by the Portuguese and Prussian court to collect samples from the provinces of Rio Grande do Sul and Uruguay were responsible for disseminating the views of Larrañaga to ever widening circles. In Larrañaga's museum, Sellow and Saint-Hilaire saw fragments of a bony coat of mail or tessellated armour, belonging to the back and tail of a very large animal. Seizing upon the clergyman's ideas, they reported to Berlin and Paris that these pieces likely belonged to *Megatherium*. Thus began a disagreement about the anatomy of this burly animal that lasted almost twenty years and reveals the impact of oral tradition on the culture of natural history. The conversations and readings went back and forth continuously to and from Montevideo, Buenos Aires or Asunción, by way of different ports and cities, where other meanings were ascribed to them before they continued on their way.[27]

4 Some Final Considerations

The history of instructions as a form of long-distance governance, as a mechanism to educate and oversee the gestures and observation of the traveller, or to assure "immutable mobiles" is not a new story.[28] It is not an innovation of the Bourbons or the Enlightenment, much less of the nineteenth century or the new American republics. Instructions, on the contrary, reveal a long history that transcended administrative reforms, revolutions, and breaks in the political order. Once the Spanish Empire had fallen, its former subjects continued behaving in the same way. Perhaps because of that, the history of the papers related to the *Megatherium* is repeated with other names and objects throughout America. Its population of former civil servants, priests, and topographers from Peru, New Grenada, New Spain and the River Plate had access to the colonial archives, worked with the data they found there, and began to write about the history and natural history of their regions. In addition, they had contact with agents who made copies or simply purchased their documents to have them translated and published in France and England, rediscovering that which, as the very papers testify, had been described decades before.[29]

---

27 On these debates, see Irina Podgorny, "Fossil Dealers: The Practices of Comparative Anatomy and British Diplomacy in Latin America, 1820–1840," *The British Journal for the History of Science* 46, no. 4 (2013).

28 Bruno Latour, "Visualisation and Cognition: Drawing Things Together," in *Knowledge and Society: Studies in the Sociology of Culture Past and Present: a Research Annual* 6 (1986): 140.

29 The colonial bureaucracy produced an enormous load of documents in original and copy. Until late in the 18th century, all the documentation regarding the overseas empire

This process of dispersal of data – which some call accumulation – happened at the same time as the establishment throughout America of institutions either as replacements for colonial ones or as brand new ones. With a difference of very few years, the various governments or collectors with the resources to do so began to establish public museums. The dates indicate trends or waves that cannot be explained by the metropolitan dispositions such as those promoted in the context of the late eighteenth-century expeditions and the Catholic – or Protestant – Enlightenment that, according to the now classic expression of José Carlos Chiaramonte, characterized the late colonial American world. Many of these collections, such as those of Franco Dávila grew out of the mix between commercial culture, the worldly passion of the Enlightenment for science and nature, and the trade expectations of the new republics. Wherever there was an important port, an initiative appeared to compile and gather the products that, in some way or another, passed through it, to exhibit and inventory them as in a storefront.

Historiography has treated the years around 1810 and the growth of new independent nations as the point of origin from which each Latin American nation began to construct its own identity and destiny. A less hasty examination reveals a political, social, and economic panorama filled with ambiguity, contradictions, confusions, and of course ruptures, but also continuities. The breaks are evident: for example, the key issue of how to rebuild finances, the foundation of any political and administrative structure.[30] However, in the majority of cases, continuity predominates: individuals continued doing what they had learned how to do; for a long time, offices continued using the same stationary, barely covering the official stamp of Ferdinand VII.

The establishment of museums and cabinets in the new nations is linked to these historical events and to the expansion of the new scientific disciplines that occurred almost in parallel. The fall of Spain, the French Revolution, and

---

was dispersed among various archives, as Simancas, Cádiz and Seville in Spain and the local administrative offices in the Americas, which housed copies and the documents needed for administration. In Spain, the Archivo General de Indias established in 1785, brought together the documentation existing in Spain. A similar initiative followed in Mexico, where the General Archive of Chapultepec was established in 1792. In the Rio de la Plata, until the creation of the Archive of Buenos Aires in 1821, papers were kept at the respective administrative offices, such as the *Cabildo*, the *Tribunal del Consulado*, and the *Contaduría General*. For a general bibliography on the topic, see Irina Podgorny, "Los archivos: Entre el síndrome de Barba Azul y los sueños de Napoleón," in *Los secretos de Barba Azul: Fantasías y realidades de los archivos del Museo de La Plata*, ed. Tatiana Kelly and Irina Podgorny (Rosario: Prohistoria, 2012).

30 Tulio Halperín Donghi, *Guerra y finanzas en los orígenes del Estado argentino, 1791–1850* (Buenos Aires: Universidad de Belgrano, 1982).

the North Atlantic War of 1812 provoked the mobilization of an enormous quantity of data that flooded the most diverse spaces. The materials gathered in the cabinets of the new Hispanic world, Madrid, or Portugal through confiscation and the traffic of drawings and documents, would shape the new disciplines of herpetology, malacology, comparative anatomy, palaeontology, and archaeology. This dispersion, combined with the access and circulation of the reports, maps, and drawings of expeditions and of colonial military engineers, implies various things: on the one hand, the rupture of an administrative order based on both the copious production of papers and documents and limited access to its archives, and on the other, the admittance of those data into the new disciplines of the nineteenth century. Thus, the materials produced in the eighteenth century, that had until that point remained in the circuit of the colonial American administration or accumulated in the office of the general Cosmographer of the Indies, are taken out of that order to enter into the so-called "spirit of the system" of nature and history, characterized by its dynamism, the reformulation of its categories, and permanent debate.

Acknowledgements

Earlier versions of this paper – written in the frame of the Research Projects PIP-CONICET 0153, and PICT 20153534 on "The Bureaucracy of Knowledge" – were discussed in several opportunities and workshops. I would like to thank María Jesús Santesmases, Mark Thurner, and Annick Lempérière, whose comments as well as the help of the ILL service from the MPIWG Berlin were much appreciated. A version of the chapter was previously published in a similar form: "Bureaucracy, Instructions, and Paperwork: The Gathering of Data about the Three Kingdoms of Nature in the Americas, 1770–1815." *Nuevo Mundo Mundos Nuevos* [Online], Paper posted on February 19, 2019. https://doi.org/10.4000/nuevomundo.75454. I would also like to thank Simona Boscani Leoni, Meike Knittel, and Sarah Baumgartner for the fruitful discussions during the Bern conference and the preparation of this volume.

Bibliography

Algorta Camusso, Rafael. *El Padre Dámaso Antonio Larrañaga: Apuntes para su biografía*. Montevideo: Barreiro y Ramos, 1922.

Barreiro, Agustín Jesús. *El Museo Nacional de Ciencias Naturales (1771–1935)*. Madrid: Doce Calles, 1992 (1944).

Bell, Stephen. *A Life in Shadow: Aimé Bonpland in Southern South America, 1817–1858*. Palo Alto: Stanford University Press, 2010.

Bourguet, Marie-Noëlle. "La collecte du monde: Voyage et histoire naturelle (fin XVIIème siècle – début XIXème siècle)." In *Le Muséum au premier siècle de son histoire*, edited by Claude Blanckaert, Claudine Cohen, Pietro Corsi and Jean-Louis Fischer, 163–96. Paris: Muséum national d'Histoire naturelle, 1997.

Brendecke, Arndt. *Imperium und Empirie: Funktionen des Wissens in der Spanischen Kolonialherrschaft*. Cologne: Böhlau, 2009.

Calatayud Arinero, María Ángeles. *Catálogo crítico de los documentos del Real Gabinete de Historia Natural (1787–1815)*. Madrid: CSIC, 2000.

Constantino, María Eugenia and Antonio Lafuente. "The Hidden Logistics of Longinos's Novohispanic Cabinet." *Nuncius* 27, no. 2 (2012): 348–70.

Di Stéfano, Roberto. *El púlpito y la plaza: Clero, sociedad y política de la monarquía católica a la república rosista*. Buenos Aires: Siglo Veintiuno, 2004.

Figueroa, Marcelo F. "Packing Techniques and Political Obedience as Scientific Issues: 18th-Century Medicinal Balsams, Gums and Resins from the Indies to Madrid." *HOST* 5 (2012): 49–67.

Gaudin, Guillaume. *Penser et gouverner le nouveau monde au XVIIe siècle: l'empire de papier de Juan Díez de la Calle, commis du Conseil des Indes*. Paris: L'Harmattan, 2013.

Gelman, Jorge. *Un funcionario en busca del Estado: Pedro Andrés García y la cuestión agraria bonaerense, 1810–1822*. Buenos Aires: Universidad Nacional de Quilmes, 1997.

Grieco, Viviana L. *The Politics of Giving in the Viceroyalty of Rio de la Plata: Donors, Lenders, Subjects, and Citizens*. Albuquerque: The University of New Mexico Press, 2014.

Halperín Donghi, Tulio. *Guerra y finanzas en los orígenes del Estado argentino, 1791–1850*. Buenos Aires: Universidad de Belgrano, 1982.

Latour, Bruno. "Visualisation and Cognition: Drawing Things Together." *Knowledge and Society: Studies in the Sociology of Culture Past and Present: a Research Annual 6 (1986): 1–40.*

Lemoine Villicaña, Ernesto. "Instrucciones para aumentar las colecciones del gabinete de historia natural de Madrid." *Boletín del Archivo General de la Nación* 2 (1961): 189–230.

Lopes, María Margaret. *O Brasil descobre a pesquisa científica. Os Museus e as ciências naturais no século XIX*. São Paulo: Hucitec, 1997.

López Piñero, Jose María. *El arte de navegar en la España del Renacimiento*. Madrid: Labor, 1979.

Lleó Cañal, Vicente. "Julian Benjamin Williams y el comercio de arte en la Sevilla del XIX." *Boletín de la Real academia Sevillana de Buenas Letras: Minervae Baeticae* 36 (2008): 187–204.

Mañé Garzón, Fernando. *El Glorioso montevideano: Vida y obra del Doctor José Manuel Pérez Castellano (1742–1815)*. Montevideo: Archivo General de la Nación, 1998–2003.

Pimentel, Juan. *Testigos del Mundo: Ciencia, literatura y viajes en la ilustración*. Madrid: Marcial Pons, 2003.

Podgorny, Irina. "Fossil Dealers, the Practices of Comparative Anatomy and British Diplomacy in Latin America, 1820–1840." *The British Journal for the History of Science* 46, no. 4 (2013): 647–74.

Podgorny, Irina "Los archivos: Entre el síndrome de Barba Azul y los sueños de Napoleón." In *Los secretos de Barba Azul: Fantasías y realidades de los archivos del Museo de La Plata*, edited by Tatiana Kelly and Irina Podgorny, 21–39. Rosario: Prohistoria, 2012.

Podgorny, Irina. "El camino de los fósiles: Las colecciones de mamíferos pampeanos en los museos franceses e ingleses." *Asclepio* 53, no. 2 (2001): 97–116.

Podgorny, Irina. "The reliability of the ruins." *Journal of Spanish Cultural Studies* 8, no. 2 (2007): 213–33.

Podgorny, Irina. "De ángeles, gigantes y megaterios. Saber, dinero y honor en el intercambio de fósiles en las provincias del Plata en la primera mitad del Siglo XIX." In *Los lugares del saber. Contextos locales y redes transnacionales en la formación del conocimiento moderno*, edited by Ricardo Salvatore, 125–57. Rosario: Beatriz Viterbo, 2007.

Podgorny, Irina. "Bureaucracy, Instructions, and Paperwork: The Gathering of Data about the Three Kingdoms of Nature in the Americas, 1770–1815." *Nuevo Mundo Mundos Nuevos* [Online], Paper posted on February 19, 2019, accessed on 21 June 2021. https://doi.org/10.4000/nuevomundo.75454.

Podgorny, Irina and Maria Margaret Lopes. *El desierto en una vitrina: Museos e historia Natural en la Argentina, 1810–1890*. Mexico: Limusa, 2008.

Podgorny, Irina and Wolfgang Schäffner. "La intención de observar abre los ojos." *Prismas*, 4 (2000): 217–27.

Ramírez Rozzi, Fernando and Irina Podgorny. "La metamorfosis del megaterio." *Ciencia Hoy* 11, no. 61 (2001): 12–19.

Safier, Neil. *Measuring the New World: Enlightenment Science and South America*. Chicago, London: The University of Chicago Press, 2008.

Sánchez Almazán, Javier I. *Pedro Franco Dávila: De Guayaquil a la Royal Society. La época y la obra de un ilustrado criollo*. Madrid: CSIC, 2012.

Shapin, Steven and Simon Schaffer. *Leviathan and the Air-Pump: Hobbes, Boyle, and the Experimental Life*. Princeton: Princeton University Press, 1985.

Schäffner, Wolfgang. "Die Verwaltung der Endlichkeit. Zur Geburt des neuzeitlichen Romans in Spanien." In *Die Endlichkeit der Literatur*, edited by Eckart Goebel and Martin von Koppenfels, 1–12. Berlin: Akademie Verlag, 2002.

Siegert, Bernhard. *Passagiere und Papiere: Schreibakte auf der Schwelle zwischen Spanien und Amerika*. Munich: W. Fink, 2006.

Socolow, Susan. *The Bureaucrats of Buenos Aires, 1769–1810: Amor al Real Servicio*. Durham, London: Duke University Press, 1987.

Stagl, Justin. *Apodemiken: Eine räsonnierte Bibliographie der reisetheoretischen Literatur des 16., 17. und 18. Jahrhunderts*. Paderborn: Schöningh, 1983.

Tanner, Jakob. "Akteure, Akten und Archive." In *Was Akten bewirken können: Integrations- und Ausschlussprozesse eines Verwaltungsvorgangs*, edited by Claudia Kaufmann and Walter Leimgruber, 150–60. Zurich: Seismo, 2008.

Trelles, Manuel R. "El Padre Manuel de Torres." *Revista de la Biblioteca Pública de Buenos Aires* 4 (1882): 439–48.

Villena Sánchez-Valero, Miguel, Javier I. Sánchez Almazán, Jesús Muñoz Fernández, and Francisco Yagüe Sánchez. *El gabinete perdido. Pedro Franco Dávila y la Historia Natural del Siglo de las Luces*. Madrid: CSIC, 2009.

CHAPTER 7

Questionnaires, Parish Registers and Prize Competitions: The Zurich Physical Society's Sources and Methods for Surveying the Territory

Sarah Baumgartner

During the early modern period, Europeans not only began to "discover" and conquer the world, but they also showed a growing interest in the usefulness of knowing the natural resources of their home territory.[1] This trend intensified in the second half of the 18th century, when rulers came to appreciate the importance of such knowledge and supported initiatives to survey their subject territories. At the same time, new methods of generating knowledge emerged. Besides, and in collaboration with, state administrations, different actors took part in such efforts, including individual researchers as well as associations. In addition, let us not forget those local informants whom the "experts" often resorted to and had to rely on.

In the Swiss city-state of Zurich, it was a private society, the "Physical Society," which took the first steps in this direction. In my article, I want to examine its motivation for doing so, as well as the methods it used and the theories that guided its activities. The focus will be on two cases where the society actively and systematically collected information: first, the censuses of the city's residents and, secondly, the Society's manifold efforts to gather information on the rural population and its subsistence strategies. Finally, this paper will show what the Society's new picture of Zurich and its territory looked like, and what conclusions the members drew from that. To begin with, however, let us provide some useful contextual information on the Society and the city where it was located.

1 Zurich in the 18th Century

Being, together with Bern, one of the political centres of the Old Swiss Confederacy, Zurich was a city-state with a "capital" of approximately 10,000

---

1 Alix Cooper, *Inventing the Indigenous: Local Knowledge and Natural History in Early Modern Europe* (Cambridge, New York: Cambridge University Press, 2007).

© SARAH BAUMGARTNER, 2022 | DOI:10.1163/9789004412477_008

inhabitants – or to be more precise about 2,000 adult male "burghers" – ruling over a considerable territory with a population that by the end of the 18th century had risen from about 120,000 to 180,000 inhabitants. The town's constitution was essentially republican, but over the centuries, political power became more and more concentrated in the hands of a few rich families.[2] Most of them owed their wealth to the export-oriented textile industry, a branch of business that sprang up in the 16th century. It was organised in the putting-out system, which meant that urban merchants – only the capital's burghers were entitled to exercise this activity – provided rural outworkers with the raw materials they would then process, and later picked up the finished goods for distribution. Hence, the putting-out industry not only strongly shaped the economic conditions and political power relations within the city, but also had far-reaching consequences for the countryside and its population. Workers earned a wage that, even when small, constituted a source of income independent of agriculture. Even poor, landless couples were now able to marry and to found families. In the second half of the 18th century, about one-third of Zurich's rural population was in some way or other occupied in this industry, spinning or weaving cotton, linen or silk.[3] This trade formed the basis of Zurich's financial wealth, but was not risk-free. A considerable proportion of the rural population were without enough land to cover their own needs, and this made them more dependent on grain imports. Poor harvests could easily result in severe subsistence crises, which is what happened, for example, in the early 1770s.

Besides its political and economic relevance, Zurich also became known as a cultural hotspot of Enlightenment Europe: indeed, the poet Johann Jacob Bodmer (1698–1783) and the theologian Johann Caspar Lavater (1741–1801) attracted visitors from all over the Continent. Zurich's citizens were interested in contemporary intellectual debate and began to get together in newly established "societies" to discuss the latest literature, as well as political ideas. This, of course, did not happen only in Zurich, but was quite a widespread phenomenon in that period. In the context of the so-called "society movement" of the Age of Enlightenment, all over Europe, a new type of social association was emerging. Although broad variations could be observed between societies, they did share some characteristics. First, usually they were voluntary

---

2 For an overview on Zurich in the 18th century, see: Conrad Ulrich, "Das 18. Jahrhundert," in *Geschichte des Kantons Zürich*, ed. Niklaus Flüeler and Marianne Flüeler-Grauwiller, vol. 2, Frühe Neuzeit. 16. bis 18. Jahrhundert, (Zurich: Werd, 1996); Clive H. Church and Randolph C. Head, *A Concise History of Switzerland* (Cambridge, New York: Cambridge University Press, 2013), 104–31.

3 For the history of industrialisation in the Zurich countryside, see: Rudolf Braun, *Industrialisation and Everyday Life, trans. Sarah Hanbury Tenison* (Cambridge, New York: Cambridge University Press, 1990).

associations of citizens; secondly, their activities pursued reform-oriented targets. In Zurich, this movement was especially important; so much, so that at least 23 societies had been founded by the end of the century.[4] One of the most remarkable was indeed the "Physical Society," the main subject of this article.

2 The Physical Society

The Zurich Physical Society (*Physikalische Gesellschaft*, later called *Naturforschende Gesellschaft*) was established in 1746 by some citizens associated with Dr Johannes Gessner (1709–1790), professor of "physicesque mathesos" at the "Carolinum," Zurich's theological academy.[5] All of them were interested in "physics," the name by which natural sciences were known in those days. What drove them to create a society was their regret that, in their city, no institution existed to promote this field of learning, which was becoming increasingly important in the 18th century. The "solution" chosen by the Zurichers, the establishment of a specialized learned association was, as already mentioned, quite a typical in the Age of Enlightenment.[6]

The famous "Royal Societies" of London, Paris and Berlin were used as a model. The founding members expressly set out to imitate these distinguished forerunners, but on a much humbler level, adjusted to their own, limited capacities. In contrast to the celebrated paragons, original research was not the main criterion or objective of the Zurich society's activities. Its main purpose was closer to what today we would call adult education, conceptualised as an agent for making new scientific findings, and especially "useful" applications of this knowledge, accessible to Zurich's burghers.[7]

4 Emil Erne, *Die schweizerischen Sozietäten: Lexikalische Darstellung der Reformgesellschaften des 18. Jahrhunderts in der Schweiz* (Zurich: Chronos, 1988), 65–157.

5 The only treatise on this society available in English is James Roger Hansen, *Scientific Fellowship in a Swiss Community Enlightenment: A History of Zurich's Physical Society, 1746–1798* (Columbus: Ohio State University, 1981).

6 James E. McClellan III, *Science Reorganized. Scientific Societies in the Eighteenth Century* (New York: Columbia University Press, 1985); Henry E. Lowood, *Patriotism, Profit, and the Promotion of Science in the German Enlightenment: The Economic Scientific Societies 1760–1815* (New York, London: Garland Pub., 1991).

7 According to the statutes, the Society's purpose was the advancement of "the knowledge of nature, insofar as it serves the comfort, utility, and necessity of human society in general, and especially of our dear fatherland." StAZ B IX 208, Erster Ursprung der Naturforschenden Gesellschaft in Zürich, Gesetze und Uebungen derselben, wie sie nun in dem Jahr 1776 geordnet und bestäthet sind, und auch für die folgenden Zeiten dienen sollen (the original 1746 document is lost).

The society was principally open to any burgher who could afford the membership fee. Its active members included medical doctors, who played an important role, but also pastors, some merchants, and a few artisans. Similar to the "Royal Societies," there was also a class of honorary members who were not obliged to take an active part, but served as patrons, ensuring that the society was anchored in the circles of the urban elites.[8]

From the very beginning, the Society began to build up its basic infrastructure, namely a library and collections of "naturalia" and scientific instruments. Every second week, and later every week, its members met to attend lectures on a wide variety of scientific subjects – the protocols mention nearly all topics of 18th-century "physics," from botanical and zoological classification to electricity – and presentations of strange "naturalia" or "monsters," which in those days would be classified as natural history. During the first decade of its existence, the Society's meetings were dominated by rather abstract subjects, while applied sciences played a minor role. This changed when, in 1759, a specialised "economic" commission was established, with the remit to focus on collecting, testing and disseminating "useful" knowledge related to agriculture.

The inspiration for widening their scope and promoting practical applications seems to have come from Bern; there, an "Economic Society" with a comparable objective had just been founded.[9] The Bern "Society" had itself been inspired by similar associations established elsewhere in Europe, which made it part of a larger movement. It was a time when elites began to develop a keen interest in agriculture; States and princes as well as private associations debated over agricultural reform programmes and tried to put them into practice. As a result, the second half of the century became the heyday of a particular type of association, the agricultural or "economic" societies.[10] Together with the Bernese society, Zurich's Economic Commission was one of the first of this kind in German-speaking Europe.

The Society's as well as the Economic Commission's activities generated very rich records, which are largely extant, preserved and now stored in the State Archives of Zurich (StAZ); they constitute the empirical basis for this study.

8 For the Society's organisational structure, see: StAZ B IX 208.

9 For this society, see: Regula Wyss and Martin Stuber, "Paternalism and Agricultural Reform: The Economic Society of Bern in the Eighteenth Century," in *The Rise of Economic Societies in the Eighteenth Century*, ed. Koen Stapelbroek and Jani Marjanen (Basingstoke: Palgrave Macmillan, 2012).

10 For an overview, see: Koen Stapelbroek and Jani Marjanen, "Political Economy, Patriotism and the Rise of Societies," in *The Rise of Economic Societies in the Eighteenth Century*, ed. Koen Stapelbroek and Jani Marjanen, (Basingstoke: Palgrave Macmillan, 2012) and the case studies in the same volume.

3 Demographic Surveys: Motivation and Theoretical Orientation

In the political discourse of that time, population was seen not only as the most important element of a state's economic and military strength, but also as an indicator of the government's ability to care for its subjects. To the contemporary eye, a growing population was infallible proof of the goodness of the authorities, and therefore of their legitimacy.[11] So it is not surprising that government administrations more and more often strove to gather information on the number and condition of their subjects. Similarly, interest in what later became known as "demography" was closely followed by growing interest in natural resources in the territory. According to the enlightened reformers' utilitarian worldview, nature provided everything for man's wellbeing – as long as its treasures were sufficiently known; for this was considered to be an essential prerequisite for any profitable exploitation. Increasingly, this sort of survey aimed not just at inventorying, but also at a systematic understanding of a country's political economy, investigating relations of dependency between factors such as farming productivity and population.

Zurich's elite, or at least its reform-friendly minds, were not unaffected by, or indifferent to, these new ideas; they were well informed of actual examples of census that had been undertaken elsewhere, for example in Sweden[12] or in Prussia. They seem to have been particularly impressed by the case of King Frederick the Great, who, thanks to an excellent knowledge of his territory's population and resources, managed to develop his relatively small country into a remarkable military power.[13]

In the Physical Society, a rallying point for such men, the opinion took hold that their hometown should not neglect these developments and that it might be useful to systematically collect information on the territory's inhabitants and nature. There was hope that this knowledge might help to improve Zurich's economy effectively and reduce its dependence on grain imports.

As the city-state lacked an adequate administrative organization able to perform a census, a complex and time-consuming operation, the Physical Society took on the responsibility for it; this is why Zurich's first censuses were not organised by the State itself, but by a private organization. It needs to be said that its members saw their undertakings as a service to their fatherland;

---

11 Edward Prince Hutchinson, *The Population Debate. The Development of Conflicting Theories up to 1900* (Boston: Houghton Mifflin, 1967), 94–109.

12 The pioneering Swedish censuses became a model for similar projects all over Europe: Peter Sköld, "The Birth of Population Statistics in Sweden," *History of the Family* 9 (2004).

13 StAZ B IX 244, Johann Heinrich Schinz, *Anmerkungen über die Abzehlungen der Einwohner zu Zürich*, 1762, 284.

they clearly emphasized that, far from being merely a way of satisfying their curiosity, these surveys were meant to be useful.

Censuses undoubtedly figure among the Physical Society's most important achievements; they were, at least for the Swiss Confederacy, pioneering; it was not until the early 19th century that the State took on the task to systematically survey its inhabitants. It was a common phenomenon that societies such as this one should develop and launch projects that would later become part of the business of the State.

The Society was well aware of the sensitive nature of its surveys; statistical data meant political power and had therefore to be treated as a state secret. It was probably because the Society's members came from families connected with the government that the authorities tolerated its activities, and even acquiesced to the publication of some of the results.[14] But in Zurich, too, statistical investigations could, unless executed with the necessary caution, have severe consequences, as the fate of Johann Heinrich Waser (1742–1780) shows. Waser, a pastor by profession and an assiduous member of the Physical Society, was an able and ardent statistician; he was absolutely fascinated by the knowledge-generating potential of numbers. He used the Society as a platform to present his results but did most of his extensive work on his own initiative. Tapping into different types of sources, such as tax registers, recruitment and parish registers, he retraced the development of Zurich's population back to the 15th century and used and drew maps to estimate the canton's surface area. Among his published or manuscript works, we also find treatises about historical changes in wages, grain prices and the value of money, as well as on life or fire insurance.[15] His confidence in himself and in his findings was such that he went as far as using them to daringly criticise the authorities – which, in those days, could end fatally. In fact, he first lost his job and livelihood and, in the end, was sentenced to death for high treason.[16]

In the 18th century, two or three different methodological approaches for gathering "statistical" information on territories can be distinguished:

14 Christian Simon, "Hintergründe Bevölkerungsstatistischer Erhebungen in Schweizer Städteorten des 18. Jahrhunderts. Zur Geschichte des demographischen Interesses," *Schweizerische Zeitschrift für Geschichte* 34, no. 2 (1984): 201–04.

15 On Waser's works, see: Albert Hauser, "Johann Heinrich Waser, Leben und Werk eines grossen Volkswirtschaftlers im Zeitalter der Aufklärung," in *Festschrift zum 70. Geburtstag von Eugen Böhler*, ed. Schweizerische Gesellschaft für Konjunkturforschung (Zurich: Polygraphischer Verlag, 1963).

16 For an overview of Waser's fate, see especially: Rolf Graber, "Der Waser-Handel. Analyse eines sozipolitischen Konflikts in der Alten Eidgenossenschaft," *Schweizerische Zeitschrift für Geschichte* 30, no.3/4 (1980).

the quantitative French and, especially, English tradition – called "political arithmetic" – on the one hand; and the decidedly qualitative and descriptive German "Universitätsstatistik" on the other.[17] All in all, the Zurich Society combined quantitative and qualitative methods, but had an intriguing interest in numerical procedures. With their readiness to use quantitative methods, the Zurich Society not only stood in contrast to the other important Swiss reform association, the already mentioned "Economic Society" of Bern,[18] but also differed from most such private reform societies.[19]

References to all the protagonists of the English "political arithmetic," from John Graunt (1620–1674), to William Petty (1623–1687),[20] Edmond Halley (1656–1742) and Gregory King (1648–1712) can be found in the societies' statistical treatises; and the Zurichers were also familiar with the work of French statisticians such as Antoine Deparcieux (1703–1768), and the Dutch Willem Kersseboom (1691–1771). Astonishingly, however, the Society just marginally adopted the descriptive "Universitätsstatistik". In its library, works by the protagonists of this science – for example the Göttingen professor Gottfried Achenwall (1719–1772), considered to be its founding father[21] – seem to be completely absent.

The Zuricher's most important theoretical reference was also of "German" origin. It originally had nothing to do with any efforts to gather information to improve statecraft, but was the result of the Prussian pastor Johann Peter Süßmilch's (1707–1767) endeavour to use statistical methods to obtain deeper insights into the principles of the divine world order. He was inspired by "physicotheology," a movement that aimed to unify naturalist research and religion by stating that closer understanding of the structure of creation would lead

17 On these differing traditions of early modern "statistics": Alain Desrosières, *Die Politik der grossen Zahlen: Eine Geschichte der statistischen Denkweise* (Heidelberg: Springer-Verlag, 2005), 19–35; Karin Johannisson, "Society in Numbers: The Debate over Quantification in 18th-Century Political Economy," in *The Quantifying Spirit in the Eighteenth Century*, ed. Tore Frängsmyr, John L. Heilbron, and Robin E. Rider (Berkeley: University of California Press, 1990).

18 The Oeconomic Society had set up a comprehensive programme for gathering descriptive data on Bern's territory: Gerrendina Gerber-Visser, *Die Ressourcen des Landes: Der ökonomisch-patriotische Blick in den Topographischen Beschreibungen der Oekonomischen Gesellschaft Bern (1759–1855)* (Baden: hier + jetzt, 2012).

19 For an overview of the activities of these societies, see Lowood, *Patriotism, Profit, and the Promotion of Science in the German Enlightenment.*

20 On Graunt and Petty: Mohamed Rassem and Justin Stagl, eds., *Geschichte der Staatsbeschreibung: Ausgewählte Quellentexte 1456–1814* (Berlin: Akademie-Verlag, 1994), 271–94.

21 On Achenwall, see: ibid., 399–424.

to a better knowledge of God himself.[22] The *Göttliche Ordnung* (the title of his work, first published in 1742 and followed by a massively enlarged second edition in 1761/62) was to become an influential and widely read work, even though the author's physicotheological inspiration was not shared by all of its readers.[23]

In order to gather demographic information necessary for his project, Süßmilch corresponded and collaborated with many researchers across Europe. One of them was President Gessner, and this probably explains why Süßmilch was so well received in Zurich.[24] Apart from that, Gessner is known to have been a keen mathematician, and his influence undoubtedly was the main reason for the society's theoretical preferences. In 1748, he published a "dissertation" entitled *De termino vitae*, where he discussed different causes of death alongside with statistical methods to calculate life expectancy. Here, he cited Süßmilch right at the beginning, and he stated that statistics was not only useful for medicine and statecraft, but especially as a means of recognizing the perfection of the Almighty's creation.[25]

Nevertheless, the influence of physicotheology began to fade over the course of the decades. The above-mentioned Waser, a former pupil of Gessner, seems not to have justified his activities in this way – even though he was a pastor! Apart from that, he criticised Süßmilch for not being experienced enough in mathematics.[26]

Another important field of study should be mentioned here for the sake of completeness, without being expanded further: medical statistics. Such efforts were undertaken by the two doctors, Johann Caspar Hirzel, father (1725–1803) and son (1751–1817), who succeeded each other in the position of Town Physician. The duties of this official role included being responsible for the city's hospitals, which allowed them to compile comprehensive lists with the causes of deaths of patients that had deceased there.[27] Here, statistical data was collected with a clear intention to support "improvement": the rate

22 Jean-Marc Rohrbasser, *Dieu, l'ordre et le nombre: Théologie physique et dénombrement au XVIIIe siècle* (Paris: Presses Universitaires de France, 2001).

23 Justus Nipperdey and Patrick Festy, "Johann Peter Süßmilch: de la loi divine à l'intervention humaine." *Population* 66, no. 3/4 (2011).

24 Herwig Birg, ed., *Ursprünge der Demographie in Deutschland: Leben und Werk Johann Peter Süßmilchs (1707–1767)* (Frankfurt a.M.: Campus, 1986), 158–59.

25 Johannes Gessner, *Dissertatio Physico-Medico-Mathematica de Termino Vitae* (Zurich: ex officina Gessneriana, 1748), 28–29.

26 StAZ B X 26.20, Johann Heinrich Waser, Rede von der politischen Rechenkunst, 12.

27 They were also presented to the Society; for example: StAZ B IX 247, Dr Johann Caspar Hirzel Jr., Relation von den merkwürdigsten Vorfällen im Spithal 1783–1788, 96–112; ibid., Ueber Krankheiten aus den Gschaubüchern 1708–1790, 140–76.

of stillborn children, for example, was seen as an indicator of how well the midwives did their job. The Hirzels also made weather observations and tried to find correspondences between the weather and certain diseases, which was popular in contemporary medical discourse.[28]

4 Alarming Trends: The Development of Zurich's Population

In 1748, President Gessner encouraged one of the society's members, the "Junker" (squire) Hans Ulrich Blaarer (1717–1793), to prepare a draft for census-tables.[29] In all probability, this was the first time that somewhere in the Old Swiss Confederacy such considerations were made – one year before the pioneering Swedish *Tabellverket* was launched, the first project to use tables for systematically surveying a country's population. That shows how well-informed and innovative the Zurichers were, even when they lacked the means to put their plans immediately into practice.

It was not until 1756 that the Society conducted Zurich's first census. By contrast with the aforementioned project, it was limited to the city's population. It proved too complex and time-consuming to extend it to the whole territory, as the Society decided not to rely on parish registers, but to perform its own survey. Moreover, Blaarer's forms were found to be too detailed, as were other proposals, for example a scheme developed by the English Royal Society. Instead, the Society chose quite a simple scheme, devised by the French engineer Sébastien Le Prestre de Vauban (1633–1707).[30] This scheme counted the inhabitants of a village per household, and only contained the categories house-father and house-mother, adult or minor children and servants; the Zurichers added two extra columns for boarders ("Tischgänger"), meaning people living as paying guests in a family – for example students from other cities – and journeymen, as they were common in the households of artisans.[31]

28 On medical meteorology and statistics in France and England: Andrea A. Rusnock, *Vital Accounts* (Cambridge, New York: Cambridge University Press, 2002), 109–36.

29 The Society was well aware of the pioneering character of these tables; even over a decade later, it thought them worth publishing: Johann Ulrich Blaarer, "Entwurf Allgemeiner Politischer Gemeind-Tafeln," *Abhandlungen der Naturforschenden Gesellschaft in Zürich* 2 (Zurich: Heidegger und Compagnie, 1764).

30 Vauban's project (not implemented) was designed to quantify the total population of the French Kingdom: [Sébastien Le Prestre de Vauban], *Projet d'une Dixme Royale* ([Rouen], 1707); for a discussion, see Michèle Virol, "Connaître et accroître les peuples du royaume: Vauban et la population," *Population* 56, no. 5 (2001).

31 StAZ B IX 85, Volkszählungen in der Stadt Zürich 1780 und 1790.

Only in some cases, but not systematically, the housefather's profession was mentioned. The census seems not to have been intended as a survey of the city's socio-economic condition – at least at the beginning. On the other hand, the distinction between burghers and non-burghers was a very important criterion: people without this status were marked in red and counted separately. As no protocols have been handed down, we know scarcely anything concerning the discussions surrounding the creation of the first census-tables; but there is little doubt that, as far as the city was concerned, the burgher-population must have occupied centre state in the Society's interests. Given the Society's social structure, this is not at all surprising; as members of the ruling class, the associates had to be concerned about changes in the ratio of their own caste and the subordinate classes. This was an important issue in order to secure political power.

For the purpose of counting, the city was divided into eleven districts, and volunteers were given two weeks' time to register every household in their area according to this prescribed scheme. Their lists were then collected, revised to eliminate errors, and finally copied into a folio volume.[32] These results were then presented to the society. According to this first, more or less reliable, census, about 11,000 people were living in Zurich, a much lower number than had been assumed until then. Some estimates went up to 20,000 people, while a careful calculation based on parish registers by President Gessner some years earlier had resulted in 12,500 inhabitants.[33]

The census was repeated in 1762, 1769, 1780, and 1790, with slightly modified tables.[34] In order to assure the uniformity and reliability of the results, a lot of details on how to complete the tables had to be prescribed; and each time, these explanations became slightly more sophisticated. For example, people staying there only on workdays – say, workers in manufactories – should be ignored. A "household" was defined as "people having their meals together."

At the beginning, the few maidservants being burghers were simply counted as daughters of their employers, as the society found it more important to identify "foreign" people than to adequately register the families. It was not until 1780 that widowers and widows were distinguished as a separate status from married.

From the second cycle onwards, pre-printed tables facilitated the work; in the same year, the member Johann Heinrich Schinz (1727–1792) took the

32 StAZ B IX 10, Volkszählung in der Stadt Zürich 1756.

33 StAZ B IX 244, Johann Heinrich Schinz, Anmerkungen über die abzählungen der Einwohner zu Zürich in den Jahren 1756 und 1762, 294.

34 The results were recorded in five volumes: StAZ B IX 10–14.

initiative and supplemented the Society's tables with excerpts from the guilds' registers in order to provide an overview of the occupational structure of the burghers and to check whether there might be an oversupply of certain crafts. As every burgher had to belong to a guild, including pastors and bailiffs living outside the city, this was also a means of determining their total number.

Apart from several technical difficulties, the Society had to cope with, at least in its early stages, some resistance from the population – even when the lack of sources makes it difficult to reconstruct how pronounced this had really been. The Society's reports mentioned two types of anxiety affecting people: first, religious qualms based on the biblical narration of the divine punishment that King David's census received; secondly, a rumour that it might be preparatory work for raising taxes.[35] It was quite common at the time for censuses to raise exactly such fears – and it is no less true that concerns about taxes were not always groundless.

From the second census onwards, the Society tried to analyse and compare the figures assembled in the tables. Three essays give us some insights on how changes in urban population were interpreted.[36] Despite the aforementioned difficulties, the Society was able to detect some changes in the number and structure of the city population. And these had, at least from its members' perspective, an alarming trend: the constant decrease in the number of inhabitants with burgher status. The society was quick in finding an answer to the question why this was happening. Changing customs, increasing laziness as well as a growing love for lavishness and "luxury" among young people – which would prevent them from marrying and founding families – were thought to be the reasons.[37] To blame any societal or economic difficulties on people's idleness was very typical of the time.[38]

Apart from that, the censuses revealed a striking anomaly, namely that, contrary to the rule that there were usually more newborn boys than girls, in

35 StAZ B IX 248, Johann Rudolf Schinz: Geschichte der von Physik. Gesellschaft in Zürich aufgenommenen Zählung der Einwohner der Stadt. Der Gesellschaft vorgelesen den 12.11.1781, 156.

36 StAZ B IX 244, Johann Heinrich Schinz, Anmerkungen über die Abzählungen der Einwohner zu Zürich in den Jahren 1756 und 1762, 282–320; StAZ B IX 248, Johann Rudolf Schinz: Geschichte der von Physik. Gesellschaft in Zürich aufgenommenen Zählung der Einwohner der Stadt. Der Gesellschaft vorgelesen den 12.11.1781, 156–71; B IX 249, Johann Jacob Pestalutz, Ueber die Volkszählung der Stadt Zürich im Jenner 1790, 160–87.

37 They especially blamed the imitation of French fashion in clothing and giving dinners: StAZ B IX 248, Johann Rudolf Schinz: Geschichte der von Physik. Gesellschaft in Zürich aufgenommenen Zählung der Einwohner der Stadt. Der Gesellschaft vorgelesen den 12.11.1781, 169–70.

38 Rusnock, *Vital Accounts*, 180.

Zurich, the male/female ratio was almost equal. This is particularly interesting, as the relative constancy of this ratio, which had been observed in the 17th century already, was considered to be proof of divine providence.[39] For this "aberrance," however, the Society's members had simply no explanation; neither did they want to speculate on possible consequences.[40]

The results were not only a shrinking urban population, but also a slowly growing proportion of non-burghers, for instance servants. According to the Society's census, at end of the 18th century about 20% of the city's inhabitants were domestic servants, most of them female.[41] Of course, a certain number of servants was accepted as indispensable, a necessary evil in a sense; but the censuses showed an increasing number of menservants, which claimed higher wages than maids and were therefore seen as the sign of a baneful predilection for luxury; in the same way, the growing proportion of maids from outside the Canton were feared to be a threat to local customs.[42] However, the foremost problem was not these foreigners integrated in families, but the growing number of non-burgher households, called "Hintersassen."[43] They were not only disliked as undesirable competitors of burgher craftsmen, but also seen as a potential burden to the city's poor relief institutions and, probably worst of all, as political rebels. A frightening negative example the Zurichers knew of was Geneva; this city had a considerable proportion of inhabitants excluded from political participation, and during the 18th century, they had expressed their discontent in several riots.[44] Nevertheless, the situation in Zurich was quite different, as the "Hintersassen" never made up more than 10% of all households. The seemingly simple solution of naturalizing some of the "Hintersassen" was refused by the urban elite, including the relatively reform-oriented members of the Physical Society – until the end of the old regime. In the late 17th century already, the acceptance of new burghers had almost completely stopped, as was the case of other cities in the Old Swiss Confederacy as well as in the Holy Roman Empire. The ruling families, even when their number steadily dropped, did not want to share their privileges with anyone who could not look back on a generations-old history of burgher ancestors.

39 Ibid., 171.

40 StAZ B IX 249, Johann Jacob Pestalutz, Ueber die Volkszählung der Stadt Zürich im Jenner 1790, 179.

41 Ibid., 170.

42 StAZ B IX 248, Johann Rudolf Schinz: Geschichte der von Physik. Gesellschaft in Zürich aufgenommenen Zählung der Einwohner der Stadt. Der Gesellschaft vorgelesen den 12.11.1781, 164–65.

43 Ibid., 168.

44 Church and Head, *A Concise History of Switzerland*, 108–09.

The changes in the urban burgher population contrasted clearly with the countryside, where the population, apart from a sharp decline caused by the famine of the early 1770s, was constantly expanding. Although it remained difficult to calculate exactly the Canton's total inhabitants, Zurich's elites were well aware of that. In 1763, President Gessner, using church registers, estimated Zurich's population at 150,000 people.[45] The famine of the 1770s also prompted the authorities to find out more about food supply in the territory. In 1771, the government "Kornkammer" (grain administration) gathered information on the population of each village, and on its grain production, estimated from the tithe. It is not clear in what way the Society was involved there, but later on its findings were also discussed in the Economic Commission.

Also, from then onwards, on a yearly basis, Dr Johann Caspar Hirzel presented baptisms and deaths of the past year, which he was able to obtain from registers assembled by the church synod, a council of Zurich's pastors. Hirzel also compared these figures to neighbouring territories and could proudly conclude that mortality was lower in his homeland than anywhere else.[46]

Hirzel and probably most of the Economic Commission were convinced of the positive significance of a growing population; the more people, the greater the supply of "working hands," the precondition for a prosperous country. Another member of the Society, pastor Waser, was rather more critical and emphasized that this would only be valid as long as population growth did not exceed farming productivity.[47] But all agreed on the importance of promoting agriculture. From 1759 onwards, this was to become the endeavour of the Economic Commission.

5 Prerequisite for Reforms: Knowing the Territory

At one of their first meetings, the members of the Economic Commission stated that comprehensive knowledge of the territory would be an absolutely essential prerequisite for planning and implementing any reforms, especially for "improving" agriculture. Therefore, they laid down that collecting information should be their commission's primary line of action. Implementing such reforms was considered a task exceeding the scope of a private society, but

45 StAZ B IX 181, Tagbuch der Naturforschenden Gesellschaft 3, meeting of 5 September 1763, 18–19.

46 StAZ B IX 59, Protokoll der Oekonomischen Kommission 2, meeting of 9 May 1772, 195.

47 StAZ B X 27.29, Johann Heinrich Waser, Bevölkerung, Sterblichkeit.

they hoped to persuade the government to establish a specialized reform committee, which then could base its efforts on the preparatory work done by the Economic Commission.[48] This plan was finally successful: two decades later, in 1779, the authorities actually established a "Landwirtschaftskommission" (Commission for agriculture). Members of the Economic Commission came to play an important role there. In one of its first meetings, the "Landwirtschaftskommission" decided to gather documents and tables from both the State Archives and those of the Physical Society.[49]

Accordingly, the Society must have been in possession of a considerable stock of information on the Canton's agriculture until then. This raises the question how the Society had managed to collect all this knowledge.

Information reached the Commission through several channels. It was in close contact with several men from the countryside, especially pastors, who supported their reformist agenda. From them, the Commission received, every now and then, more or less detailed descriptions of certain regions. In addition, some of its members undertook journeys all around the Canton and recorded their observations in travel journals. These contributions were discussed in the Commission's meetings and usually, copies of the manuscripts were archived.

Information collected systematically played an even more important role. To do so, the Economic Commission had used a set of different tools and techniques, some of them traditional, others quite innovative. As for demography, quantification was important, but the Commission also began to be interested in rather descriptive cameralist methods. An example of this was a treatise by Carolus Linnaeus on the usefulness of travelling in the homeland, that was read and discussed in a meeting.[50]

The basic and most important tools were tables and questionnaires. Applying methods similar to those used to conduct city censuses, a village's inhabitants were counted by the households they belonged to. In addition, details of their land ownership as well as of the livestock they kept were collected as well. Interestingly, some years later, the Commission's members began to consider the numerical data gathered by means of the tables as insufficient,

48 StAZ B IX 58, Protokoll der Oekonomischen Kommission 1, meeting of 21st February 1759, 7.

49 StAZ B III 155, Protokoll der Landwirtschaftlichen Commission 1, meeting of 28th April 1779, 1.

50 StAZ B IX 183, Tagbuch der Naturforschenden Gesellschaft 5, meeting of 27th February 1769, 25.

and introduced "qualitative" questionnaires to complement this information.[51] In 82 detailed questions, aggregated in ten groups, information concerning the buildings in a village, the exact use and condition of different kinds of land – pastures, fields, commons and forests – as well as details of cattle, horses and small livestock had to be given. These tables and questionnaires, although they do not exist for every village, today are an important source for reconstructing Zurich's economic history.[52] From 1762 onwards, usually once or twice a year, a couple of them were sent to the pastors of some villages; the clergyman had to fill them in; afterwards, farmers from these villages were invited to the city in order to discuss some details. These meetings, called "Bauerngespräche" (conversations with farmers), were unusual – to many contemporaries, urban gentlemen and simple farmers talking together seemed to be a daring way of overstepping class barriers. But, needless to say, the commission took care not to upset the traditional hierarchy; so that farmers were not allowed to talk as they felt like, but only to answer questions.[53] Nonetheless, although it is rather debatable how much "Bauerngespräche" effectively contributed to an actual exchange of information, they were a clear sign that members of the urban elite had begun to give credit to their rural subjects for possessing potentially relevant and useful knowledge.

The Commission was interested not only in the number of the Canton's inhabitants and the condition of Zurich's fields and forests, but also in the everyday work of farmers. To get hold of this information, concerning problems that many of the urban members were not so familiar with, the Commission resorted to prize competitions. Admittedly, such competitions were quite a common way used by reform societies to reach out to a broader audience. Yet, the way the Economic Commission adapted them to its own purposes was particular. Whereas, usually, such questions were intended to be answered by well-educated men, the Zurichers' questions were directly addressed to local farmers and asked them questions about their everyday work in a very detailed manner. Between 1762 and 1803, a total of 48 questions were advertised,

---

51 StAZ B IX 47a, Kurze Instruction oder Anleitung, vor Diejenige, so den ökonomischen Zustand […], in die von Lobl. Physikalischer Gesellschaft in Zürich gedruckten Tabellen eintragen wollen.

52 StAZ B IX, 86–91, Oekonomische Tabellen, 1762–1789; for a comprehensive analysis of this material, see: Ulrich Pfister, *Die Zürcher Fabriques. Protoindustrielles Wachstum vom 16. zum 18. Jahrhundert* (Zurich: Chronos-Verlag, 1992).

53 For a critical discussion of the "Bauerngespräche", see Rolf Graber, "Die Züricher Bauerngespräche: Innovation der Volksaufklärung oder Instrument der Herrschaftssicherung?" in *Die Entdeckung von Volk, Erziehung und Ökonomie im europäischen Netzwerk der Aufklärung*, ed. Holger Böning, Werner Greiling, and Reinhart Siegert (Bremen: edition lumière, 2011).

covering subjects from forestry, to so-called "artificial meadows,"[54] from potato crops, to the drainage of wetlands, winegrowing, and livestock.

The number and quality of the answers differed from year to year, reaching from just five participants up to more than 30.[55] Even when there were a handful of very interested farmers who seem to have participated on their own initiative, one can observe that often, the village's pastor's commitment played an important role whether farmers were involved or not.

Thanks to these questions, over the years, the Commission gathered a lot of information on farming practices – and their diversity throughout the Canton. This knowledge did not remain in the society's archives, but was compiled into manuals, which were later distributed among the farmers.[56]

We know relatively little about how the Society processed and interpreted the material collected; this observation, incidentally, is valid also for many other states and institutions that were collecting "statistical" data in that period. Besides political concerns – "false" interpretations might lead to trouble with the authorities – the eschewal of detailed evaluations was often simply due to the lack of skills for processing the data, or the incompleteness of the information collected. Tables were undoubtedly an important invention for securing uniformity of information, but they could not absolutely guarantee it. This was especially the case when data collecting was not performed by a small group of well-trained men; therefore, the ways in which the *Oeconomische Tabellen* were completed varies from village to village – in contrast to the much more systematic city census tables.

The most sophisticated efforts to evaluate the data were not undertaken by the Society as such, but by some dedicated members on their own. A very comprehensive attempt to make sense of the Commission's data was Dr Johann Caspar Hirzels's 1788 treatise entitled *Beantwortung der Frage: Ist die Handelschaft,* [...] *unserem Lande schädlich oder nützlich* [...][57] [Answering the question: Is commerce useful for our country or not?]. In this essay, Hirzel used

---

54 On this subject, see Sarah Baumgartner, "'Nützliche Gras-Arten und Kräuter.' Die Zürcher Ökonomische Kommission und das Wissen vom Klee- und Wiesenbau," in *Wer das Gras wachsen hört. Wissensgeschichte(n) der pflanzlichen Ressourcen vom Mittelalter bis ins 20. Jahrhundert*, ed. Simona Boscani Leoni and Martin Stuber, Jahrbuch für Geschichte des ländlichen Raumes 14 (Innsbruck: Studien-Verlag, 2017).

55 A considerable proportion of them, more than 500 items, can still be consulted: StAZ B IX 18–24, Preisabhandlungen von Landleuten, 1762–1784.

56 Even when their titles always begin with "Anleitung für die Landleute" [Instructions for Farmers], they were often read by urban agronomists or pastors rather than farmers.

57 Johann Caspar Hirzel, *Beantwortung der Frage: Ist die Handelschaft, wie solche bey uns beschaffen, unserm Lande schädlich oder nützlich in Absicht auf den Feldbau und die Sitten des Volkes?* ([Zurich], 1788).

statistical data collected by the Commission, but also a good deal of information that he had gathered on his walks through the Canton to assess the effects of the putting-out industry on the Canton's agriculture. The treatise reveals a new view of the territory. For example, Hirzel introduces an innovative subdivision of the Canton, following agro-ecological criteria, not the political subdivision in bailiwicks. Even when he complained that some probably important information was still wanting, he was able to give a detailed assessment on how the impact of industry differed from region to region. Whereas, with regard to corn producing regions in the lowlands, the concern that it was taking away workforce from agricultural occupations could be confirmed, Hirzel showed that the impact of industry on mountainous regions was rather positive. Here, the additional source of income enabled people to care better for their fields and, in addition, offered a chance of employment during the winter.

6 Conclusion

Around the middle of the 18th century, the members of the reform-oriented Zurich Physical Society, inspired by the discourse in the literature, began to deliberate on how to gather information for a better knowledge of the city-state's territory. At first, their activities seem to be motivated by rather theoretical "scientific" interests, where physicotheological ideas played an important role. Subsequently, however, the society began to collect information, towards a more vigorous support of practical improvement in agriculture. A broad range of authors on these subjects seem to have been familiar to the Zurichers – at least nominally, as only a few of the members were able to really delve in statistical methods and problems. To achieve its goals, the Society tested and combined different techniques and sources of information, qualitative as well as quantitative. The Society conducted its own surveys, but it also relied on older records, such as Church registers, which it used for its own new purposes. The intention to use data for analysing economic or social issues in order to implement reforms distinguishes the Society's surveys from traditional data collections by the Church or the State, which was always done for specific, limited purposes, for example raising taxes. They not only tried to adapt methods from the contemporary discourse, but also devised innovative approaches for getting in contact with the rural population, like the "Bauerngespräche".

The perspectives were equally diverse; on the one hand, there was a "rational" interest in understanding the Canton's economic system and its inhabitants' productivity, but traditional criteria for categorizing people remained important – as the city-census' emphasis on the distinction between burghers and non-burghers shows. It is also interesting to see how the application of systematic and quantitative methods resulted in the discovery of the territory's actual diversity. The Physical Society's records are an interesting source for modern historians, but it is difficult to reconstruct what conclusions the contemporaries drew from this material. This would not be unusual, since typically, at the time, quite impressive records were collected, but then no appropriate means were available for processing them. Institutionalization also was in its infancy. Even though the Society and its Economic Commission took on certain tasks, the personal initiative of several members remained crucial, as the examples of Hirzel's and Waser's work shows. The Society already collaborated with the authorities, but it was not until the 19th century that statistical surveys became part of the State's obligations.

Acknowledgements

This contribution is based on research funded by the Swiss National Science Foundation and carried out in the context of the project "Kulturen der Naturforschung" (Cultures of Natural History), Project-Nr. 144731, led by Simona Boscani Leoni at the University of Bern (Switzerland) from 2014–2018.

Bibliography

Manuscripts

StAZ (Staatsarchiv Zurich) B III 155, Protocoll der Landwirtschaftlichen Commission 1, meeting of 28 April 1779, 1.

StAZ B IX 10, Volkszählung in der Stadt Zürich 1756.

StAZ B IX 18–24, Preisabhandlungen von Landleuten, 1762–1784.

StAZ B IX 47a, Kurze Instruction oder Anleitung, vor Diejenige, so den ökonomischen Zustand […], in die von Lobl. Physikalischer Gesellschaft in Zürich gedruckten Tabellen eintragen wollen.

StAZ B IX 58, Protokoll der Oekonomischen Kommission 1, meeting of 21 February 1759, 7.

StAZ B IX 59, Protokoll der Oekonomischen Kommission 2, meeting of 9 May 1772, 195.

StAZ B IX 85, Volkszählungen in der Stadt Zürich 1780 und 1790.

StAZ B IX 86–91, Oekonomische Tabellen, 1762–1789.

StAZ B IX 181, Tagbuch der Naturforschenden Gesellschaft Zürich 3, meeting of 5 September 1763, 18–9.

StAZ B IX 183, Tagbuch der Naturforschenden Gesellschaft Zürich 5, meeting of 27 February 1769, 25.

StAZ B IX 208, Erster Ursprung der Naturforschenden Gesellschaft in Zürich, Gesetze und Uebungen derselben, wie sie nun in dem Jahr 1776 geordnet und bestäthet sind, und auch für die folgenden Zeiten dienen sollen.

StAZ B IX 244, Johann Heinrich Schinz, Anmerkungen über die Abzählungen der Einwohner zu Zürich in den Jahren 1756 und 1762, 282–320.

StAZ B IX 248, Johann Rudolf Schinz: Geschichte der von Physik. Gesellschaft in Zürich aufgenommenen Zählung der Einwohner der Stadt. Der Gesellschaft vorgelesen den 12.11.1781, 156–71.

StAZ B IX 249, Johann Jacob Pestalutz, Ueber die Volkszählung der Stadt Zürich im Jenner 1790, 160–87.

StAZ B IX 247, Dr. Johann Caspar Hirzel jun., Relation von den merkwürdigsten Vorfällen im Spithal 1783–1788, 96–112; ibid., Ueber Krankheiten aus den Gschaubüchern 1708–1790, 140–76.

StAZ B X 26.20, Johann Heinrich Waser, Rede von der politischen Rechenkunst.

StAZ B X 27.29, Johann Heinrich Waser, Bevölkerung, Sterblichkeit.

Printed Sources

[Vauban, Sébastien Le Prestre de]. Projet d'une Dixme Royale. [Rouen], 1707.

Blaarer, Johann Ulrich. "Entwurf Allgemeiner Politischer Gemeind-Tafeln." *Abhandlungen der Naturforschenden Gesellschaft in Zürich* 2 (Zurich: Heidegger und Compagnie, 1764): 277–308.

Gessner, Johannes. *Dissertatio Physico-Medico-Mathematica de Termino Vitae*. Zurich: ex officina Gessneriana, 1748.

Hirzel, Johann Caspar. *Beantwortung der Frage: Ist die Handelschaft, wie solche bey uns beschaffen, unserm Lande schädlich oder nützlich in Absicht auf den Feldbau und die Sitten des Volkes?* [Zurich], 1788.

Literature

Baumgartner, Sarah. "'Nützliche Gras-Arten und Kräuter.' Die Zürcher Ökonomische Kommission und das Wissen vom Klee- und Wiesenbau." In *Wer das Gras wachsen hört. Wissensgeschichte(n) der pflanzlichen Ressourcen vom Mittelalter bis ins 20. Jahrhundert*, edited by Simona Boscani Leoni and Martin Stuber, 81–95. Jahrbuch für Geschichte des ländlichen Raumes 14. Innsbruck: Studien-Verlag, 2017.

Birg, Herwig, ed. *Ursprünge der Demographie in Deutschland: Leben und Werk Johann Peter Süßmilchs (1707–1767)*. Frankfurt am Main: Campus, 1986.

Braun, Rudolf. *Industrialisation and Everyday Life*. Translated by Sarah Hanbury Tenison. Cambridge, New York: Cambridge University Press, 1990.

Church, Clive H. and Randolph C. Head. *A Concise History of Switzerland*. Cambridge, New York: Cambridge University Press, 2013.

Cooper, Alix. *Inventing the Indigenous: Local Knowledge and Natural History in Early Modern Europe*. Cambridge, New York: Cambridge University Press, 2007.

Desrosières, Alain. *Die Politik der grossen Zahlen: Eine Geschichte der statistischen Denkweise*. Heidelberg: Springer-Verlag, 2005.

Erne, Emil. *Die schweizerischen Sozietäten: Lexikalische Darstellung der Reformgesellschaften des 18. Jahrhunderts in der Schweiz*. Zurich: Chronos, 1988.

Gerber-Visser, Gerrendina. *Die Ressourcen des Landes: Der ökonomisch-patriotische Blick in den Topographischen Beschreibungen der Oekonomischen Gesellschaft Bern (1759–1855)*. Archiv des historischen Vereins des Kantons Bern 89. Baden: hier + jetzt, 2012.

Graber, Rolf. "Der Waser-Handel. Analyse eines soziopolitischen Konflikts in der Alten Eidgenossenschaft." *Schweizerische Zeitschrift für Geschichte* 30, no. 3/4 (1980): 321–56.

Graber, Rolf. "Die Züricher Bauerngespräche: Innovation der Volksaufklärung oder Instrument der Herrschaftssicherung?" In *Die Entdeckung von Volk, Erziehung und Ökonomie im europäischen Netzwerk der Aufklärung*, edited by Holger Böning, Werner Greiling, and Reinhart Siegert, 43–58. Philantropismus und populäre Aufklärung – Studien und Dokumente 1. Bremen: edition lumière, 2011.

Hansen, James Roger. *Scientific Fellowship in a Swiss Community Enlightenment: A History of Zurich's Physical Society, 1746–1798*. Columbus: Ohio State University, 1981.

Hauser, Albert. "Johann Heinrich Waser, Leben und Werk eines grossen Volkswirtschaftlers im Zeitalter der Aufklärung." In *Festschrift zum 70. Geburtstag von Eugen Böhler, edited by Schweizerische Gesellschaft für Konjunkturforschung*, 43–58. Zurich: Polygraphischer Verlag, 1963.

Hutchinson, Edward Prince *The Population Debate. The Development of Conflicting Theories up to 1900*. Boston: Houghton Mifflin, 1967.

Johannisson, Karin. "Society in Numbers: The Debate over Quantification in 18th-Century Political Economy." In *The Quantifying Spirit in the Eighteenth Century*, edited by Tore Frängsmyr, John L. Heilbron, and Robin E. Rider, 343–62. Berkeley: University of California Press, 1990.

Lowood, Henry E. *Patriotism, Profit, and the Promotion of Science in the German Enlightenment: The Economic Scientific Societies 1760–1815*. New York, London: Garland Pub., 1991.

McClellan III, James E. *Science Reorganized. Scientific Societies in the Eighteenth Century*. New York: Columbia University Press, 1985.

Nipperdey, Justus and Patrick Festy. "Johann Peter Süßmilch: de la loi divine à l'intervention humaine." *Population* 66, no. 3/4 (2011): 699–716.

Pfister, Ulrich. *Die Zürcher Fabriques. Protoindustrielles Wachstum vom 16. zum 18. Jahrhundert.* Zurich: Chronos-Verlag, 1992.

Rassem, Mohamed and Justin Stagl, eds. *Geschichte der Staatsbeschreibung: Ausgewählte Quellentexte 1456–1814*. Berlin: Akademie-Verlag, 1994.

Rohrbasser, Jean-Marc. *Dieu, l'ordre et le nombre: Théologie physique et dénombrement au XVIIIe siècle.* Paris: Presses Universitaires de France, 2001.

Rusnock, Andrea A. *Vital Accounts.* Cambridge, New York: Cambridge University Press, 2002.

Simon, Christian. "Hintergründe Bevölkerungsstatistischer Erhebungen in Schweizer Städteorten des 18. Jahrhunderts. Zur Geschichte des demographischen Interesses." *Schweizerische Zeitschrift für Geschichte* 34, no. 2 (1984): 186–205.

Sköld, Peter. "The Birth of Population Statistics in Sweden." *History of the Family* 9 (2004): 5–21.

Stapelbroek, Koen and Jani Marjanen. "Political Economy, Patriotism and the Rise of Societies." In *The Rise of Economic Societies in the Eighteenth Century*, edited by Koen Stapelbroek and Jani Marjanen, 1–25. Basingstoke: Palgrave Macmillan, 2012.

Ulrich, Conrad. "Das 18. Jahrhundert." In *Geschichte des Kantons Zürich*, edited by Niklaus Flüeler and Marianne Flüeler-Grauwiller, vol. 2, Frühe Neuzeit. 16. bis 18. Jahrhundert, 364–511. Zurich: Werd, 1996.

Virol, Michèle. "Connaître et accroître les peuples du royaume: Vauban et la population." *Population* 56, no. 5 (2001): 845–75.

Wyss, Regula, and Martin Stuber. "Paternalism and Agricultural Reform: The Economic Society of Bern in the Eighteenth Century." In *The Rise of Economic Societies in the Eighteenth Century*, edited by Koen Stapelbroek and Jani Marjanen, 157–81, Basingstoke: Palgrave Macmillan, 2012.

CHAPTER 8

Social Anthropology *avant la lettre*: The Economic Enlightenment Perspective on Traditional Uses of Wetlands

Martin Stuber

1 Introduction

The period from the late 18th to the mid-19th century is a distinctive era for scientific ethnographic writing.[1] The research literature on the history of this discipline connects these essential origins of social anthropology mainly to the great, global scientific expeditions of James Cook (1728–1779) and Louis Antoine de Bougainville (1729–1811).[2] By contrast, I have adopted a complementary approach, along the lines of Alix Cooper's *Inventing the Indigenous*.[3] My research grew out of the exact descriptions of the traditional economy, which today continue to be a main empirical interest of the disciplinary fields of "cultural anthropology," "social anthropology," "ethnology" and "folklife studies" ("Volkskunde, Völkerkunde").[4] Such records of traditional economic practices were kept not only in the context of so-called global expeditions but on a larger scale also in the context of Economic Enlightenment ("Ökonomische Aufklärung") which, since the mid-18th century, gradually covered all of Europe.

A constitutive element of Economic Enlightenment is the systematic development of knowledge stocks for the purposeful exploitation of resources in

1 This chapter is based on: Martin Stuber and Matthias Bürgi, *Vom "eroberten Land" zum Renaturierungsprojekt. Geschichte der Feuchtgebiete in der Schweiz seit 1700* (Bern, Stuttgart, Wien: Haupt Verlag, 2018), 19–43.

2 Harry Liebersohn, "Scientific Ethnography and Travel, 1750–1850," in *History of Science*, vol. 7, *The Modern Social Sciences*, ed. Theodore M. Porter and Dorothy Ross (Cambridge, New York: Cambridge University Press, 2003), 100.

3 Alix Cooper, *Inventing the Indigenous: Local Knowledge and Natural History in Early Modern Europe* (Cambridge, New York: Cambridge University Press, 2007).

4 Adam Kuper, "Anthropology," in *History of Science*, vol.7, *The Modern Social Sciences*, ed. Theodore M. Porter and Dorothy Ross (Cambridge, New York: Cambridge University Press, 2003), 354.

view of a comprehensive modernisation of agriculture and forestry.[5] However, Economic Enlightenment cannot be reduced to the modernistic view. Indeed, the reformers also describe the rural economy negatively, as an economy too heavily based on tradition, in order to make their own modernisation goals look brighter. Furthermore, reformers were also interested in the exact knowledge of the rural economy to improve the coherence of their reforms. These attempts to view rural society from the inside reveal these reformers as social anthropologists *avant la lettre*. In many respects, their observations are in line with ethnographic reports and lay the foundations for a contemporary reconstruction of the traditional economy of rural society.

The German historian Rainer Beck developed the concept of "natural economy" ("Naturale Ökonomie") providing examples from Bavaria in the 18th and early 19th centuries: rural societies were dependent on what their efforts produced within their regions year after year, in accordance with the metabolism of nature, and on the amount of energy they were able to reproduce. According to Beck's micro-historical analyses, in this context we must focus not only on areas of intensive agriculture, but also on fringe and outside areas. For example, wetlands, which we are going to draw attention later, were anything but an underused resource in the "natural economy." On the contrary, they were an indispensable complement to more intensively cultivated fields near the villages.[6]

Among the many economic-patriotic societies of Switzerland, the Economic Society of Bern, founded by patricians in 1759, was considered the most important one already at the time. This was due mostly to the quality of its bilingual publication organ, which was read both in the German- and the French-speaking regions.[7] The Economic Society saw the exact and spatially

---

5 Marcus Popplow, ed., *Landschaften agrarisch-ökonomischen Wissens. Strategien innovativer Ressourcennutzung in Zeitschriften und Sozietäten des 18. Jahrhunderts* (Münster: Waxmann, 2010), 3–5; Koen Stapelbroek and Jani Marjanen, eds., *The Rise of Economic Societies in the Eighteenth Century: Patriotic Reform in Europe and North America* (Basingstoke: Palgrave Macmillan, 2012).

6 Rainer Beck, *Naturale Ökonomie. Unterfinning: Bäuerliche Wirtschaft in einem oberbayerischen Dorf des frühen 18. Jahrhunderts* (Munich, Berlin: Deutscher Kunstverlag, 1986); Beck, *Ebersberg oder das Ende der Wildnis. Eine Landschaftsgeschichte* (Munich: C. H. Beck, 2003); Rita Gudermann, *Morastwelt und Paradies. Ökonomie und Ökologie in der Landwirtschaft am Beispiel der Meliorationen in Westfalen und Brandenburg (1830–1880)* (Paderborn: Schöningh, 2000), 74–78.

7 Martin Stuber, Peter Moser, Gerrendina Gerber-Visser, and Christian Pfister, eds., *Kartoffeln, Klee und kluge Köpfe. Die Oekonomische und Gemeinnützige Gesellschaft des Kantons Bern OGG (1759–2009)* (Bern, Stuttgart, Wien: Haupt Verlag, 2009); Regula Wyss and Martin Stuber,

differentiated mapping of the country as its crucial starting point.[8] Firstly, the Society organised a continuous and standardised network of meteorological stations;[9] secondly, it drew up inventories of the current and the potential plant resources of the territory;[10] thirdly, by means of a total of 48 Topographic Descriptions covering each region it recorded the current state of the population, of agriculture, of trades as well as commerce and, based on this, estimated the existing potential for development.[11] How comprehensive the recording of the regionally differentiated agricultural landscapes designed to be becomes obvious from the work programme of the Economic Society of Bern (1762) which, on 54 printed pages, lists a total of 416 research questions structured into five categories, namely natural history, population, agriculture, trade, and commerce.[12]

"Paternalism and Agricultural Reform. The Economic Society of Bern in the Eighteenth Century," in *The Rise of Economic Societies in the Eighteenth Century*, ed. Stapelbroek and Marjanen, 157–81.

8 See Alix Cooper, "'The Possibilities of the Land': The Inventory of 'Natural Riches' in the Early Modern German Territories," in *Oeconomies in the Age of Newton*, ed. Margret Schabas and Neil de Marchi (Durham, London: Duke University Press , 2003), 129–53; *Inventing the Indigenous*, 116–51 (chapter 4: *The Nature of the Territory*); Torsten Meyer and Marcus Popplow, "'To employ each of Nature's Products in the most Favorable Way possible' – Nature as a Commodity in Eighteenth Century German Economic Discourse," *Historical Social Research* 29, no. 4 (2004).

9 Christian Pfister, *Agrarkonjunktur und Witterungsverlauf im westlichen Schweizer Mittelland zur Zeit der ökonomischen Patrioten 1755–1797. Ein Beitrag zur Umwelt- und Wirtschaftsgeschichte des 18. Jahrhunderts* (Bern: Lang Druck, 1975), 160–62.

10 Martin Stuber and Luc Lienhard, "Nützliche Pflanzen. Systematische Verzeichnisse von Wild- und Kulturpflanzen im Umfeld der Oekonomischen Gesellschaft Bern," in *Nützliche Wissenschaft und Ökonomie im Ancien Régime. Akteure, Themen, Kommunikationsformen*, ed. André Holenstein, Martin Stuber, and Gerrendina Gerber-Visser (Heidelberg: Palatina, 2007).

11 Gerrendina Gerber-Visser, *Die Ressourcen des Landes: Der ökonomisch-patriotische Blick in den Topographischen Beschreibungen der Oekonomischen Gesellschaft Bern (1759–1855)* (Baden: hier + jetzt, 2012), 360–61.

12 "Entwurf der vornehmsten Gegenstände der Untersuchungen, die zur Aufnahme des Feldbaues, des Nahrungsstandes und der Handlung abzielen sollen," in *Abhandlungen und Beobachtungen durch die ökonomische Gesellschaft zu Bern gesammelt*, 1. Stück (1762); see Martin Stuber, "Die Entdeckung der Landschaft als territoriale Ressource," in *Geschichte der Landschaft in der Schweiz: Von der Eiszeit bis zur Gegenwart*, ed. Jon Mathieu, Norman Backhaus, Katja Hürlimann, and Matthias Bürgi (Zurich: Orell Füssli, 2016); Simona Boscani Leoni, "Queries and Questionnaires. Collecting Local and Popular Knowledge in 17th and 18th Century Europe," in *Wissenschaftsgeschichte und Geschichte des Wissens im Dialog – Connecting Science and Knowledge*, ed. Kaspar von Greyerz, Silvia Flubacher, and Philipp Senn (Göttingen: V&R unipress, 2013).

This economic-patriotic work programme includes a number of items on wetlands. Basically, one sought to establish kinds and origin of the various wetlands. How did these wetlands develop, in which ways were they supplied with water, and where did the water drain off? Furthermore, one was interested in their current economic significance: How could they be exploited? What would be the best way of cultivating wet meadows? Which trees or bushes grow best on wet grounds, each according to a different climate?[13]

A number of research questions concerned the conversion of wetlands into farmland, which could be used for agriculture or forestry. How could, "by way of draining the bogs, and by way of improving low or wet grounds, efforts and expenses be reduced?" Are there local gravel pits "for the improvement of wet grounds?" When laying out the area of the "drained bogs" for cultivation for the first time, what must be taken into consideration if they are supposed to be turned into artificial pastures, vegetable or herb gardens or are meant for any other kind of "field crops"?

In the economic-patriotic work programme, wetlands, though not explicitly mentioned, belong to the superior context of the desired population growth, which was considered a sign of good government. Because of demographic interests, one did not want to know only the essential demographic variables for each region – marriages, births and deaths – but also the ratio of cultivated ("kultiviertes") to fallow ("ungebautes") land. If one is aware of the fact that the wetlands accounted for a significant portion of "fallow" land, it becomes immediately obvious how relevant the following research questions are for our topic. What is the ratio of population numbers to acreage of cultivated land? Of what kind is the fallow land in each individual district? Could it not be cultivated? And which crops could best be grown there? What is the price of fallow land?[14] All this refers to the fact that an increase in cultivated land by draining wetlands was believed to have an immediately positive impact on the desired population growth – and consequently on the desired growth of the national economy.

The many observations and functional connections of the "natural economy" on wetlands recorded by the Economic Society of Bern shall now be structured and presented according to the various landscapes. Through the Topographic Descriptions ("Topographische Beschreibungen") of the Economic Society of Bern, I reach back to a stock of sources which were opened up and analysed by Gerrendina Gerber-Visser in the context of her PhD thesis.[15] The records

13 "Entwurf der vornehmsten Gegenstände der Untersuchungen," 20–21.

14 Ibid., 17–19.

15 Gerber-Visser, *Die Ressourcen des Landes*.

were made first by local pastors. The quality of their observations rests on the fact that, on the one hand, they had been educated in the cities; and on the other hand that their official position as rural priests made them were well acquainted with the local situations.[16]

2 Bernese Oberland

Vicar Albrecht Stapfer (1722–1798) is a case in point. In his award-winning *The Best Way of Turning all Kinds of wetlands into Cultivated Land* (1761), he proves to be a good observer of the specific "natural economy" of Saanenland. He explains the logic of the local exploitation system in detail, which keeps many rural people from draining their boggy meadows. For, only if they leave their wetlands in the natural state can they be provided, without effort, with sufficient fodder for their young cattle and horses also in the future. In addition, the draining of the Saanen bogs as well as of the boggy areas near Zweisimmen and Boltigen would affect the meadows. This, he said, was because in their current state wetlands grew an astonishing amount of reed and bad timothy, which – since grain is not grown there – is indispensable as a bed for the cattle which then, together with dung, also serves for fertilising the grassland.[17]

We find additional hints in the Topographic Descriptions of other valleys of the *Bernese Oberland*. In Grindelwald, the pastor notes that people use sacks stuffed with litter as mattresses.[18] The Vicar of St Stephan observes that, apart from maple leaves, most of all litter is used for beds in the stables.[19] The Vicar of Sigriswil comments critically on the many wetland areas, although just a few drainage ditches would easily help; on the common land, about half an hour

16 Regula Wyss and Gerrendina Gerber-Visser, "Formen der Generierung und Verbreitung nützlichen Wissens. Pfarrherren als lokale Mitarbeiter der Oekonomischen Gesellschaft Bern," in *Nützliche Wissenschaft und Ökonomie im Ancien Régime. Akteure, Themen, Kommunikationsformen*, ed. André Holenstein, Martin Stuber, and Gerrendina Gerber-Visser (Heidelberg: Palatina, 2007).

17 [Albrecht Stapfer], "Abhandlung von Auströcknung der Möser, Sümpfe, Moräste und Nuzbarkeit derselben," in *Der Schweitzerischen Gesellschaft in Bern Sammlungen von landwirthschaftlichen Dingen*, 2. Stück (1761); see Martin Stuber and Matthias Bürgi, *Hüeterbueb und Heitisträhl. Traditionelle Formen der Waldnutzung in der Schweiz 1800 bis 2000* (Bern, Stuttgart, Wien: Haupt Verlag, 2011), 89–93.

18 [Friedrich Kuhn and Bernhard Friedrich Kuhn], "Versuch einer Beschreibung des Grindelwaldthales [1. Teil]," *Magazin für die Naturkunde Helvetiens* 1 (1787).

19 Johann Heinrich Schmid, *Beschreybung des Kirchspiehls St. Stephan im Simmenthal*, Manuscript (1783): Burgerbibliothek Bern: GA Oek. Ges. 123 (5). Transcription Gerrendina Gerber-Visser.

FIGURE 8.1 River Mouth of the Aare, near Interlaken, 1812 – H. Rieter, Schweizerische Nationalbibliothek, Bern, Sammlung Gugelmann

uphill from the village, nobody bothers to drain so many wide, boggy places with the help of ditches, to prevent creeks and ditches being swept away.[20]

The Vicar of Meiringen, too, is critical of the common land. Regarding the Haslital, he notes pointedly that there were so many acres of wet ground where only litter grew that the horses there would not take to a better quality, so used they were to the existing bad fodder. People, he claimed, would at once be motivated to improve this situation if the common lands, most of which were wet ground, were equally distributed among the people as their private property.[21]

Vicar Johann Rudolf Nöthiger (1739–1826) in his *Physisch-topographische Beschreibung der Kirchgemeinde Gsteig bei Interlaken* (1783) argues in a similar way. The large bog area, he says, would turn into exceptionally fertile lands producing abundant crops if it were privately owned and everybody could improve their own plot and use it for their own benefit as they liked. However, as the privatisation of the common lands had not started yet, Nöthiger states,

20 Gottlieb Jakob Kuhn, "Versuch einer oekonomisch-topographischen Beschreibung der Gemeinde Sigriswyl," *Alpina* 3 (1808).

21 [Johannes Sprüngli], "Beschreibung des Hasle-Lands im Canton Bern [1. Teil]," in *Der Schweitzerischen Gesellschaft in Bern Sammlungen von landwirthschaftlichen Dingen*, 4. Stück (1760); "Von dem Haslethal [2. Teil]," in *Abhandlungen und Beobachtungen durch die ökonomische Gesellschaft zu Bern gesammelt*, 4. Stück (1762).

with regret, that the various parts of the bogs are subject to a variety of different, partly overlapping, rights. For example, the "Auermoos" is subject to the common rights of the village community of Aarmühle which, however, as the right to graze horses, may be exercises only in spring, from early May to the end of June, as well as in autumn from September 14th to October 16th. What grows in the meantime is the property of several private owners from the communities in the vicinity and is divided into plots of land or "Meder," which may be inherited, purchased, subdivided into smaller plots. The size of these "Meder" may vary greatly and yield between one and thirty loads of litter fodder. The *Wilderswilermoos*, on the other hand, is jointly owned by the two village communities of Matten and Wilderswil, only the latter being entitled to grazing in spring and autumn. The Monastery of Interlaken had likewise acquired ownership of two "Meder."[22]

3 Aare and Gürbe Valleys

The great significance of the wetlands for the local "natural economy" becomes also obvious from Vicar Samuel Massé's (1704–1782) manuscript on the parish of Belp, which owns a large wetland area.[23] According to Massé, the bog area is flooded twice a year by the Gürbe river, which is swollen by rain and meltwater, making the fodder inedible because of evil-smelling morass and sand. At its southern end, it provides meagre grass and litter for horses and oxen, only in relatively dry periods. The bog area along the Aare river provides fodder of better quality, which is good for all kinds of horses and cattle. On the question of how the population is affected by the wetlands, Massé's conclusion is ambivalent. On the one hand, making up most of the common land they are an important economic resource most of all for the poor. On the other hand, Massé connects them to a number of different diseases which he believes to be caused by the wet state of the bog and the dwelling places sinking deeply into

22 Johann Rudolf Nöthiger, *Physisch-topographische Beschreibung der Kirchgemeinden Unterseen, Habkern und Beatenberg*, Manuscript (1783): Burgerbibliothek Bern: GA Oek. Ges. 123 (11). Transcription Gerrendina Gerber-Visser.

23 Massé, Samuel, *Antwortlicher Bericht über die vorgeschlagenen Fragen der Oeconomischen Gesellschaft, soweit dieselben dem Bezirk der Kirchgemeinde Belp betreffen mögen*, Manuscript (1762): Burgerbibliothek Bern: GA Oek. Ges. 123 (1). Transcription Gerrendina Gerber-Visser; siehe Melanie Salvisberg, *Der Hochwasserschutz an der Gürbe: Eine Herausforderung für Generationen (1855–2010)* (Basel: Schwabe Verlag, 2017), 68–75.

FIGURE 8.2 Plan of the Belp Bog (Grundriss des Belp-Mooses), 1709 – Staatsarchiv des Kantons Bern: AA IV 2157

the ground. The bog air, he says, causes scurvy and frequent toothache, bad hearing, and weakness of limbs.[24]

According to the description given by theologian and naturalist Samuel Jakob Wyttenbach (1748–1830), also in the community of Gurzelen, about 15 kilometres upriver, most of the lands are of a bog nature. Whereas the meadows of the share owners are on grass land, all common lands and common property are on bog or peat lands. Where the bog is too deep and the grass is bad the litter growing there is auctioned every year.[25] For a long time the boggy land at Gurzelen was only of little value, until a few years ago one acre did not sell for more than 20 Crowns. Very recently, however, precisely the prices for these grounds started rising, a trend that Wyttenbach connects to the fact that cultivatable boggy areas are perfectly suitable for cabbage gardens, that the

24 Massé, *Antwortlicher Bericht über die vorgeschlagenen Fragen der Oeconomischen Gesellschaft*; see Sinja Clavadetscher, "Die Topographischen Beschreibungen zweier Moosgebiete: Belp und Kerzers" (Proseminararbeit University of Bern, 2015), 9–10, 15–18.

25 Jakob Samuel Wyttenbach, *Beschreibung der Pfarrgemeinde Gurzelen*, Manuscript (1776): Burgerbibliothek Bern: Mss. Hist. Helv. XX 9 C 20. Transcription Gerrendina Gerber-Visser.

"Thurnen cabbage" growing there is sold to Bern in large amounts and generates considerable profits for farmers.[26]

4 Seeland

In the *Oekonomische Beschreibung des Kirchspieles Kerzers* (1763) by the Vicar David Albrecht Bolz (1713–1782), the longest section of the whole work, is devoted to the wetlands. This is in line with their significance for local cattle farming, which Bolz explains by comparing the five village communities. As they are located farther away from the *Grosses Moos*, in Golaten, Gurbrü and Wileroltigen, there is not much horse breeding and only little cattle husbandry; horses and cattle must be grazed in the little fertile woods and meadows. Kerzers and Fräschels, on the other hand, being located near the *Grosses Moos*, are capable of large-scale horse breeding and can have large cattle herds. As most meadows are not privately owned, one makes use of the rights to "common meadows" from whose crops even a poor man can feed at least two heads of cattle over the winter.

However, the bog meadow is a health concern, in particular at the height of summer. According to Bolz, the water in the ditches, which starts smelling in the heat, is a major cause of widespread cattle disease in the area. In the hot summer of 1761, near Fräschels many cows, oxen and horses died from anthrax, which must doubtlessly be blamed on the poisonous mists that often rise from the ground like rot, and on the rotten water. Bolz, however, does not mention any negative effects on people. On the contrary, he classifies the vast bog area under the "hardenings" that make the local people healthy, strong, cheerful, hard-working. And when in a cold, wet spring they often struggle to chase their cattle on the bog meadows while standing in waist-deep morass and water, they simply change their clothes, which happens every day, and they remain in good health while doing so.[27]

For the sake of completeness, here we may refer to two other Topographic Descriptions of Seeland written at the same time. At Bözingen – today a quarter of the City of Biel – in the direction of Pieterlen there is a bog of about one hour's length which has developed where the waters of Leber- and Büttenberg

26 Ibid.

27 David Albrecht Bolz, "Oekonomische Beschreibung des Kirchspieles Kerzers," in *Abhandlungen und Beobachtungen durch die ökonomische Gesellschaft zu Bern gesammelt*, 1. Stück (1763); see Clavadetscher, "Die Topographischen Beschreibungen zweier Moosgebiete," 11–15.

collect and cannot drain. One part of the bog is let to the community people for a yearly rent, as hay rights; the other part is the bog gardens where barley, cabbage and other vegetables are grown. The author, the book printer Niklaus Heilmann (1739–1817) from Biel, views the forgoing of collective rights as the key to an improved cultivation of bog land. For two reasons. To begin with, by doing so the community people would no longer lose the urgently needed dung, which is lost during the eight months of grazing their cattle on the common meadows, but could collect it for their fields; secondly, if so, owners would no longer be satisfied with what nature provides voluntarily but would of their own accord diligently and carefully improve their share.[28]

Even 70 years later, wetlands are omnipresent in the *Beschreibung der Kirchgemeinde Gampelen* written by Judge Gottlieb Stauffer (1786–1882). Whereas in most places the ground of the forests, fields and vineyards – all of them located in a hilly area – is clay, in single cases also sandy, the meadows on low ground are mostly "mossy and boggy" and about one-fifth of the area is periodically flooded. If these floods last only for a short time, they have a "beneficial" effect, however they cause extraordinary damage in those years when they last longer. After the two-year flood of 1816/17, two decades later both the fodder yield and grass for grazing were still lower than before.

Similarly, at Gampelen, the utilisation of the *Grosses Moos* goes beyond the use of meadows and litter. For example, the garden plants for the weekly markets in Neuchâtel are not only grown around the houses but also in the bog gardens, which results in constant losses for many people. Particularly after wet years a considerable number of frogs are caught; whereas the thighs of the hind legs are sold to merchants at Muntelier, the other parts of the frogs make good swine fodder a small bucket is sold for about 2.5 to 3 Batzes. Until recently, leeches ("Blutegeln") were collected in great numbers and sold to Neuchâtel; unfortunately, now they are almost extinct. This seems to have been caused by their high prices, no spawn being spared. Moreover in the *Grosses Moos*, during the periodical floods, often considerable amounts of fish are caught

28 [Heilmann, Niklaus], "Topographisch und Oekonomische Beschreibung der Landschaft um Biel gelegen," in *Abhandlungen und Beobachtungen durch die ökonomische Gesellschaft zu Bern gesammelt*, 4. Stück (1766); similarly: Abraham Pagan: "Versuch einer Oekonomischen Beschreibung der Graffschaft oder Landvogtey Nidau im Canton Bern," in *Abhandlungen und Beobachtungen durch die ökonomische Gesellschaft zu Bern gesammelt*, 4. Stück (1761): The best yield from the "Möösern oder Morästen" is the "Beünden" (small gardens) where kitchen herbs, cabbage, hemp and flax are grown. Most of them are harvested or mown after a certain time of the year. Of them we must positively distinguish the single shares separated from the common lands and meadows: "These plots torn from the service of grazing illustrate well the different yield, and what unlimited diligence may contribute to the fertility of lands."

FIGURE 8.3 General Map of the Jura Water Bodies (General Charte der Jura Gewässer), 1816/17, F. Trechsel; J. Oppikofer – Staatsarchiv des Kantons Bern (Atlanten 21)

with the help of nets or six-tined forks, which are then sold to the merchants of Muntelier or Lüscherz who sell them on in Solothurn, Bern, Fribourg and Neuchâtel.[29]

5 Conclusion

For the reformers, the stocktaking of the rural economy served, on the one hand, as a negative backdrop for their criticism, and on the other hand as a basis for working out reform strategies. This double motivation yielded early social-anthropologic observations, which may be summarised in three points:

29 Gottlieb Stauffer, *Beschreibung der Kirchgemeinde Gampelen und des in derselben liegenden ehemaligen Klosters St. Johannsen*, Manuscript (1835/1839): Burgerbibliothek Bern: GA Oek. Ges. 125 (18); GA Oek. Ges. 125 (20). Transcription: Gerrendina Gerber-Visser.

1. The economic significance of the wetlands was observed and recorded in detail: being an extensive way of land use, it allowed for some minimum animal husbandry even for poorer households, which were not dependent on cash and individual land property.
2. The observations are not limited to the traditional economy; indeed the reformers also perceived dynamic elements, such as the harvesting of litter used to meet the growing demand for beds as a result of increasing stable feeding in summer, a key element of the modernisation of agriculture. Frogs and leeches caught in the wetlands were supplied the growing urban markets of the cities of Bern, Fribourg, Neuchâtel and Solothurn; and so were vegetables increasingly cultivated on formerly boggy ground.
3. The "natural economy" was not considered in isolation but as embedded in institutions. Most wetlands were common property, which the reformers criticized as being an obstacle for innovation. From their point of view, only private interests would motivate the necessary improvements. In the long run, this criticism is somewhat ironic: from an ecological point of view, today it is precisely in the absence of those collective methods of exploitation – mowing and grazing – that wetlands silt up. Today, these ways of utilisation must be imitated by complex – and appropriately costly – environmental protection measures.[30]

Acknowledgements

This contribution is based on: Martin Stuber and Matthias Bürgi, *Vom "eroberten Land" zum Renaturierungsprojekt. Geschichte der Feuchtgebiete in der Schweiz seit 1700* (Bern, Stuttgart, Wien: Haupt Verlag, 2018), 19–43. I wish to thank Gerrendina Gerber-Visser (Bern) for kindly handing over the transcriptions of the unpublished Topographical Descriptions to me; and to Mirko Wittwar (Morsbach, D), for his careful translation in English.

30 See Meinrad Küchler, *Moore der Schweiz. Zustand, Entwicklung, Regeneration* (Bern, Stuttgart, Wien: Haupt Verlag, 2018); Meinrad Küchler and Helen Küchler, "Entwicklung der Vegetation in geschützten Moorflächen, 1995–2007," in *Vom "eroberten Land" zum Renaturierungsprojekt*, ed. Martin Stuber and Matthias Bürgi (Bern, Stuttgart, Wien: Haupt Verlag, 2018).

Bibliography

Manuscripts

Massé, Samuel. *Antwortlicher Bericht über die vorgeschlagenen Fragen der Oeconomischen Gesellschaft, soweit dieselben dem Bezirk der Kirchgemeinde Belp betreffen mögen*. Manuscript (1762): Burgerbibliothek Bern: GA Oek. Ges. 123 (1). Transcription Gerrendina Gerber-Visser.

Nöthiger, Johann Rudolf. *Physisch-topographische Beschreibung der Kirchgemeinden Unterseen, Habkern und Beatenberg*. Manuscript (1783): Burgerbibliothek Bern: GA Oek. Ges. 123 (11). Transcription Gerrendina Gerber-Visser.

Schmid, Johann Heinrich. *Beschreybung des Kirchspiehls St. Stephan im Simmenthal*. Manuscript (1783): Burgerbibliothek Bern: GA Oek. Ges. 123 (5). Transcription Gerrendina Gerber-Visser.

Stauffer, Gottlieb. *Beschreibung der Kirchgemeinde Gampelen und des in derselben liegenden ehemaligen Klosters St. Johannsen*. Manuscript (1835/1839): Burgerbibliothek Bern: GA Oek. Ges. 125 (18); GA Oek. Ges. 125 (20). Transcription: Gerrendina Gerber-Visser.

Wyttenbach, Jakob Samuel, *Beschreibung der Pfarrgemeinde Gurzelen*. Manuscript (1776): Burgerbibliothek Bern: Mss. Hist. Helv. XX 9 C 20. Transcription Gerrendina Gerber-Visser.

Printed Sources

Bolz, David Albrecht. "Oekonomische Beschreibung des Kirchspieles Kerzers." In *Abhandlungen und Beobachtungen durch die ökonomische Gesellschaft zu Bern gesammelt*. 1. Stück (1763), 69–89.

"Entwurf der vornehmsten Gegenstände der Untersuchungen, die zur Aufnahme des Feldbaues, des Nahrungsstandes und der Handlung abzielen sollen." In *Abhandlungen und Beobachtungen durch die ökonomische Gesellschaft zu Bern gesammelt*. 1. Stück (1762), 1–54.

Kuhn, Gottlieb Jakob. "Versuch einer oekonomisch-topographischen Beschreibung der Gemeinde Sigriswyl." *Alpina* 3 (1808): 116–69.

[Heilmann, Niklaus]. "Topographisch und Oekonomische Beschreibung der Landschaft um Biel gelegen." In *Abhandlungen und Beobachtungen durch die ökonomische Gesellschaft zu Bern gesammelt*. 4. Stück (1766), 53–82.

[Kuhn, Friedrich and Bernhard Friedrich Kuhn]. "Versuch einer Beschreibung des Grindelwaldthales [1. Teil]." *Magazin für die Naturkunde Helvetiens* 1 (1787): 1–28.

Pagan, Abraham. "Versuch einer Oekonomischen Beschreibung der Graffschaft oder Landvogtey Nidau im Canton Bern." In *Abhandlungen und Beobachtungen durch die ökonomische Gesellschaft zu Bern gesammelt*. 4. Stück (1761), 785–859.

[Sprüngli, Johannes]. "Beschreibung des Hasle-Lands im Canton Bern [1. Teil]." In *Der Schweitzerischen Gesellschaft in Bern Sammlungen von landwirthschaftlichen Dingen.* 4. Stück (1760), 859–85.

[Stapfer, Albrecht]. "Abhandlung von Auströcknung der Möser, Sümpfe, Moräste und Nuzbarkeit derselben." In *Der Schweitzerischen Gesellschaft in Bern Sammlungen von landwirthschaftlichen Dingen.* 2. Stück (1761), 243–306.

"Von dem Haslethal [2. Teil]." In *Abhandlungen und Beobachtungen durch die ökonomische Gesellschaft zu Bern gesammelt.* 4. Stück (1762), 130–43.

Literature

Beck, Rainer. *Naturale Ökonomie. Unterfinning: Bäuerliche Wirtschaft in einem oberbayerischen Dorf des frühen 18. Jahrhunderts.* Munich, Berlin: Deutscher Kunstverlag, 1986.

Beck, Rainer. *Ebersberg oder das Ende der Wildnis. Eine Landschaftsgeschichte.* Munich: C. H. Beck, 2003.

Boscani Leoni, Simona. "Queries and Questionnaires. Collecting Local and Popular Knowledge in 17th and 18th Century Europe." In *Wissenschaftsgeschichte und Geschichte des Wissens im Dialog – Connecting Science and Knowledge*, edited by Kaspar von Greyerz, Silvia Flubacher, and Philipp Senn, 187–208. Göttingen: V&R unipress, 2013.

Clavadetscher, Sinja. "Die Topographischen Beschreibungen zweier Moosgebiete: Belp und Kerzers." Proseminararbeit, University of Bern, 2015.

Cooper, Alix. "'The Possibilities of the Land': The Inventory of 'Natural Riches' in the Early Modern German Territories." In: *Oeconomies in the Age of Newton*, edited by Margret Schabas, Neil de Marchi, 129–53. Durham, London: Duke University Press, 2003.

Cooper, Alix. *Inventing the Indigenous: Local Knowledge and Natural History in Early Modern Europe.* Cambridge, New York: Cambridge University Press, 2007.

Gerber-Visser, Gerrendina. *Die Ressourcen des Landes: Der ökonomisch-patriotische Blick in den Topographischen Beschreibungen der Oekonomischen Gesellschaft Bern (1759–1855).* Archiv des historischen Vereins des Kantons Bern 89. Baden: hier + jetzt, 2012.

Gudermann, Rita. *Morastwelt und Paradies. Ökonomie und Ökologie in der Landwirtschaft am Beispiel der Meliorationen in Westfalen und Brandenburg (1830–1880).* Paderborn: Schöningh, 2000.

Küchler, Meinrad and Helen Küchler. "Entwicklung der Vegetation in geschützten Moorflächen, 1995–2007." In *Vom "eroberten Land" zum Renaturierungsprojekt. Geschichte der Feuchtgebiete in der Schweiz seit 1700*, edited by Martin Stuber and Matthias Bürgi, 242–54. Bern, Stuttgart, Wien: Haupt Verlag, 2018.

Küchler, Meinrad et al. *Moore der Schweiz. Zustand, Entwicklung, Regeneration.* Bern, Stuttgart, Wien: Haupt Verlag, 2018.

Kuper, Adam. "Anthropology." In *History of Science,* vol.7, *The Modern Social Sciences,* edited by Theodore M. Porter and Dorothy Ross, 354–78. Cambridge, New York: Cambridge University Press, 2003.

Liebersohn, Harry. "Scientific Ethnography and Travel, 1750–1850." In *History of Science,* vol. 7, *The Modern Social Sciences,* edited by Theodore M. Porter and Dorothy Ross, 100–12. Cambridge, New York: Cambridge University Press, 2003.

Meyer, Torsten and Marcus Popplow. "'To employ each of Nature's Products in the most Favorable Way possible' – Nature as a Commodity in Eighteenth Century German Economic Discourse." *Historical Social Research* 29, no. 4 (2004): 4–40.

Pfister, Christian. *Agrarkonjunktur und Witterungsverlauf im westlichen Schweizer Mittelland zur Zeit der ökonomischen Patrioten 1755–1797. Ein Beitrag zur Umwelt- und Wirtschaftsgeschichte des 18. Jahrhunderts.* Bern: Lang Druck, 1975.

Popplow, Marcus, ed. *Landschaften agrarisch-ökonomischen Wissens. Strategien innovativer Ressourcennutzung in Zeitschriften und Sozietäten des 18. Jahrhunderts.* Münster: Waxmann 2010.

Salvisberg, Melanie. *Der Hochwasserschutz an der Gürbe. Eine Herausforderung für Generationen (1855–2010).* Basel: Schwabe Verlag, 2017.

Stapelbroek, Koen and Jani Marjanen, eds. *The Rise of Economic Societies in the Eighteenth Century: Patriotic Reform in Europe and North America.* Basingstoke: Palgrave Macmillan 2012.

Stuber, Martin and Luc Lienhard. "Nützliche Pflanzen. Systematische Verzeichnisse von Wild- und Kulturpflanzen im Umfeld der Oekonomischen Gesellschaft Bern." In *Nützliche Wissenschaft und Ökonomie im Ancien Régime. Akteure, Themen, Kommunikationsformen,* edited by André Holenstein, Martin Stuber, and Gerrendina Gerber-Visser, 65–106. Heidelberg: Palatina, 2007.

Stuber, Martin, Peter Moser, Gerrendina Gerber-Visser and Christian Pfister, eds. *Kartoffeln, Klee und kluge Köpfe. Die Oekonomische und Gemeinnützige Gesellschaft des Kantons Bern OGG (1759–2009).* Bern, Stuttgart, Wien: Haupt Verlag, 2009.

Stuber, Martin and Matthias Bürgi. *Hüeterbueb und Heitisträhl. Traditionelle Formen der Waldnutzung in der Schweiz 1800 bis 2000.* Bern, Stuttgart, Wien: Haupt Verlag, 2011.

Stuber, Martin. "Die Entdeckung der Landschaft als territoriale Ressource." In *Geschichte der Landschaft in der Schweiz. Von der Eiszeit bis zur Gegenwart,* edited by Jon Mathieu, Norman Backhaus, Katja Hürlimann, and Matthias Bürgi, 91–104. Zurich: Orell Füssli, 2016.

Stuber, Martin and Matthias Bürgi. *Vom "eroberten Land" zum Renaturierungsprojekt. Geschichte der Feuchtgebiete in der Schweiz seit 1700.* Bern, Stuttgart, Wien: Haupt Verlag, 2018.

Wyss, Regula and Gerrendina Gerber-Visser. "Formen der Generierung und Verbreitung nützlichen Wissens. Pfarrherren als lokale Mitarbeiter der Oekonomischen Gesellschaft Bern." In *Nützliche Wissenschaft und Ökonomie im Ancien Régime. Akteure, Themen, Kommunikationsformen*, edited by André Holenstein, Martin Stuber, and Gerrendina Gerber-Visser, 41–64. Heidelberg: Palatina, 2007.

Wyss, Regula and Martin Stuber. "Paternalism and Agricultural Reform. The Economic Society of Bern in the Eighteenth Century." In *The Rise of Economic Societies in the Eighteenth Century*, edited by Stapelbroek and Marjanen, 157–81. Basingstoke: Palgrave Macmillan, 2012.

PART 3

Defining Territories

CHAPTER 9

Divergent Perception: Deserts and Mountains in Transition to Modernity, Seen through Alexander von Humboldt's *Views of Nature*

Jon Mathieu

1 Introduction

This chapter deals with the historical perception of two major ecosystems – deserts and mountains – in transition to Western modernity. I will argue that, from 1750 to 1850, the perception underwent a process of divergence between the two systems, which is significant in its own right and can offer some clues as to general factors underlying environmental perception. Alexander von Humboldt's book *Ansichten der Natur mit wissenschaftlichen Erläuterungen* ("Views of Nature with Scientific Annotations") is used for illustrative and certain analytical purposes. The book, published by the author three times in different editions (1808, 1826 and 1849), belongs to the famous pieces of nature writing of the 19th century. When it was first published, however, it received less-than-flattering reviews. A critic in the leading *Göttingische Gelehrte Anzeigen* stated that reading Humboldt's book had been both pleasant and instructive. Nevertheless, the gratification would have been greater if its language had been simpler. "The poetic prose seems to be unfortunate in several respects," namely for the learned audience and for the treated subjects "which are inapt to provoke poetic enthusiasm." Moreover, the author tries to establish relationships between totally different things. "To our mind, the true researching genius manifests itself not so much through the establishment of resemblances between unequal objects as through the disclosure of the characteristic features of both single objects and entire species." Instead of familiarising the audience with the plains of South America, observed in his travels, he brings them together with deserts in Asia, Africa, and even with heaths of European coastlands. This will certainly cause some readers to consider these very different surfaces as similar.[1]

1 *Göttingische Gelehrte Anzeigen unter der Aufsicht der königl. Gesellschaft der Wissenschaften*, 28 December 1809, 2049–2050, my translation from German. I use published translations when available; for *Views of Nature* there is a careful new American version: Alexander von

DOI:10.1163/9789004412477_010

English and French commentators believed that the reception of *Ansichten der Natur* was "encouraging" or even "a very flattering success" in Germany. It is not clear how the public, and different parts thereof, reacted precisely. In the long run, however, the book was read more and more as a piece of creative writing and became a literary classic.[2] In the following chapter, I would like to take up the criticised points of the first moment: the form and language of the book, and the problem of a global approach to nature writing. They give relevant context information to the main question of this chapter: the general perception of deserts and mountains in the 18th and 19th centuries. *Views of Nature* provides a good entrance to environmental perception for several reasons. Humboldt absorbed many influences of the older literature, put them in a specific context, transformed and bequeathed them to numerous other people. His audience included important "multipliers" like Charles Darwin, Henry David Thoreau, Ernst Haeckel, George Perkins Marsh, and John Muir.[3] As for the perception of deserts, my reading of Humboldt is inspired by Diana K. Davis and her thought-provoking survey *The Arid Lands: History, Power, Knowledge*, published in 2016. On the side of mountain perception, my own research experience goes back quite some time.[4]

The first section offers a succinct bibliographical inquiry of *Views of Nature*. The next two sections deal with the perception of deserts and mountains respectively as inferred from Humboldt and other sources. The conclusions try to bundle and analyse the historical change of environmental perception in a summary form.

Humboldt, *Views of Nature*, trans. Mark W. Person, ed. Stephen T. Jackson and Laura Dassow Walls (Chicago, London: The University of Chicago Press, 2014).

2 *Monthly Review* 59 (1809), 473; Alexander von Humboldt, *Tableaux de la nature, ou considérations sur les déserts, sur la physionomie des végétaux, et sur les cataractes de l'Orénoque*, trans. J. B. B. Eyriès, 2 vols. (Paris: F. Schoell, 1808), I.

3 These personalities are mentioned by Wulf in her 2015 biography; there are quite a few biographies of Humboldt, but no standard academic one: Andrea Wulf, *The Invention of Nature: The Adventures of Alexander von Humboldt, the Lost Hero of Science* (London: John Murray, 2015); a detailed chronology is provided by Kurt-R. Biermann and Ilse Jahn, *Alexander von Humboldt: Chronologische Übersicht über wichtige Daten seines Lebens* (Berlin: Akademie-Verlag, 1983).

4 Jon Mathieu, "Alpenwahrnehmung: Probleme der historischen Periodisierung," in *Die Alpen! Zur europäischen Wahrnehmungsgeschichte seit der Renaissance – Les Alpes! Pour une histoire de la perception européenne depuis la Renaissance*, ed. Jon Mathieu and Simona Boscani Leoni (Bern, Berlin, Bruxelles: Peter Lang, 2005); Mathieu, *The Third Dimension: A Comparative History of Mountains in the Modern Era* (Cambridge: White Horse Press, 2011).

2 *Views of Nature* 1808–1849: Publishing History

Alexander von Humboldt (1769–1859) was an intellectual celebrity in natural history from early on. It is well known that his voyage to South America from 1799 to 1804 boosted that fame in many countries. In the first three decades of the 19th century, he received more than eighty awards from learned societies and other institutions. The number of published articles, books, and illustrations is impressive. It was not only the result of Humboldt himself but also of a kind of Humboldt industry with scientific and technical assistants, independent publishing agents, translators, and intermediaries of all kinds. To date, getting an overview of the complete work and its ramifications remains difficult. In Humboldt's extensive account of the South American journey, according to his bibliographers Horst Fiedler and Ulrike Leitner, it is not even possible always to identify the affiliation to the various series of his work. The publication started in 1814 and was discontinued in 1831, with only a third of the travel route described. For the breach of contract Humboldt had to pay a heavy penalty to his publisher in Paris, whom he called "my tyrant" in a letter to his German publisher Georg von Cotta.[5]

Views of Nature belonged to the decidedly popularising parts of Humboldt's work, and Cotta – the publisher of Johann Wolfgang Goethe, Friedrich Schiller and other foremost poets and intellectuals – was its cornerstone. While the editions of 1808, 1826 and 1849 varied in size and contents, the Cotta family remained unchanged. Possibly the very idea came from that side. The 1808 edition consisted of the publication of three academic lectures about arid lands, plant physiognomy, and a particular South American waterfall, delivered by Humboldt to the Royal Prussian Academy of Sciences in Berlin in 1807 and 1808. He completed the text of these oral presentations with endnotes, called "annotations and additions" ("Erläuterungen und Zusätze"). One of the essays was a reprint from a prior separate publication. The small format booklet was designated as "first volume" on the title page, and the last page announced the

5 Horst Fiedler and Ulrike Leitner, *Alexander von Humboldts Schriften: Bibliographie der selbständig erschienenen Werke* (Berlin: Akademie Verlag, 2000), XVI, 70; a quantitative chronological approach to Humboldt's publications and awards in Jon Mathieu, "Von den Alpen zu den Anden: Alexander von Humboldt und die Gebirgsforschung," in *Wissenschaft – Berge – Ideologien. Johann Jakob Scheuchzer (1672–1733) und die frühneuzeitliche Naturforschung – Scienza – montagna – ideologie. Johann Jakob Scheuchzer (1672–1733) e la ricerca naturalistica in epoca moderna*, ed. Simona Boscani Leoni (Basel: Schwabe, 2010), 308; Oliver Lubrich, ed., *Alexander von Humboldt: Das graphische Gesamtwerk* (Darmstadt: Lambert Schneider, 2014) offers a complete edition of the illustrative work; his low figures of the book production (p. 750) do not include the numerous derivative prints.

"end of the first volume." A second volume, however, did not appear. Probably there had been a plan to include other essays, which was then abandoned.[6] The realisation came nearly twenty years later in the 1826 edition. Without a new preface, the three original texts were now rearranged and completed with two new ones: a lecture about volcanoes, read in 1823 to the same Academy in Berlin, and a narration about the "Life Force" published back in 1795 in a literary journal attended by Schiller. Since 1846, then, Cotta and Humboldt discussed plans for a third edition. Humboldt should redo the annotations. Surprisingly, after that extensive revision work, the busy elderly naturalist came up with two new essays about his South America travel almost half a century ago, one about the nocturnal wildlife in the rainforest, the other one about the Andes.[7] Table 9.1 gives a survey of the titles and the formal organisation of these three first-hand German editions.

TABLE 9.1 First-hand German editions of *Views of Nature*: bibliographical indications

Edition Publisher	Essays Title (pages)	Endnotes (pages)	Number	Volumes Total pages
1808 Tübingen: J. G. Cotta	0. Preface (V–VIII)	–	–	1 vol. designated as "first volume" 334 pages
	1. Concerning the Steppes and Deserts (1–46)	(46–155)	50	
	2. Ideas for a Physiognomy of Plants (157–204)	(205–278)	36	
	3. Concerning the Waterfalls of the Orinoco near Atures and Maypures (281–330)	(331–334)	7	

6 In the French version of the 1808 edition, the first essay was augmented by a letter and published as volume 1, the other two essays as volume 2. The translation was concluded only half a year after the German manuscript (Fiedler and Leitner, *Alexander von Humboldts Schriften*, 39, 51); this could mean that it was rather quickly clear that the announced second volume in German would not follow.

7 Fiedler and Leitner, *Alexander von Humboldts Schriften*, 37–42.

TABLE 9.1 First-hand German editions of *Views of Nature*: bibliographical indications (*cont.*)

Edition Publisher	Essays Title (pages)	Endnotes (pages)	Number	Volumes Total pages
1826 Stuttgart, Tübingen J. G. Cotta	0. Preface to the First Edition (V–VI)	–	–	2 vols. vol. 1: essays 1–2 vol. 2: essays 3–5 434 pages
	1. Concerning the Steppes and Deserts (1–45)	(46–180)	50	
	2. Concerning the Waterfalls of the Orinoco near Atures and Maypures (183–230)	(231–234)	6	
	3. Ideas for a Physiognomy of Plants (1–47)	(48–125)	36	
	4. Concerning the Structure and Action of Volcanoes in Various Regions of the Earth (126–178)	(179–186)	2	
	5. The Life Force, or The Rhodian Genius (187–200)	–	–	
1849 Stuttgart, Tübingen: J. G. Cotta	0. Preface of the First Edition (VII–X)	–	–	2 vols. vol. 1: essays 1–4 vol. 2: essays 5–8 744 pages
	0. Preface to the Second and Third Edition (XI–XVIII)	–	–	
	1. Concerning the Steppes and Deserts (1–38)	(39–248)	52	
	2. Concerning the Waterfalls of the Orinoco near Atures and Maypures (249–286)	(287–316)	12	
	3. The Nocturnal Wildlife of the Primeval Forest (317–337)	(338–340)	6	
	4. Hypsometric Addenda (341–350)	–	–	
	5. Ideas for a Physiognomy of Plants (1–41)	(42–248)	37	
	6. Concerning the Structure and Action of Volcanoes in Various Regions of the Earth (249–289)	(290–296)	3	

TABLE 9.1 First-hand German editions of *Views of Nature*: bibliographical indications (*cont.*)

Edition Publisher	Essays Title (pages)	Endnotes (pages)	Number	Volumes Total pages
	7. The Life Force, or The Rhodian Genius (297–308)	(309–314)	1	
	8. The Plateau of Cajamarca, the Old Residential City of Inca Atahualpa; First Sight of the Pacific from the Ridge of the Andes Chain (315–367)	(368–394)	20	

SOURCES: Humboldt, *Ansichten der Natur mit wissenschaftlichen Erläuterungen* (1808, 1826, and 1849); English titles of the essays according to the 2014 translation; in the originals, the essay about the "Life Force" has a subtitle: "Eine Erzählung" (a narration).

From the bibliographical indications, it is evident that *Views of Nature* experienced notable growth from 1808 to 1849. It grew from three essays on 334 small format pages to eight essays on 744 bigger pages. Growth was most noticeable at the annotation level. The proportion of endnotes in the total essay text increased from 71 to 85 percent in the piece on arid lands, from 61 to 83 percent in the botanical essay, and from 8 to 44 percent in the waterfall essay. Interestingly, the number of endnotes did not go up very much. The three original essays received only between two and five supplementary notes. This indicates that the "annotations and additions" had fixed entry points, which would then be filled in with new results from ongoing research. There was no one-to-one, flexible relation between the two levels of text. The essays, originally written for oral presentation, remained relatively stable, while the annotations changed from one edition to the next. In the essay on arid lands, for example, the annotations offered about sixty references in 1808. In the 1826 edition there were more than ninety, particularly new self-references to Humboldt's main publications. The 1849 edition, then, brought another huge increase in references, both to cutting edge research and in the self-referencing mode. The new supplementary notes deserve special attention. In essays 5 and 7, they defend a field of research considered outdated (Physiognomy of Plants), and an older combination of literature and science (Life Force). The supplementary notes in the arid lands essay offer new information about the Sahara and the steppe vegetation, revealing by contrast that Humboldt did not consider the preceding editions strong in these matters.[8]

A good indicator of the public attention attracted by *Views of Nature* is provided by its translation history. The 1808 edition was rendered completely into French and Dutch, while the 1826 edition had only a complete French translation. Very different was the situation in 1849 when the possibility of selling non-German versions of the illustrious scholar's popularising work prompted fierce competition between publishers both in France and in Britain. It resulted in two rival translations respectively. Perhaps a similar pattern manifested itself a bit later in Russia with one translation in 1853 (except for essay 7) and another one in 1855.[9] Table 9.2 presents the complete and partial translations during Humboldt's lifetime in a survey format.

TABLE 9.2 Translations of *Views of Nature* from the German editions until Humboldt's death 1859

Edition	Language: Translator	Year	Index	Illustration	Comments
1808	French: J. B. B. Eyres	1808	–	–	Includes additional item
	Italian: anonymous	1808	–	–	Only part of essay 1
	Dutch: G. Troost	1808	–	–	
	Russian: anonymous	1818	–	–	Only essay 1
1826	French: J. B. B. Eyres	1828	–	–	
	Polish: anonymous	1828/43	–	–	Only essays 4 and 1
	Russian: anonymous	1829	–	–	Only essay 5
1849	English: E. J. Sabine	1849	DC, ID	–	Rival translation to Otté
	English: E. C. Otté	1850	DC, ID	1 IM	Rival translation to Sabine
	French: F. Hoefer	1850/51	DC	4 IM, 2 MP	Rival translation to Galusky
	French: Ch. Galusky	1851	DC	5 MP, 2 FG	Rival translation to Hoefer

8 Humboldt, *Views of Nature*, 77–82, 238–41, 264–66.

9 Fiedler and Leitner, *Alexander von Humboldts Schriften*, 61; being unable to read Russian, I cannot check that point.

TABLE 9.2 Translations of *Views of Nature* from the German editions (*cont.*)

Edition	Language: Translator	Year	Index	Illustration	Comments
	Dutch: E. M. Beima	1850	DC	–	
	Russian: N. Ch. Ketčer	1853	DC	–	Without essay 7
	Russian: A. Nazymov	1855	(?)	IM	Complete translation
	Czech: Vojtěch Šafařík	1859	–	–	Only part of essay 3

DC: detailed table of contents; ID: alphabetical index; IM: image; MP: map; FG: figure; (?): unsettled. The numbers of the essays correspond to the respective edition, see table 9.1. SOURCE: Fiedler and Leitner, *Alexander von Humboldts Schriften*, 50–63 and available online versions; I thank Oliver Lubrich for providing me the Russian version of 1853.

The table includes indications on the make-up of the books and shows that the translations of the 1849 editions differed quite significantly from the earlier ones. Some of them were now illustrated with images, maps, and figures. And most, if not all, possessed a detailed, summarising table of contents which helped the readership through the various and sometimes surprisingly diverse essays. The translations were taken from the German 1849 edition, which offered – in contrast to earlier editions – an extensive survey of about a dozen pages in each volume. The British translations went a step further and included an alphabetical index. Humboldt did not believe that *Views of Nature* could receive much attention with an English audience. In a letter to a friend, he noted that the book, although he had enriched it with a lot of science, was completely aimed to the "German way of feeling" ("deutsche Gefühlsweise") and that the English were more rigorous in this matter. Shortly after, however, he remarked that the English version was better received than expected. Indeed, the translation by E. J. Sabine was reprinted in three further editions after 1849.[10] A notable feature of later editions concerns the relationship between the essays and the scholarly annotations. While the latter had been extended considerably during the first half of the 19th century, in the 20th century they were scaled back. A popular German edition, first published in 1969, included an extremely flattering postscript on the prominent compatriot and completely removed the annotations in which the very same person

10 Alexander Humboldt, *Briefe von Alexander von Humboldt an Christian Carl Josias Freiherr von Bunsen* (Leipzig: Brockhaus, 1869), 115–16, 119; Fiedler and Leitner, *Alexander von Humboldts Schriften*, 46–47.

had invested a great deal of effort. The latest American edition is more respectful skipping the scholarly part only in the title, not in the contents.[11]

Our main purpose here is to employ *Views of Nature* as a starting point to shed light on the perception of deserts and mountains. The particular form of two parallel texts is very appropriate. Whereas the essay level leans towards subjectivity, emotional expression and solemn literary language, almost cleansed from references, the annotation level is on the objective side with a more value-neutral outward look. Together, the two levels reflect a broad spectrum of environmental discourses and attitudes. Humboldt was quite conscious of the difficulties of the "dual direction of this text" which he sketched in the last preface as "a painstaking effort to heighten the enjoyment of Nature through living depictions, while simultaneously increasing insight into the harmonious cooperative effect of forces according to the state of scientific understanding of the time." In the first preface the difficulties were highlighted with an ostentatious self-critique. "In appealing to feeling and fancy," Humboldt wrote in 1808, "style easily degenerates into poetic prose," of which the following pages, regrettably, would offer manifold examples.[12] As we have seen, this point was taken up by a German reviewer criticising the high-tone literary style. An English reviewer disliked the dual organisation. According to him, Humboldt should have "moulded his materials into one uniform and consistent mass, instead of detaching such a large and valuable portion of them into the form of supplementary annotations."[13]

For Humboldt's career, playing two games at the same time was probably helpful, particularly in Germany. Besides showing his qualities as a naturalist, it positioned him near the distinguished world of poets. Moreover, the use of his native tongue instead of the dominant French language was advisable for political reasons. For a Prussian and indeed a chamberlain of the Prussian King, the creation period of *Views of Nature* was troubled. On 27 October 1806, Napoleon entered Berlin after his victory in two battles. On 29 January 1807, Humboldt read the first lecture about arid lands leading to his book. When the book was in print one year later, the *Grande Armée* was still in town. Humboldt's preface reflected the situation in an allusive yet unmistakable way: "To embattled minds particularly, these pages are dedicated. 'Who saves himself from

11 Alexander von Humboldt, *Ansichten der Natur*, ed. Adolf Meyer-Abbich (Stuttgart: Reclams Universal-Bibliothek, 1969); Humboldt, *Views of Nature*.

12 Humboldt, *Views of Nature*, 25, 27; see also Thomas Richter, *Alexander von Humboldt: "Ansichten der Natur". Naturforschung zwischen Poetik und Wissenschaft* (Tübingen: Stauffenburg, 2009), 58–66.

13 *Göttingische Gelehrte Anzeigen*, 2049–50; *Monthly Review*, 474–75 (about the 1808 French version).

life's stormy wave' will follow me gladly into the thickets of the forest, into the immeasurable steppes, and out upon the spine of the Andes range."[14]

3 Perception of Deserts

In the first essay of *Views of Nature*, Humboldt recalled travel impressions of the steppe-like Llanos in Venezuela, which he visited in 1800, and then proceeded to global comparisons. While the preface promised his readers "enjoyment of Nature," here they had to be content with dark scenery and dramatic animal fights. The entrance to the semi-arid lands is depicted as a transition to another world: "From the luxuriant fullness of organic life, the astonished wanderer comes to the barren edge of a sparse and treeless desert. No hill, no cliff rises as an island in this incalculable space." The unbounded space reminds the wanderer of an ocean, and both steppes and oceans provide a feeling of infinity. "But while the clear ocean surface in which ripples the graceful, softly foaming wave is a friendly sight, dead and stiff lies the steppe, stretched out like the naked rocky crust of a desolate planet." This entrance sets the tone for the rest of the essay. Arid lands are dismal areas with few or no people. They constrict the mood of the visitor. Everywhere drought announces death. The Sahara, in particular, is a "terrible sea of sand." And shepherd populations can be dangerous. The Mongols and other peoples have brought "calamity and devastation across the globe." More precisely: "Over the course of the centuries, whenever early intellectual culture has travelled like revitalizing sunlight from east to west, so too have barbarism and rawness of custom subsequently threatened to creep over Europe like a fog."[15]

Humboldt's 1800 journey across the Llanos lasted two and a half weeks. Probably this short "desert" experience was one of the reasons that his essay was comparative and based on literature to a large degree. The method was described as both productive and demanding: "It is a rewarding (if difficult) exercise in general regional geography to compare the natural properties of remote regions to one another and to describe the results of this comparison in concise terms." The German reviewer of 1808, quoted above, believed that the exercise was misleading, not rewarding, in this case. The arid or semi-arid areas across the globe, included in *Views of Nature*, were too diverse for useful analysis.[16] These methodological tensions between generalising and individualising

14 Humboldt, *Views of Nature*, 25.

15 Humboldt, *Views of Nature*, 29–31, 34, 36–38.

16 Humboldt, *Views of Nature*, 33; *Göttingische Gelehrte Anzeigen*, 2049–50.

approaches accompanied Humboldt throughout his life. It is known that the global perspective remained central to his thinking and his success. When speaking of a specific scholarly style – a "Humboldtian Science" – historians insist on his planetary research strategy.[17] With the booming production of world atlases involving Humboldt's support, or only his name, that aspect received a visual expression. For the problem of aridity, for example, readers could refer a global map of precipitation in Berghaus's *Physikalischer Atlas*.[18] The *Views of Nature* allude repeatedly to the methodological tension. In the essay on volcanoes, added to the 1826 edition, Humboldt shows how the scholarly focus on the small volcanoes Vesuvius and Etna had produced erroneous assumptions – to be corrected by the inclusion of the huge volcanoes in South America and other parts of the world. In the essay on nocturnal wildlife in the rainforest, added to the 1849 edition, he reflects on the relationship between language and "natural truth" ("Naturwahrheit"). In contrast to the other strategy, a "limiting individualization" ("beschränkende Individualisierung") is presented as suitable to achieve natural truth.[19]

In her recent survey on *The Arid Lands*, desert ecologist and historian Diana K. Davis points out that Humboldt, from the first to the second edition, made a seemingly small yet significant change in *Views of Nature*. When describing the African situation, he added a further element to the conditions of the "terrible sea of sand," highlighted here with italics: "Associated with the effect of hot land breezes in Africa, to the extent that we are familiar with it, is the lack of large rivers, *of forests that exhale water vapor and create a cooling effect*, and of high mountains." Humboldt did not change the text level of his book easily. In this case, after 1808, he must have seen that forests should be mentioned in the desert context as a counteracting force.[20] For explanation, Davis refers to a generalising passage in his travel account about Venezuela, published in 1819: "By felling the trees that cover the tops and the sides of mountains," Humboldt asserts, "men in every climate prepare at once two calamities for future generations; the want of fuel and a scarcity of

17 See e.g. Susan Faye Cannon, *Science in Culture: The Early Victorian Period* (New York: Dawson, 1978), 73–100.

18 Heinrich Berghaus, *Physikalischer Atlas zu Alexander von Humboldt, Kosmos. Entwurf einer physischen Weltbeschreibung*, ed. Ottmar Ette and Oliver Lubrich (Frankfurt am Main: Eichborn, 2004, 1st edition 1845–1848), 20–21.

19 Humboldt, *Views of Nature*, 141–42, 243–46.

20 Diana K. Davis, *The Arid Lands: History, Power, Knowledge* (Cambridge MA: The MIT Press, 2016), 82–86; Humboldt, *Views of Nature*, 34, with Alexander von Humboldt, *Ansichten der Natur mit wissenschaftlichen Erläuterungen* (Tübingen: J. G. Cotta'scher Verlag, 1808, 1st edition; Stuttgart: J. G. Cotta'scher Verlag, 1826, 2nd ed.), 18.

water." Trees cool down the climate and diminish the evaporation of soils by sheltering them from the sun. When forests are destroyed the springs dry up or become less abundant. The dry beds of rivers are converted into torrents whenever great rainfalls commence which, in turn, causes erosion and inundations. Hence, the destruction of forests, the want of permanent springs and the existence of torrents are three closely connected phenomena. For proof, Humboldt refers to the Alps and the study of a director of the *École nationale des ponts et chaussées* about floods in the Po Valley.[21]

Davis puts the Humboldtian account, and its change, into a long-term approach to the perception of deserts in the West. Her centrepiece is the politically driven colonial assumption that arid regions had been laid waste by "traditional" forms of land use. Originally – it was assumed – the regions had been covered and cooled down by forests. However, "barbarous" mobile forms of pastoralism turned them into the "desolate" arid state which the colonial officers found in North Africa and elsewhere, and which should be recovered by "improvement" through European style agriculture or "reforestation." In the Christian tradition, deserts had been ambivalent spaces of temptation and deception, or of perfection as with the monastic movement of the "Desert Fathers" of the 3rd and 4th centuries. But they were mostly considered a simple fact of life on earth – not a degenerated part – in continuation of classical views. During the early modern period, when European powers expanded and scholarly discourse intensified, the association between changes of landscape and climate grew in importance and enhanced the assumed role of human action in environmental transformation. By the mid-18th century, the desiccation belief formed a rather coherent ensemble of thought: arid lands have been abused through deforestation producing a decrease in rainfall and hence the necessity of restoration. During the 19th century, then, this belief dominated policies in Europe and the colonies to a considerable degree. Some of the policies were socially repressive and ecologically damaging. Recent arid area research has shown that the variability of precipitation and the long-term sustainability of pastoralism have been clearly underrated by enlightened, colonial and postcolonial scholarship.[22]

---

21 Alexander von Humboldt and Aimé Bonpland, *Personal Narrative of Travels to the Equinoctial Regions of the New Continent During the Years 1797–1804*, vol. 4 (London: Longman, Hurst, Rees, Orme, and Brown, 1819), 142–43; the footnoted French author was Gaspard de Prony.

22 Davis, *The Arid Lands: History, Power, Knowledge*, chapters 2–4; see also Diana K. Davis, *Resurrecting the Granary of Rome: Environmental History and French Colonial Expansion in North Africa* (Athens OH: Ohio University Press, 2007); Davis, "Deserts," in *The*

When Humboldt put forward a desolate image of deserts in 1808, and added a forest element thereto in 1826, he was in line with ideas of his time, particularly in France. During his long stay in Paris (1808–1827), he maintained close contact with leading persons in academia and administration. It is quite probable that he knew François Antoine Rauch, chief of service in the department *Ponts et chaussées*. Rauch's work gives a telling example of the desiccationist approach in this milieu.[23] In 1792, the engineer and geographer, inspired by the Revolution, published his first plan for reform including elements of deforestation fear and sent it to politicians in Paris and all over France. Ten years later he came up with a two-volume work programmatically entitled *Hydro-Vegetal and Meteorological Harmony, or Research into the Means of Recreating with our Forests the Strength of the Temperatures, and the Regularity of the Seasons, by Reasoned Plantations* (my translation from French). In 1818, Rauch published an augmented version of his work. While the 1802 edition had 675 pages and was dedicated to the "Citoyen Premier Consul" (Napoleon), the latter version extended to 900 pages with a dedication to "all the Sovereigns and all the Governments." The First Consul was addressed with the following words: "Rome has only be great and illustrious under its consuls. France, restored by your military genius, shining of glory, from the ruins of the most astonishing Revolution ever happened in the annals of humankind, enters the consulate lustres in view of a most brilliant destiny!" After a few sentences, then, Rauch came to the point and stated that "our beautiful mountains" ("nos belles montagnes") have lost their majestic forests that protected them from devastation. Chapter 1 instructed the readers in general terms on the benign effects of woods. Chapter 2 presented a panorama of the sad effects of deforestation observed in Asia and Africa and of the alarming signs in Europe.[24]

Oxford Handboook of Environmental History, ed. Andrew C. Isenberg (New York: Oxford University Press, 2014).

23 Rauch's 1802 book is discussed by Davis, *The Arid Lands: History, Power, Knowledge*, 64–69; a direct link to Humboldt cannot be shown so far.

24 François Antoine Rauch, *Harmonie hydro-végétale et météorologique, ou recherches sur les moyens de recréer avec nos forêts la force des températures et la régularité des saisons, par des plantations raisonnées*, 2 vols. (Paris: Levrault 1802); Raphaël Larrère, "Les utopies de François Antoine Rauch: ou comment sortir de la physiocratie, tout en maintenant l'homme dans la nature," in *Nouvelles sciences: modèles techniques et pensée politique de Bacon à Condorcet*, ed. Franck Tinland (Seyssel: Champ Vallon, 1998).

4 Perception of Mountains

Nos belles montagnes – Rauch was one of many administrators in France who wanted to restore them by extending their forest cover. Drawing on earlier discourses, during the first half of the 19th century, the general concept emerged that the mountains of the nation should belong to forests as the plains to agriculture. According to the concept, mountain torrents damaged the plains, and were caused by using slopes for pasturing animals, not for cultivating trees (studies of our time would show that extraordinary rainfalls played the dominant role). Non-forested mountain stretches were considered a recent phenomenon, produced by an ignorant, predatory peasantry (in reality they went back to early modern and medieval population processes). An alarmist atmosphere called for state intervention by pushing the peasants off their communal lands and claiming them for the national territory. In a first period, things took their course by legislation and by trying to enforce laws with a military-style administration. Afterwards active reforestation in often rough mountain territory was institutionalised. A national forest code passed in 1827, and a special Alpine reforestation law in 1860. Administrative wisdom (in contrast to contemporary botanical knowledge) did not suggest that reforestation could be a problem above a certain altitude. It seemed possible everywhere, except for the rocky summits. According to a detailed study of Tamara L. Whited, mountain peasants tried to resist these inroads right from the beginning in many ways. She compares the Alpine situation to French colonial politics in Algeria. The battles between state power, obsessed with forest ideology, and dry-land pastoralists had striking parallels, yet the intervention was much harsher in subdued North Africa.[25]

As mentioned above, Humboldt was in line with many enlightened ideas about forests in desert and mountain contexts. However, he did not blame pastoralists for deforestation in *Views of Nature*, and perhaps would have been opposed to the crude policies of the late 19th century. But he was not interested in the two ecosystems in the same way: the arid zones attracted him for some time, whereas he had a lifelong affair with mountains.[26] It started with a study at the mining academy in Saxony (1791), continued with the appointment as a mining inspector (1792–1796) and several visits to the Alps (1792, 1795, 1797–1798), and climaxed with his attempt to reach the top of the Chimborazo

---

25 Tamara L. Whited, *Forests and Peasant Politics in Modern France* (New Haven: Yale University Press, 2000), particularly 53–67, 199–211; see also Davis, *Resurrecting the Granary of Rome*, 72–79.

26 Mathieu, "Von den Alpen zu den Anden"; Mathieu, *The Third Dimension*, 23–32.

volcano in Ecuador, believed to be the highest summit of the world (1802). In spite of Humboldt's varied interests, the upland image stuck to him for decades. A famous portrait, painted in the year of his death, shows him as an old man in front of Chimborazo.[27]

Interestingly, in *Views of Nature*, the principal mountain plot is to be found at the annotation level and is very much centred on measuring elevation above sea level. The reader can participate in the expanding quantitative description of the world's third dimension. In the 1849 edition, this speciality of Humboldt remained no longer confined to endnotes, but emerged also at the upper essay level as *Hypsometric Addenda* (see table 9.1). During the forty years between the first and the third editions, the Chimborazo, step by step, lost its former position as the highest summit. After his return from South America, Humboldt announced by personal letter to the King of Prussia that he had climbed "the highest mountain ranges of the world, the Cordilleras de los Andes."[28] In 1808, in spite of opposite views, he still defended that opinion. He described how he had measured the Andean summit, and maintained that the "Himalah" mountains in Asia could perhaps be only slightly higher than the Etna in Sicily. By 1826, it was clear that the Asian mountain ranges surpassed the Andes. Humboldt, based on more information, distinguished four of them; perhaps "Pic Dhawalagiri" was their highest summit. In 1849, he included North America in his account, and for South America he had now more comprehensive reports, showing that even there, things had changed. According to Robert Fitz-Roy and Charles Darwin, the explorers on HMS Beagle, Aconcagua was the top Andean summit.[29] Thus globalisation was both a blessing and a curse for Humboldt. His dreams of measuring the planet were realised with the work of followers, relativising by the same action his own physical achievement of 1802.

On the side of environmental "enjoyment," Humboldt's book offers two mountain views. The first was rather conventional and adopted the emphatic language of the European enlightenment and romanticism. "In the mountains is freedom! The breath of the tomb / Cannot climb up to the purest air's home, /

27 By Julius Schrader, today in the Metropolitan Museum of Art, New York.

28 Otto Krätz, *Alexander von Humboldt: Wissenschaftler – Weltbürger – Revolutionär* (Munich: Callwey Verlag, 1997), 182.

29 Humboldt, *Ansichten der Natur mit wissenschaftlichen Erläuterungen*, 75–77, 208–12 (1808); 10–96 (1826, vol. 1); 60–72, 149–53 (2014 (1849)); at the end, Humboldt did not find mountaineering really important anymore: "Unfortunately, as I develop more completely in another work, these mountain expeditions beyond the line of everlasting snow (as much as they capture the imagination of the public) are of very little scientific value!" (ibid., 172).

The world is perfect anywhere, / If Humanity's anguish has not entered there." Humboldt borrowed these verses from a drama of Schiller, published five years earlier, and offered them in the preface to the "embattled minds" of the readers to distract them from political sorrows (see section 2). When appealing to landscape painting or to personal experience of his audience, the conventionality of this kind of mountain imagery becomes particularly clear. Humboldt refers to the "Swiss Nature" model of landscaping, and presents, in another work, a picture of the Chimborazo with hints to the Alps: "The travellers who have seen from near the summits of Mont-Blanc and of Mont-Rose are the only ones capable of appreciating the character of this impressive, calm and majestic scenery."[30] The second view went beyond that restricted European experience and consciously stressed the restriction. Humboldt believed that the "enjoyment of Nature" was enhanced by biological and physical diversity. Steppes and deserts were mostly monotonous, while the highest degree of diversity was offered by tropical mountains, his favourite ecosystem. Vegetation in this zone was most lush, and the vertical botanical succession provided further enrichment.[31]

The social conventions of decidedly positive attitudes towards mountains, adopted and enlarged by Humboldt, emerged in the second half of the 18th century. The new attitudes did not come out of nowhere. In earlier periods the mountain imagery was often ambivalent, with shifting mixtures of negative and positive elements. After 1750, the rise of enlightened ideas in leading sections of the European societies prompted a relative increase of positive elements, and above all, a massive increase in the attention devoted to mountains. They were now seen as an embodiment of nature with which society should reconcile. In a first period the move to the mountains, and particularly to the European Alps, had been sustained by naturalists. Later on, other travellers and cultural protagonists joined them, leading to the birth of nature tourism. The current vocabulary for mountain scenery offered a panoply of adjectives, ranging from "picturesque" and "romantic" to "grand" and "sublime." The discourse about grand and sublime nature often had spiritual overtones. A certain

30 Humboldt, *Views of Nature*, 160; Alexander von Humboldt, *Vues des Cordillères, et Monumens des Peuples indigènes de l'Amérique* (Paris: F. Schoell, 1810), 200 (it is the picture reproduced in the English edition, figure 3).

31 See e.g. Humboldt, *Views of Nature*, 167–69; in his late work *Cosmos*, Humboldt integrated this view in an approach to a theory of environmental perception, see Alexander von Humboldt, *Kosmos: A General Survey of the Physical Phenomena of the Universe*, vol. 1 (London: H. Baillière, 1845), 6–16.

sacralisation was one of the tendencies within this modern aesthetic, emotional view of mountains.[32]

How far did the ennoblement of nature by "sublimity" extend? Mountains were included from the beginning; and so, certainly, were seas and oceans.[33] But what about arid zones – did they deserve the title? Diana K. Davis mentions the restricted possibility of the desert sublime. In the Christian tradition, certain positive opinions went back to the "Desert Fathers" and their monastic asceticism (see section 3). According to circumstances, positive images could appear occasionally also after the spread of the desiccationist paradigm. In the early 20th century, a group of idealistic British, taken in by the mysterious world of "Arabia," for a while romanticised the local nomads instead of blaming them for a barbarous and damaging way of life.[34] Another interesting case in point concerns the integration of the Far Western landscape of the United States in the repertoire of her national culture. In a first moment, environmental perception in North America depended heavily on European aesthetic ideals. The parched, arid regions of the Southwest did not fit into these categories. Explorers spoke of "dreary," "dismal," even "revolting" regions. It was only around the turn of the 20th century when perception changed and some of these lands, judged by new standards, became popular and celebrated.[35]

5 Conclusions

Environmental perceptions are often formed by an overlay of older and newer images. They are mostly of different intensity. Some of them can be considered social conventions or fashions, some leave more space for individual attitudes. Thus, complexity seems to be the rule in many regards. Nonetheless, it is important to delineate the main historical trajectories in the understanding of nature. This chapter has argued that the perception underwent a process of divergence between deserts and mountains in the transition to Western

32 Jon Mathieu, "The Sacralization of Mountains in Europe during the Modern Age," *Mountain Research and Development* 26, no. 4 (2006); there are quite a few recent accounts of the history of mountain imagery; most of them dismiss the earlier black-and-white approach (change from absolute negative to absolute positive images).

33 John Mack, *The Sea: A Cultural History* (London: Reaktion Books, 2011), 95–99.

34 Davis, *The Arid Lands: History, Power, Knowledge*, 38–39, 115–16; see also Yi-Fu Tuan, "Desert and Ice: Ambivalent Aesthetics," in *Landscape, Natural Beauty and the Arts*, ed. Salim Kemal and Ivan Gaskell (Cambridge, New York: Cambridge University Press, 1993), 145–46.

35 Anne Farrar Hyde, *An American Vision: Far Western Landscape and National Culture, 1820–1920* (New York: New York University Press, 1990).

modernity. Up to the mid-18th century, the two major ecosystems, in public discourse, were both considered ambivalent spaces with no truly generalised values attached. Later, when "modernity" took command, deserts were clearly devalued and mountains clearly elevated. The process of divergence can be roughly fixed to the period between 1750 and 1850 or 1900.

For analytical reasons, in this discourse, we can discern technical (or technocratic) and aesthetic (or culturalist) directions and arguments. They were driven by population and economic growth, state-building, political expansion, and intensification of research and education in the 18th and 19th centuries. The two directions can be considered analogous to the two directions of Humboldt's *Views of Nature*. This book, published in a rather improvised manner in 1808 and augmented and revised in 1826 and 1849 by the author, comprised two parallel texts: the essays were written in an aesthetic literary style with numerous valuing adjectives, whereas the endnotes included long and rather technical information about the landscapes, plants, animals and cultural monuments described. This double organisation was first criticised by naturalists (rejecting the high-tone style) and later by literary scholars (cancelling the annotation level). For our purpose, however, it is a good starting point.

The distinction between technical and aesthetic discourses can help to explain the divergence between desert and mountain perception. The key notion for important sections of the contemporary elite was "improvement." Deserts had to bloom through European style agriculture or through "restoration" of forests, falsely assumed to have covered the regions before pastoralism had "devastated" them. The same technocratic, forest-centric ideology aimed at "restoring" the mountains of the nation, "damaged" by upland pastoralism. This concerned particularly France with its emphasis on territorial homogenisation, yet it permeated other countries as well.[36] In the mountain case, the negative technical discourse was overshadowed by a much broader aesthetic discourse celebrating the glory of the summits – a central location of sublimity. In the desert case, sublimity, to a certain degree, was a possibility, too. But it was strongly overshadowed by the negative technical view.

36 For Germany, on the Bavarian example: Richard Hölzl, *Umkämpfte Wälder: Die Geschichte einer ökologischen Reform in Deutschland 1760–1860* (Frankfurt am Main: Campus, 2010), 16–17, 44–48, 473–85, 496.

Bibliography

Printed Sources

Göttingische Gelehrte Anzeigen unter der Aufsicht der königl. Gesellschaft der Wissenschaften, 28 December 1809, 2049–2056 (Review of Humboldt 1808).

Humboldt, Alexander von. *Ansichten der Natur mit wissenschaftlichen Erläuterungen*. Tübingen: J. G. Cotta'scher Verlag, 1808, 1st ed.; Stuttgart: J. G. Cotta'scher Verlag, 1826, 2nd ed.; Stuttgart: J. G. Cotta'scher Verlag, 1849, 3rd ed.

Humboldt, Alexander von. *Tableaux de la nature, ou considérations sur les déserts, sur la physionomie des végétaux, et sur les cataractes de l'Orénoque*. Translated by J. B. B. Eyriès, 2 vols. Paris: F. Schoell, 1808.

Humboldt, Alexander von. *Vues des Cordillères, et Monumens des Peuples indigènes de l'Amérique*. Paris: F. Schoell, 1810.

Humboldt, Alexander von. *Kosmos: A General Survey of the Physical Phenomena of the Universe*. Vol. 1. London: H. Baillière, 1845.

Humboldt, Alexander von. *Briefe von Alexander von Humboldt an Christian Carl Josias Freiherr von Bunsen*. Leipzig: Brockhaus, 1869.

Humboldt, Alexander von. *Ansichten der Natur*, edited by Adolf Meyer-Abbich. Stuttgart: Reclams Universal-Bibliothek, 1969.

Humboldt, Alexander von. *Views of Nature*. Translated by Mark W. Person, edited by Stephen T. Jackson and Laura Dassow Walls. Chicago, London: University of Chicago Press, 2014.

Humboldt, Alexander von and Aimé Bonpland. *Personal Narrative of Travels to the Equinoctial Regions of the New Continent During the Years 1797–1804*. Vol. 4. London: Longman, Hurst, Rees, Orme, and Brown, 1819.

Monthly Review 59 (1809), 473–83 (Review of Humboldt 1808).

Rauch, François Antoine. *Harmonie hydro-végétale et météorologique, ou recherches sur les moyens de recréer avec nos forêts la force des températures et la régularité des saisons, par des plantations raisonnées*. 2 vols. Paris: Levrault, 1802.

Literature

Berghaus, Heinrich. *Physikalischer Atlas zu Alexander von Humboldt, Kosmos. Entwurf einer physischen Weltbeschreibung*, edited by Ottmar Ette and Oliver Lubrich. Frankfurt am Main: Eichborn, 2004; 1st ed. 1845–1848.

Biermann, Kurt-R. and Ilse Jahn. *Alexander von Humboldt. Chronologische Übersicht über wichtige Daten seines Lebens*. Berlin: Akademie-Verlag, 1983.

Cannon, Susan Faye. *Science in Culture: The Early Victorian Period*. New York: Dawson, 1978.

Davis, Diana K. *Resurrecting the Granary of Rome: Environmental History and French Colonial Expansion in North Africa*. Athens: Ohio University Press, 2007.

Davis, Diana K. "Deserts." In *The Oxford Handboook of Environmental History*, edited by Andrew C. Isenberg, 108–32. New York: Oxford University Press, 2014.

Davis, Diana K. *The Arid Lands: History, Power, Knowledge*. Cambridge MA: The MIT Press, 2016.

Fiedler, Horst and Ulrike Leitner. *Alexander von Humboldts Schriften. Bibliographie der selbständig erschienenen Werke*. Berlin: Akademie Verlag, 2000.

Hölzl, Richard. *Umkämpfte Wälder: Die Geschichte einer ökologischen Reform in Deutschland 1760–1860*. Frankfurt am Main: Campus, 2010.

Hyde, Anne Farrar. *An American Vision: Far Western Landscape and National Culture, 1820–1920*. New York: New York University Press, 1990.

Krätz, Otto. *Alexander von Humboldt: Wissenschaftler – Weltbürger – Revolutionär*. Munich: Callwey Verlag, 1997.

Larrère, Raphaël. "Les utopies de François Antoine Rauch: ou comment sortir de la physiocratie, tout en maintenant l'homme dans la nature." In *Nouvelles sciences: modèles techniques et pensée politique de Bacon à Condorcet*, edited by Franck Tinland, 143–57. Seyssel: Champ Vallon, 1998.

Lubrich, Oliver, ed. *Alexander von Humboldt. Das graphische Gesamtwerk*. Darmstadt: Lambert Schneider, 2014.

Mack, John. *The Sea: A Cultural History*. London: Reaktion Books, 2011.

Mathieu, Jon. "Alpenwahrnehmung: Probleme der historischen Periodisierung." In *Die Alpen! Zur europäischen Wahrnehmungsgeschichte seit der Renaissance – Les Alpes! Pour une histoire de la perception européenne depuis la Renaissance*, edited by Jon Mathieu and Simona Boscani Leoni, 53–72. Bern, Berlin, Bruxelles: Peter Lang, 2005.

Mathieu, Jon. "The Sacralization of Mountains in Europe during the Modern Age." *Mountain Research and Development* 26, no. 4 (2006): 343–49.

Mathieu, Jon. "Von den Alpen zu den Anden: Alexander von Humboldt und die Gebirgsforschung." In *Wissenschaft – Berge – Ideologien. Johann Jakob Scheuchzer (1672–1733) und die frühneuzeitliche Naturforschung–*, edited by Simona Boscani Leoni, 293–308. Basel: Schwabe, 2010.

Mathieu, Jon. *The Third Dimension: A Comparative History of Mountains in the Modern Era*. Cambridge: White Horse Press, 2011.

Richter, Thomas. *Alexander von Humboldt: "Ansichten der Natur". Naturforschung zwischen Poetik und Wissenschaft*. Tübingen: Stauffenburg, 2009.

Tuan, Yi-Fu. "Desert and Ice: Ambivalent Aesthetics." In *Landscape, Natural Beauty and the Arts*, edited by Salim Kemal and Ivan Gaskell, 139–57. Cambridge, New York: Cambridge University Press, 1993.

Whited, Tamara L. *Forests and Peasant Politics in Modern France*. New Haven: Yale University Press, 2000.

Wulf, Andrea. *The Invention of Nature: The Adventures of Alexander von Humboldt, the Lost Hero of Science*. London: John Murray, 2015.

CHAPTER 10

Alpine Landscapes of Health: The Swiss Whey Cure and Therapeutic Tourism between 1750 and 1870

Barbara Orland

1 Introduction: Alps, Health, and Knowledge

The notion that landscapes influence human health can be traced back at least to the Hippocratic treatise *On Airs, Waters, and Places*.[1] The author explained how the environment could cause certain illnesses, and he assumed that people inherit certain physical and mental features from their environment, e.g. differences in character or complexion. People who "inhabit a country which is mountainous, rugged, elevated, and well-watered, and where the changes of the seasons are very great" are likely to be tall, and have character qualities of endurance, ferociousness and courage.[2] Apart from this, all living beings should keep a constant balancing of the so-called six *Res Non-Naturales* (air; food and drink; excretions; sleep; movement and rest; and emotions) in order to serve their health and well-being. Prevention was one of the cornerstones of ancient medicine.[3]

This Hippocratic association between environment and health, further elaborated through the medical and natural philosophical discourses of more than a millennium,[4] was still accepted by numerous descriptions of the Alpine territory of the late 18th and early 19th century. An unnamed author of the German *Morgenblatt für gebildete Leser* wrote, "that at an elevation everything is finer and better than in the lowlands. He [Hippocrates, BO] proves that as a rule the mountain-dwellers are of stronger and nobler character, and that also

1 Hippocrates, *On Airs, Waters, and Places*, trans. Francis Adams, accessed on 3 July 2018. http://classics.mit.edu//Hippocrates/airwatpl.html.

2 Ibid.

3 Sandra Cavallo and Tessa Storey, eds., *Conserving Health in Early Modern Culture. Bodies and Environments in Italy and England* (Manchester: Manchester University Press, 2017); Albrecht Classen, ed., *Bodily and Spiritual Hygiene in Medieval and Early Modern Literature: Explorations of Textual Presentations of Filth and Water* (Berlin, Boston: De Gruyter, 2017).

4 David Cantor, ed. *Reinventing Hippocrates* (Ashgate: Aldershot, 2002); Frederick II Sargent, *Hippocratic Heritage: A History of Ideas about Weather and Human Health* (New York: Pergamon Press, 1982).

the air, water, plants, and all other objects in nature are purified and strengthened by the influence of elevation."[5] So strong, in fact, was the link between environment (particularly meteorology) and medicine that the term "character" occasionally retained its ancient medical significance, and – in the case of the Alpine herdsmen – was transformed into a kind of racial trope. The British journalist Frederic Shoberl wrote:

> The herdsmen of Appenzell, accustomed to live continually in mountain solitudes, have something wild in their appearance. A jacket and trousers constitute almost the whole of their dress, and milk is their chief subsistence. The purity of the air and water of their mountains, the aromatic qualities of the milk on which they live, the liberty which they enjoy, and the share which they have in the government of their country, all contribute to that liveliness, which distinguishes them from the rest of the Swiss, and which frequently inspires them with original repartees.[6]

In this paper, I will elaborate how such opinions developed at least since the middle of the 18th century.[7] Not only physicians, but also innkeepers and herdsmen of the canton Appenzell were among the first to become acquainted with the neo-Hippocratic enthusiasm of the day, and assessed the mountainous

5 "Schon Hippokrates macht die Bemerkung, daß auf den Höhen alles feiner und besser sey als in den Ebenen. Er beweist, daß die Bergbewohner in der Regel kräftigere und edlere Naturen sind und daß auch auf die Luft, auf das Wasser, auf die Pflanzen und alle Gegenstände in der Natur die Höhen eben so einen läuternden und kräftigenden Einfluß üben. Auch in den Alpen hat man vielfach Gelegenheit, diese alte hippokratische Behauptung bestätigt zu finden." *Morgenblatt für gebildete Leser*, no. 3, January 3 (1849): 5.

6 Frederic Shoberl, *The World in Miniature. Switzerland, Containing a Description of the Characters, Manners, Customs, Diversions, Dress &c. of the People of that Country in General, and of the Inhabitants of the twenty-two Cantons in particular* (London: R. Ackermann, 1827), 140.

7 Such arguments could already be found in the works of Johann Jakob Scheuchzer. See e.g. Simona Boscani Leoni, "Tra Zurigo e le Alpi: le 'Lettres des Grisons' di Johann Jakob Scheuchzer (1672–1733). Dinamiche della comunicazione erudita all'inizio del Settecento," in *Die Alpen! Zur europäischen Wahrnehmungsgeschichte seit der Renaissance – Les Alpes! Pour une histoire de la perception européenne depuis la Renaissance*, ed. J. Mathieu and S. Boscani Leoni (Bern, Berlin, Bruxelles: Peter Lang, 2005). See further e.g. Daniela Vaj, "Le Bon air des Alpes et le développement des stations climatiques d'altitude," in *Bon air des Alpes: des stations climatiques au tourisme de bien-être: actes du colloque, Sierre, 7–8.10.2004 – Die gute Alpenluft: von den Luftkurorten zum Wellnesstourismus: Tagungsband, Siders, 7.–8.10.2004*, ed. HES-SO Valais (Sierre: HES-SO Valais, 2007); Daniela Vaj, "La géographie médicale et l'immunité phtisique des altitudes aux sources d'une hypothèse thérapeutique – [Medical Geography and Phthisic Immunity in the High Altitudes: The Origins of a Therapeutic Hypothesis]," *Revue de géographie alpine – Journal of Alpine Research* 93, no. 1 (2005).

environment as a reservoir of health-promoting sources. They developed the so-called *Gaisschottenkur* (goat's whey cure) that made a name for itself for about a century between roughly 1750 and the end of the 19th century. They transformed a rural environment into fashionable health resorts that became much *en vogue* among the European bourgeoisie. In his benevolent report on Swiss milk diets, one British physician reflected in 1869: "There is a general feeling in favour of Swiss whey; and if they are not in Switzerland, patients are gratified to learn that the whey is made by a native of Appenzell, and still better pleased, if the said native shows himself in his national costume."[8] Yet what was Swiss about the whey cure? How were representations of the landscape and the local habits meshed with neo-Hippocratic ideas about health and medicine? Referring to folk cultures and rural practices of milk consumption, travellers to the Swiss mountains most often emphasized the importance of local knowledge in their search for therapeutic landscapes.[9] However, as will be shown in this paper, the whey cure that laid the foundation for the establishment of Swiss health resorts was in fact an invention of the late 18th century. Previously whey had been considered an unattractive substance for nourishment and dairy business.

Until now historians have mentioned the Swiss whey cure as an anecdote in the history of Alpine tourism.[10] When the Swiss mountains became a tourist destination in the late 18th and early 19th century, very soon the territory and its climate were detected as a "therapeutic landscape" with fresh air, healing herbs, and milk as an "extraordinary fluid."[11] Historians have described how the interest in the sublime mountain landscapes attracted visitors on their Grand Tour, and as a romantic condemnation of urban living conditions in cities that

8 John Macpherson, *The Baths and Wells of Europe: Their Action and Uses, with Hints on Change of Air and Diet Cures* (London: Macmillan and Co, 1869), 320.

9 See e.g., Gabriel Rüsch, *Heiden und seine Molkenkuranstalt im Freihof* (Trogen: Schläpfer, 1854), 17.

10 Susan R. Barton, *Healthy Living in the Alps: The Origins of Winter Tourism in Switzerland, 1860–1914* (Manchester: Manchester University Press, 2008), 8–12; Hans Peter Treichler, "Molken, Milch und Traubenberge. Neue Ansätze im Kurwesen," in *Zauberberge. Die Schweiz als Kraftraum und Sanatorium*, ed. Felix Graf and Eberhard Wolff (Baden: hier + jetzt, 2010).

11 Heini Hofmann, *Gesundheits-Mythos St. Moritz: Sauerwasser, Gebirgssonne, Höhenklima* (St. Moritz: Montabella-Verlag, 2011), 32. Therapeutic landscape is a concept mainly used by geographers and anthropologists, see e.g. Ulrich Gebhard and Thomas Kistemann, eds., *Landschaft, Identität und Gesundheit. Zum Konzept der therapeutischen Landschaften* (Wiesbaden: Springer VS, 2016).

were said to be breeding grounds for diseases.[12] It is a commonly held view that some places in the Alps already in the 15th and 16th century had a reputation as places of healing waters, firstly as an aristocratic life style and then as medical places that should attract the ill and convalescents of all classes.[13] The spa of the later 18th century played an important role in the erosion of the barriers between the classes, historian Karl Wood argued, it was influential in the development of a new *Bürgertum*.[14]

And yet, despite many details, most narratives on the Alpine bathing traditions and health resorts tend to treat the landscape as a kind of precondition of businesses that had an independent existence beyond the realm of human meanings. Primarily, class and social relations anchor the spa histories, and are combined with enquiries into the spa as a commercial enterprise. Few studies mention that early modern bathing manuals conventionally start their content with a description of the location, or ask how the healing waters were generated from the surrounding soil, received the "*visceribus terrae*" and the properties of minerals.[15] Little interest has been directed to the authority of natural history descriptions of Alpine nature, the medical discourses and scientific investigations of Alpine plants, minerals, waters etc., and particularly the Alps as a "space of healing and recreation."[16] By contrast, geographers and

12 Andrew Beattie, *The Alps: A Cultural History* (Oxford: Oxford University Press, 2006), 124–44.

13 Werner Vogler, "Heilbäder in den Alpen im Spätmittelalter und in der Renaissance am Beispiel von Pfäfers," in *Die Alpen als Heilungs – und Erholungsraum. Historikertagung in Meran – Le alpi: luogo di cura e riposo : convegno storico a Merano : 19.–21.X.1988*, ed. Josef Nössing (Bozen: Verlagsanstalt Athesia, 1994); Richard Palmer, "'In this our lightye and learned tyme': Italian Baths in the Era of the Renaissance," in *The Medical History of Water and Spas*, ed. Roy Porter (London: Wellcome Institute for the History of Medicine, 1990); Margrit Wyder, *Kräuter, Kröpfe, Höhenkuren: die Alpen in der Medizin – die Medizin in den Alpen; Texte aus zehn Jahrhunderten* (Zurich: Verlag NZZ, 2003), 43.

14 Karl E. Wood, *Health and Hazard: Spa Culture and the Social History of Medicine in the Nineteenth Century* (Newcastle upon Tyne: Cambridge Scholars Publishing, 2012), 19–21.

15 Frank Fürbeth, "Adaptionen gelehrten Wissens für laikale Zwecke in der Bäderheilkunde der frühen Neuzeit," in *Wissenschaftsgeschichte und Geschichte des Wissens im Dialog – Connecting Science and Knowledge: Schauplätze der Forschung*, ed. Kaspar von Greyerz, Silvia Flubacher, and Philipp Senn (Göttingen: V&R unipress, 2013), 228.

16 Christa Habrich, "Medizin- und naturwissenschaftshistorische Aspekte der Alpen als Heilungs- und Erholungsraum," in *Die Alpen als Heilungs- und Erholungsraum. Historikertagung in Meran – Le alpi : luogo di cura e riposo : convegno storico a Merano : 19.–21.X.1988*, ed. Josef Nössing (Bozen: Verlagsanstalt Athesia, 1994), 11; Simona Boscani-Leoni, ed. *Wissenschaft – Berge – Ideologien. Johann Jakob Scheuchzer (1672–1733) und die frühneuzeitliche Naturforschung – Scienza – montagna – ideologie. Johann Jakob Scheuchzer (1672–1733) e la ricerca naturalistica in epoca moderna* (Basel: Schwabe, 2010); Claude Reichler, *La montagne réinventée: Géographes, naturalistes et sociétés (XVIIIe–XXe*

anthropologists interested in the historical changes of Alpine landscapes, commonly view the mountainous landscapes as cultural constructions. To them, changes in the Alps are not merely a by-product of social, economic and political developments, but a complex system of regional experiences, practices, and symbols of identities.[17] The focus is mainly on the relation between the natural environment, economy, work, and identity – the "nationalisation of nature" that transformed the Alps into "a national mass symbol," as Oliver Zimmer put it,[18] with tourism as a rare issue in this respect.[19] Histories of Alpine agriculture in turn taught us to distinguish between the land created by agricultural work and the popular imagery of an area that tends to freeze the existing landscape as natural scenery.[20]

There remains scope for further research to complete the picture of the Alps as a landscape of health and healthcare, not only in a strict geographic sense.[21]

---

siècles) = *Reinventing Mountain Areas: Geographers, Naturalists and Society* (*18th–20th centuries*) (Grenoble: Revue de géographie alpine (Institut de Géographie Alpine), 1994); Reichler, *La découverte des Alpes et la question du paysage* (Chêne-Bourg: Georg, 2002); Silvia Flubacher, "Alpen-Tiere. Lokale Wissenswelten in der schweizerischen Naturgeschichtsschreibung," in *Wissenschaftsgeschichte und Geschichte des Wissens im Dialog – Connecting Science and Knowledge: Schauplätze der Forschung*, ed. Kaspar von Greyerz, Silvia Flubacher, and Philipp Senn (Göttingen: V&R unipress, 2013). The chemical analysis of mineral waters has attracted some interest by historians of science. See e.g. Noel G. Coley, "Physicians, Chemists and the Analysis of Mineral Waters: 'The most difficult part of Chemistry'," in *The Medical History of Water and Spas*, ed. Roy Porter (London: Wellcome Institute for the History of Medicine, 1990); Matthew Daniel Eddy, "The Sparkling Nectar of Spas, or, Mineral Water as a Medically Commodifiable Material in the Province, 1770–1805," in *Materials and Expertise in Early Modern Europe. Between Market and Laboratory*, ed. Ursula Klein and Emma C. Spary (Chicago, London: The University of Chicago Press, 2010).

17 Among a voluminous literature, see the different issues of the journal: Association Internationale pour l'Histoire des Alpes, ed. *Histoire des Alpes* (Zurich: Chronos, 1996–2018); Jon Mathieu, Norman Backhaus, Katja Hürlimann, and Matthias Bürgi, eds., *Geschichte der Landschaft in der Schweiz. Von der Eiszeit bis zur Gegenwart* (Zurich: Orell Füssli, 2016); Werner Bätzing, *Die Alpen: Geschichte und Zukunft einer europäischen Kulturlandschaft*, 4th. rev. ed. (Munich: Beck, 2015).

18 Oliver Zimmer, "In Search of Natural Identity: Alpine Landscape and the Reconstruction of the Swiss Nation," *Comparative Studies in Society and History* 40, no. 4 (1998).

19 Laurent Tissot, "From Alpine Tourism to the 'Alpinization' of Tourism," in *Touring Beyond the Nation: A Transnational Approach to European Tourism History*, ed. Eric G. E. Zuelow (Farnham: Ashgate, 2011).

20 Jon Mathieu, *Eine Agrargeschichte der inneren Alpen. Graubünden, Tessin, Wallis 1500–1800* (Zurich: Chronos-Verlag, 1992); Barbara Orland, "Alpine Milk: Dairy Farming as a Pre-modern Strategy of Land Use," *Environment and History* 3 (2004).

21 Historical research on Alpine spas concentrates on Swiss and German spas, more or less on Italy, Austria and the Eastern Alps, if authors do not just explore just one place (e.g.

Notable research on other regions highlights the extent to which scholarly debates about the cause and effect of directly observed natural phenomena influenced hygienic, dietetic and therapeutic practices.[22] As Albrecht Classen put it, throughout the centuries there existed much discussion about the clean water or fresh and healthier air of some regions, making medical experts to early environmentalists, who offered very practical advice that could be easily combined with folklore prescriptions.[23] Johann Jakob Scheuchzer's consideration of the bath of Pfäfers relied on an eclectic collection of knowledge from different sources, as Philipp Senn discovered,[24] religion contributed to the relation between popular culture and healing practices,[25] and knowledge could been imported from other regions.[26]

Taking inspiration from these works, this chapter will consider how diverse actors emphasized the constitutive force of landscape, agriculture and diet in order to allure urban visitors and patients to enjoy a drinking cure in a rural village in Switzerland. I will start my description with the case of the village of Gais in the Swiss canton Appenzell, which was said to be the first place in Switzerland where whey cures were invented in 1749. My intention is to offer a description of the methods and rhetoric used to define an Alpine health resort

Davos). Christian Schürer, Der Traum von Heilung. Eine Geschichte der Höhenkur zur Behandlung der Lungentuberkulose (Baden: hier + jetzt, 2017).

22 Raingard Esser and Thomas Fuchs, eds., *Bäder und Kuren in der Aufklärung: Medizinaldiskurs und Freizeitvergnügen* (Berlin: Berliner Wissenschafts-Verlag, 2003); Elizabeth Neswald, "Asserting Medical Identities in Mid-Nineteenth Century Ireland: the Case of the Water Cure in Cork," in *Science and Technology in Nineteenth-Century Ireland*, ed. Janina Adelman and Eadaoin Agnew (Dublin: Four Courts Press, 2010).

23 Albrecht Classen, "Introduction: Bathing, Health Care, Medicine, and Water in the Middle Ages and Early Modern Age," in *Bodily and Spiritual Hygiene in Medieval and Early Modern Literature: Explorations of Textual Presentations of Filth and Water*, ed. Albrecht Classen (Berlin, Boston: De Gruyter, 2017), 76.

24 Philipp Senn, "Forscher vor Ort. Johann Jakob Scheuchzer (1672–1733), Bündner Gönner und die Balneologie," in *Wissenschaftsgeschichte und Geschichte des Wissens im Dialog – Connecting Science and Knowledge: Schauplätze der Forschung*, ed. Kaspar von Greyerz, Silvia Flubacher, and Philipp Senn (Göttingen: V&R unipress, 2013), 276.

25 Ute Lotz-Heumann, "Finding a Cure. Representations of Holy Wells and Healing Waters in Early Modern Germany," in *Wissenschaftsgeschichte und Geschichte des Wissens im Dialog – Connecting Science and Knowledge: Schauplätze der Forschung*, ed. Kaspar von Greyerz, Silvia Flubacher, and Philipp Senn (Göttingen: V&R unipress, 2013), 233–54; Alexandra Walsham, *The Reformation of the Landscape. Religion, Identity, and Memory in Early Modern Britain and Ireland* (Oxford: Oxford University Press, 2011).

26 Vladimir Jankovic, "The Last Resort: A British Perspective on the Medical South, 1815–1870," *Journal of Intercultural Studies* 27, No. 3 (2006). See also the different case studies in: John K. Walton, ed. *Mineral Springs Resorts in Global Perspective: Spa Histories* (London, New York: Routledge 2014).

and particularly whey as a local agricultural product of therapeutic significance. The connection between Alpine landscape and health as materialized in the whey cure did, however, slowly cease after the turn of the 19th century, although businesses still went quite well for some decades. The following section considers how science affected attitudes associated with the nature of the mountainous environment and altered assumptions about the medical value of animal milk. I will close this paper by bringing some of the arguments into the foreground that were responsible for the decline of the whey cure business during the second part of the 19th century.

2 The Story of Gais and Its Invention of the Swiss Whey Cure

By the early 1800s, a story circulated throughout the European medical community of how the first whey cure had been invented in Switzerland. The physician Johann Heinrich Heim (1802–1876), a native of the village that stood at the heart of the narrative, disseminated the story like this:[27] In the year 1749, a Dr Meyer of Arbon in canton Thurgau advised his brother-in-law, a Mr Steinbrüchel of Zurich, to try a therapy with milk. The patient, who suffered from a chest disease, was *advised* to undertake the cure in the little village Gais in the Swiss canton of Appenzell. The innkeeper Hans-Ulrich Heim (1720–1814) of the *Gasthof zum Ochsen* [and grandfather of Johann Heinrich Heim] instructed local herdsmen to deliver fresh and warm goat's whey every morning. After several weeks the man was cured and returned home. News of this cure rapidly spread around medical circles, so that other doctors started to send patients to Gais. During the next decades, Gais – "a village of cheerful appearance"[28] – became a renowned centre for the so-called *Gaisschottenkur* (goat's whey cure). Further establishments were built up during the next decades, firstly in the region, among them in Weissbad, Trogen, Appenzell and Heinrichsbad near Herisau. From there, the cure – most often connoted with Swiss Alpine nature – spread to other mountain regions, mainly

27 Joh. Heinrich Heim, *Die Heilkräfte der Alpenziegen-Molken und der Molkenkurort Gais* (Zurich: Schultheß, 1844), 96, 97. See also Fr.[iedrich] K. v. Kronfels, *Gais, Weisbad und die Molkenkuren im Canton Appenzell* (Constance: W. Wallis, 1826), 39–40. Herman Weber, a physician at the German Hospital in London, disseminated the story in Great Britain: Herman Weber, "Notes on the Climate of the Swiss Alps, and on some of their Health Resorts and Spas," *The Dublin Quarterly Journal of Medical Science* 37, no. 2 (1864): 333–34.

28 Weber, "Notes on the Climate," 362.

in Switzerland, Austria, Bavaria, but Gais remained the place where the story of the overwhelming success of a specific health resort had its beginning.[29]

Before I can discuss this fashionable story of an invention, we have to rethink some of the preconditions of the spa culture. From a modern perspective, Gais needed three basic components to become a whey spa – the natural resources of sufficient pasture, animals and milk, in addition "spa" facilities that do not necessarily depend on natural resources, and a therapeutic system based on milk/whey. To begin with the last aspect, it is well known in medical history that at least since antiquity milk has been recognized as having a healthy effect on people. Physicians, healers and lay people used milk of all sorts in the regular treatment and dietary care for patients.[30] Old schemes of *materia medica* found empirically in the earliest time of humankind were largely used into the 18th century, and particularly enlightened medicine with its emphasis on Hippocratic theory and practice had become a master in the formulation of past traditions.[31] Occasionally, contemporary authors labelled milk cures as a controlled empirical therapy of the past or even quoted from Plinius, Hippocratic and Galenic texts to verify their prescriptions.[32]

---

29 There exists a wealth of printed sources on the whey spa culture of the 19th century, see, for instance, Gabriel Rüsch, *Anleitung zu dem richtigen Gebrauche der Bade- und Trinkcuren ueberhaupt, mit besonderer Betrachtung der schweizerischen Mineralwasser und Badeanstalten*, 2 vols. (Ebnat, St.Gallen: Abraham Keller, 1825, 1826); F. B. Zeller, *Die Molkenkur in Verbindung der Mineral-Bronnenkur* (Würzburg: Etlinger, 1826); H. Rheiner, *Das Moosberger oder Heinrichs-Bad im Kanton Appenzell, historisch, chemisch und topographisch beschrieben* (St. Gallen: Huber und Compagnie, 1833); Friedrich Garlichs, *Über den medizinischen Gebrauch der Milch und der Molke: Eine Inaugural-Abhandlung* (Würzburg: Becker, 1837); Vincenz Müller, *Specielle Beschreibung der Molkenkur-Anstalten des Königreichs Bayern* (Munich: Eigenverlag, 1843); J. J. Strasser, *Interlaken im Berner Oberlande als Luft- und Molkenkur-Ort, vom therapeutischen Standpunkte aus betrachtet.* (Interlaken: Kurhaus-Verwaltung, 1863); Félix Roubaud, *Les Cures de Petit-Lait en Suisse, en Allemagne, dans le Tyrol et la Styrie.* (Paris: Adrien Delahaye, 1867).

30 See Barbara Orland, "Le régime suisse du petit lait (18/19ème siècle)," in *Allaiter. Histoire(s) et cultures d'une pratique*, ed. Yasmina Foehr-Janssens and Daniela Solfaroli Camillocci (Turnhout: Brepols Publisher 2022), forthcoming. Eugen Bircher, "Die Molkenkur, ihre Geschichte und geographische Verbreitung. Eine geschichtliche Studie," *Schweizerische Medizinische Wochenschrift* 39 (1953): 937–41. See more general Jutta Gisela Sperling, ed., *Medieval and Renaissance Lactations. Images, Rhetorics, Practices* (Burlington: Ashgate, 2013).

31 See Cantor, *Reinventing Hippocrates.*

32 Georg Heinrich Behr, *Zwey Bücher von der Materia Medica, oder vollständige Beschreibung aller und jeder Arzeney-Mittel: Samt beygefügter wohl-eingerichteter und höchst-nutzbarer Therapie* (Straßburg: Johannes Beck, 1748), 241–49.

If the prescription of milk was an at least age-old practice, there is no evidence to believe that people in all European regions or living conditions used milk or its components as a *materia medica* or even as a food. In the Alpine dairy business, for instance, whey was judged as a negligible waste material, and the use of *Schotte* (whey) as a remedy was by no means self-evident. Because of its low nutritional value, the residues from the production of butter and cheese were in general not seen as human nutrition, but were merely used as animal feed. In 1706, Johann Jakob Scheuchzer argued in his *Beschreibung der Natur-Geschichten des Schweizerlands*, while the "exquisite pastures" supply man with milk, butter, and cheese, pigs live from the whey. Only to make use of the remains from making cheese do Swiss dairymen take some pigs to the summer meadows on the Alps.[33] In other words, there existed no local therapeutic regime or infrastructure that could be adapted to the organization of a whey cure intended to treat foreign people.

Years before the sensational report of the successful recovery of the patient from Zurich, however, another Appenzeller, Laurenz Zellweger (1692–1764), a trained physician and enlightened savant of this region, created the conditions sustaining the proliferation of the use of whey cures in his homeland.[34] Even at a time when the European elites had not yet learned about the healthy places of canton Appenzell, Zellweger, who lived in the village Trogen, established the goat's whey cure among his urban friends in Zurich and abroad. At least since 1734, some of his closest friends – among them several prominent writers like Johann Jakob Bodmer (1698–1783), Salomon Gessner (1730–1788), or learned physicians like Hans Caspar Hirzel (1725–1803) – visited him occasionally for uplifting conversations, mountain hikes in the Alps, and weeklong whey cures. Zellweger and his friends became known as the *Whey Brothers* or *Whey Society*, and there is some evidence that only because of this illustrious group a doctor from abroad got the idea to send a patient into a village of Appenzell. However, when the whey cure business gained momentum, the fame of the *Whey Brothers* had already faded.

---

33 Johann Jakob Scheuchzer, *Beschreibung der Natur-Geschichten des Schweizerlands*, vol. 1, no. 8 (Zurich: Michael Schaufelb.[erger]s. E.[rben] und Christoff Hardmeier, 1706), 30–31.

34 I analyse Zellweger's motivation and scientific context in: Orland, "Le régime suisse du petit lait". See also Heidi Eisenhut, "Gelehrte auf Molkenkur: Laurenz Zellweger und sein Kreis in Trogen," in *Heilkunst und schöne Künste: Wechselwirkungen von Medizin, Literatur und bildender Kunst im 18. Jahrhundert*, ed. Heidi Eisenhut, Anett Lütteken, and Carsten Zelle (Göttingen: Wallstein, 2011).

3 The Pastoral Economy of Gais and Canton Appenzell

In Appenzell[35] as everywhere else, the use of milk heavily depended on the economic value of the different dairy products, and this in turn was dependent on the relationship between agriculture and livestock production. Gais, located in the centre of Ausserrhoden, was the only village (among six others of the half-canton) that lies in a mountainous area and possessed different levels of elevation (around 900–1300m above sea level). Due to the topographical conditions, a scattered settlement emerged very early and surprised many travellers. While at the end of the 18th century visitors commented on the luscious green of the pastureland,[36] half a century before they would have found a much more diverse land use.[37] In a chronicle of 1740 it is written that in Ausserrhoden people planted a lot of grain, wheat, barley, peas, oats and flakes. Beans and other fruits would thrive there too, "so beautiful and perfect, as in the most distinguished countries of Europe."[38] Fifty years later, the cereal crop had nearly disappeared; a radical conversion of arable land into meadows and pastures had taken place. Pastures made up the bulk of the habitable surface, and the export of livestock, skins, butter and cheese dominated peasant activities. Grazing management and pastoral economy with summer Alpine farming on higher elevations had replaced all mixed modes of farming. But because there existed no market for animal feed and animals could not be fed through the winter, farmers only partly engaged in cattle breeding.[39] Rather, they fattened imported and rented livestock over the summer and sold

35 Since 1597, the canton Appenzell was divided into two subdivisions Innerrhoden and Ausserrhoden, as a result of the Swiss Reformation. Gais was one of six *Landsgemeinden* belonging to the north or half-canton Ausserrhoden, which in many aspects was quite different to its counterpart.

36 See e.g. Heinrich von Malten, *Beschreibung aller berühmten Bäder in der Schweiz. Nebst einer allgemeinen Uebersicht der Bäder zweiten Ranges und der unbenutzten Heilquellen* (Aarau: Sauerländer, 1830), 21, 48.

37 For the whole paragraph, see Hanspeter Ruesch, *Lebensverhältnisse in einem frühen schweizerischen Industriegebiet. Sozialgeschichtliche Studie über die Gemeinden Trogen, Rehetobel, Wald, Gais, Speicher und Wolfhalden des Kantons Appenzell Ausserrhoden im 18. und frühen 19. Jahrhundert*, Basler Beiträge zur Geschichtswissenschaft, vol. 139/140 (Basel, Stuttgart: Helbing & Lichtenhahn, 1979); Walter Schläpfer, *Wirtschaftsgeschichte des Kantons Appenzell Ausserrhoden bis 1939* (Herisau: Appenzell-Ausserrhodische Kantonalbank, 1984).

38 Quoted in Ruesch, *Lebensverhältnisse*, 95.

39 About the development of cattle breeding in Appenzell, see Matthias Weishaupt: "'Viehveredelung' und 'Rassenzucht'. Die Anfänge der Appenzellischen Viehschauen im 19. Jahrhundert," in *Appenzeller Viehschauen*, ed. Mäddel Fuchs (St. Gallen: Typotron AG, 1998).

it in autumn, while farmers, mainly interested in dairy farming, depended on the possibility to manage a year-round stall-feeding. Yet with the exception of those herdsmen who managed about 20 till 24 cows on the Alpine summer meadows, most farms did not have more than two or three cows, which were milked only for self-sufficiency.[40]

However, the transition from agriculture to livestock farming in the second half of the 18th century was not caused by the invention of the whey cure. The main reason lay in the difficult weather conditions of the region. Crop failures have been repeatedly *blamed on* this phenomenon.[41] The chroniclers of the Appenzeller Land reported on frosts and droughts, cloudbursts, devastating storms and hailstorms, untimely snowfall, which forced the shepherds with their herds into the valley. Apart from that, livestock farming was based on cattle breeding, oxen, meat and cow milk. Goats played a useful role, less as a poor man's cow than as good feeders on rough Alpine pastures, albeit to a lesser extent than in the neighbouring canton. The available amounts of goat's milk would not have anyway been enough to build a business on it. As Laurenz Zellweger wrote in his *Versuch einiger physicalisch und medicinischer Betrachtungen*, a book length description of the living conditions in canton Appenzell, there was not even enough cows' milk to serve the needs of the local population.[42] Whole milk is for the rich, skimmed milk for the poor, he observed. His fellow citizens commonly processed butter from the available milk. Only on feast days, would people not use the available cream for butter production but drink it as so-called *Lobmilch*. The residual skimmed milk (buttermilk), with the addition of sour whey, was used to make a small and low-fat cheese. But while butter was produced the whole year round, only on their summer meadows on the Alps, Appenzellers produced a fatty cheese from the whole milk. In doing so, the remains (called whey, cheese water or *Schotte*) were heated several times, depending on the proportion of cheesy matter and acidity. What was left was of little use, the watery part of the milk, the drag-out swill, good enough for the pigs. Appenzellers would only drink this *Schotte* in the absence of water, Zellweger explained, and they would never

40 More on the system of Alpine dairy farming can be found in Orland, "Alpine Milk".

41 Ruesch, *Lebensverhältnisse*, 86.

42 Between the years 1714 and 1723, Zellweger drove all over canton Appenzell to familiarize himself with the habits of the people. Laurenz Zellweger, "Versuch einiger physicalisch und medicinischer Betrachtungen," in *Abhandlungen der Naturforschenden Gesellschaft in Zürich* 2 (Zurich: Heidegger und Compagnie, 1764). In reference to Scheuchzer, he discussed rural commons like Alpine grazing rights, the establishment of rules concerning the access of resources etc. For over ten years, he tried to persuade the herdsmen to let him in on some of their customs, but without success.

consume milk in times of illness, despite being prescribed by a doctor/healer in the cases of specific diseases. Like Galenic medicine advised, they did not appreciate whey.[43]

Visitors, who quite often came from cities, did not gain much insight into the pastoral economy of the Alpine regions. They were mainly delighted by the location. Many praised the fresh green of the hillsides and meadows, the healthy cattle and the local customs.[44] Equally clueless, outsiders saw the working and living conditions of the local population, recognizing only seldom that most peasant families could not live from the land, but had to earn their living as weavers.[45] Appenzell Ausserrhoden, from the 1720s onwards, was known for its linen and cotton spinning, and the fabrication of calico and mousseline, which were mostly exported to Italy, Spain and Germany. "The canvas production in Trogen and Herisau employed almost all the workers of the country, and only those who had not yet decided to leave their fragrant mountain meadow and their Alpine hut inherited from the fathers remained for the care of the cattle and for the cultivation of the Alps," the French Comte de Walsh summed up in 1823.[46] He further observed that because of their cottage industry, the Protestant half-canton Ausserrhoden supposed to be much more prosperous than the Catholic Innerrhoden.

This positive impression changed rapidly when Switzerland came under French occupation after 1797. Now the dependency of peasant families on home industry became highly visible. Chroniclers considered the disturbing health developments as direct results of an industrialized way of living and nutrition. If the diet was monotonous even in normal times, one author wrote, alongside with bad weather conditions, followed by crop losses and grain export barriers of foreign countries, it now quickly changed into famine.[47] Another French gentleman, who travelled through Appenzell in 1798, wrote about the local food crisis:

---

43 On the medical use of whey, see Orland, "Le régime suisse du petit lait".

44 See e.g. Kronfels, *Gais*, 35–8; [Johann Heinrich] Ernst, " Nachricht von Gaiss, und von dem daselbst üblichen Gebrauch der Ziegenmolken," *Museum der Heilkunde* 3 (1795).

45 On Appenzell home industry, see Ruesch, *Lebensverhältnisse*, 138–40.

46 "Die Leinwandfabrikation in Trogen und Herisau beschäftigte fast alle Arbeitskräfte des Landes, und zur Pflege des Viehs und zur Bewirtschaftung der Alpen blieben nur noch die, welche sich noch nicht entschliessen konnten, ihre duftende Bergwiese und ihre von den Vätern ererbte Alphütte zu verlassen." Quoted in F. Hunziker, "Das Appenzellerland nach früheren französischen Reisebeschreibungen 1750–1840," *Appenzellische Jahrbücher* 63 (1936): 29.

47 Hanspeter Ruesch, "Medizinhistorisches aus Appenzell Ausserrhoden (1800–1830)," *Gesnerus* 36 (1979): 24.

> The number of beggars, especially women and children, is appalling; they come out of their laughing houses with hungry eyes and pleading voice, to gather nettles along the way, from which they live. They are no longer able to procure bread and live almost entirely at the expense of the community and the few affluent people. One distributes oats soups with some meat in them, which are probably donated by the cities (Trogen, Herisau).[48]

In fact, during the years 1770–1772, 1798–1799 and 1816–1818 heavy food crises hit the country.[49]

4 The Development of a Spa Infrastructure

Any nascent spa tourism would have been affected by such crises, but Gais and its neighbouring villages faced a further problem – their remote location. When around 1760 the legendary healing success with goat's whey drew more patients to Gais, there existed no special infrastructure or transport system. Like any other traveller, the health seekers stayed at the only inn, the *Zum Ochsen* (the oxen).[50] The journey to Gais was arduous and could only be done on foot or on horseback from St. Gallen. It was not until 1784 that a guest dared to travel at breakneck speed on a rocky road in a coach – admired by the villagers as a world event.[51] As late as 1789, guests were advised not to go without a guide, as the paths branched and ramified, intersected and suddenly disappeared in the rocks; a labyrinth, which in winter and in bad weather could not be attempted. Around the same time, a traveller circumscribed the rural

48 "Die Anzahl der Bettler, vor allem Frauen und Kinder, ist entsetzlich; sie kommen aus ihren lachenden Häusern mit hungrigem Blick und flehender Stimme, um längs des Weges die Nesseln zu sammeln, von denen sie leben. Sie können sich das Brot nicht mehr verschaffen und leben fast ganz auf Kosten der Gemeinde und der wenig wohlhabenden Leute. Man teilt Hafersuppen mit etwas Fleisch darin aus, die wahrscheinlich von den Städten (Trogen, Herisau) gestiftet werden." Quoted in Hunziker, "Das Appenzellerland", 31.

49 Ruesch, "Medizinhistorisches", 24. The last Swiss food crisis of 1816/1817 is analysed in detail by Daniel Krämer, *"Menschen grasten nun mit dem Vieh": Die letzte grosse Hungerkrise der Schweiz 1816/17* (Basel: Schwabe 2015). Krämer describes Appenzell and the eastern part of Switzerland as the most affected regions.

50 Still in 1825, Kronfels described the *Ochsen* as the most important inn of Gais, although there existed two others. See Kronfels, *Gais*, 45.

51 The story is mentioned in the obituary "Altdistriktstatthalter Samuel Heim von Gais, der älteste Landmann," *Appenzellische Jahrbücher* 4 (1860): 73.

seclusion of Gais: "Is it possible, that people from different cantons of Switzerland and southern Germany come here to banish themselves in this sad loneliness? There is no tree, only meadows and pastures, and bare, wildly rugged rocks. This is the darkest place I have ever encountered."[52]

Cut off from the main thoroughfares of the Ancien Regime, Gais needed to develop its own transport and supply system in order to provide accessibility, and this required time.[53] And because, at the beginning, there was no comfortable accommodation or improved transport, the only unique features that could be highlighted to attract visitor's attention were the inimitable goat's whey and the healthy environment. This discursive strategy seems to have worked quite well, as more guests reportedly began coming to Gais since the 1760s.[54] Yet one can estimate that during the first two decades no more than a handful of people stayed for about four to six weeks during summer in Gais. The *Ochsen* was described as a wooden farmhouse with a dining room, the host family apartment, servant chambers and seven small and simple rooms under the roof for guests.[55]

Then, in 1780, a bad fire promoted innovations. A large part of the village had to be rebuilt; the inn was affected too. Despite economic difficulties, the innkeeper Hans-Ulrich Heim managed to rebuild the tavern larger and more spacious. Now trading under the name of *Zum alten Ochsen* (the old oxen), it offered space for significantly more guests in 25 comfortable rooms.[56] The number of *Schottatrinker* or *Schottaherren*, as they were called, increased. Spa guests from other Swiss cities (Aarau, Basel, Bern, St. Gallen, etc.) and from the southern part of Germany came, and despite the devastating fire, Gais was able to consolidate its reputation as one of the healthiest whey resorts in Switzerland. Nevertheless, the atmosphere remained relaxed as the village pastor remembered in 1791: "If the spa guests were not too distinguished," one was allowed to sit with sleepyhead, night coat and in slippers at the table.[57] But

---

52 "Ist es möglich, dass aus verschiedenen Kantonen der Schweiz und aus Süddeutschland Leute hierher kommen, um sich in diese traurige Einsamkeit zu verbannen? Man sieht keinen Baum, nur Matten und Weiden, und darüber kahle, wild zerklüftete Felsen. Dies ist der düsterste Ort, den ich je angetroffen habe." Quoted in Hunziker, "Das Appenzellerland", 5.

53 Heim, *Die Heilkräfte*,183. Hermann Grosser, "Der Fremdenverkehr und seine Entwicklung vom 17. Jahrhundert bis zur Gegenwart im Appenzellerland: unter besonderer Berücksichtigung von Appenzell Innerrhoden," *Innerrhoder Geschichtsfreund* 22 (1977–1978).

54 There exist only rough data, see ibid.

55 Kronfels, *Gais*, 41–42.

56 Ibid., 42.

57 "Altdistriktstatthalter Samuel Heim", 73.

as the number of guests increased from year to year, in 1796 the building was again expanded, and from then on the inn was renamed *Kurhaus* (spa hotel), and Gais a *Molkenkurort* (whey spa resort).[58] Already two years later, however, the thriving whey business was interrupted again. The French army occupied Switzerland, and a division started building the first settlement and headquarter in Gais. Officers and soldiers bordered at the *Kurhaus*, and although the generals of the French and Austrian armies, who were stationed in Switzerland between 1798 and 1803 (Helvetic Republic), were not interested in a whey cure, they took the opportunity to relax in Gais. Famous guests like Louis Napoleon, who visited Gais with his wife in 1812, or several renowned aristocrats like Pauline of Württemberg or the Markgraf Wilhelm von Baden did their part to increase the fame of Gais.[59]

The business of the innkeeper family, who dominated the health resort, did well for some decades.[60] In 1791, Samuel Heim (1764–1860) took over the inn *Zum Ochsen* from his father Hans-Ulrich Heim (1720–1814), who had begun the whey business and had expanded the house after the fire in 1780. It was Samuel, who in 1796 converted the inn into a spa centre, famous for its goat products. The herdsmen also played an important role, particularly Anton Josef Inauen (1725–1791), called *Schottensepp*. During the summer meadow season (between May and August), he brought the fresh whey from the alp Oberer Mesmer to Gais. Due to growing demand, Alpine dairymen from Ebenalp and Meglisalp also supplied goat milk to Gais. Up to the 1780s, the *Schottensepp* mainly supplied Gais, but the successful expansion of Gais as a health resort encouraged him to introduce whey cures to the nearby village Weissbad as well.[61] His son Karl Jakob (1755–1811) realized this plan in 1790 and founded the *Weissbad Molkenkuranstalt*. This new *Kurhaus* developed into the largest domestic hotel company under the grandson of the *Schottensepp*, Ignaz Johann Anton Inauen (1794–1864), and housed up to 400 guests from all over Europe during its best days.[62]

58 Ibid., 73.

59 Grosser, "Der Fremdenverkehr", 47.

60 Schläpfer, *Wirtschaftsgeschichte*, 205–13; Albert Nägeli, "Die Molkenkuren im Appenzellerland," *Appenzeller Kalender* 220 (1941): n.p.

61 Ibid.

62 The Kurhaus remained under the direction of the Inauen family for five generations up to 1898.

5 The Therapeutic Regime

In 1822, when with Johann Heinrich Heim (1802–1876) the next generation took over the *Kurhaus*, the health resorts of Appenzell were well established, and the businesses consolidated. Accordingly, Johann Heinrich Heim, the son of Samuel, studied medicine and could officiate as the first spa doctor in Gais. With his dissertation of 1824, in which he described the healing power of goat's whey,[63] the *Gaisschottenkur* received a scientific legitimation. Heim represented both, the local inventors in a health food, and the knowledge and language of medicine and science. He frequently referred to the extensive use of the mountains as an aesthetic element of the health resort, arguing for the acceptance of a therapeutic landscape.[64] Finally, he integrated social and environmental considerations; for instance, in the annex to his treatise on the healing power of whey he published some literary works and letters of recommendation of famous guests.[65]

As mentioned at the beginning, recent spa histories have illuminated how traveling was related to the widespread and growing neo-Hippocratic discussions of a therapeutic regimen that involved bathing, exposure to fresh air, personal and domestic hygiene, clothing reform and many other aspects of living. Heim too draw on Hippocrates and explained health and disease by looking at the role of specific agents, namely air, sites, soil and food. He reconfirms that discriminating "unhealthy places" from the opposite, the healthy environment, became an obsession of the elite population in Europe, who sought "to determine what aspects of the environment might be modified to weaken or eliminate their capacity to cause disease."[66] Quite the same fascination with the health aspects of the environment can be found with respect to the Alpine whey cure. First, goat's whey was transformed into a health food, which was not self-evident, as I said before. Whey usually was not used as a human food and only seldom as a laxative.[67] Not as a precondition but as a result of the whey boom, secondly, the goat became the unique animal of Gais and the

63 See Johann Ulrich Heim, *Ueber den medicinischen Gebrauch der Molken.* Eine Inaugural-Abhandlung (St. Gallen: Wegelin und Rätzer, 1824).

64 Ibid., 84.

65 Heim, *Die Heilkräfte*, 132–82.

66 James C. Riley, *The Eighteenth-Century Campaign to Avoid Disease* (New York: Knopf, 1987), ix.

67 In my opinion, it was probably a coincidence that the success story of 1749 happened using goat's instead of cow's whey.

Appenzell, and herdsmen developed a flourishing business with the white Appenzell goat breed.[68]

A further innovation was the sweat whey recipe. The drink served by the health resorts of the Appenzell villages should deliberately not be acidic in order to be easier to digest. Heim described the production like this:[69] A good whey could only be preserved when the goats had already spent fourteen days "in full freedom" on the Alps, where they enjoyed the aromatic Alpine herbs. Basically, it is possible to produce sweet whey in the valley too, he argued, but the best is the milk of mountain goats. Although the goat commonly was seen as an inferior animal, to Heim goat's whey was in fact as good as the cow's whey, only a bit less nutritious. After all, however, this would be a negligible disadvantage, compared to the many benefits of the Alpine sources of this juice.

Most crucially, doctors described whey no longer as the inferior water of milk, but as the juice of herbs. A Dr. Ernst in his enthusiastic report about the Appenzell whey business quoted the renowned physician Samuel Auguste Tissot from Lausanne, who had argued that because milk is nothing else than the more or less unmodified food of the animal, and because the fodder of the Alpine cattle essentially consists of herbs, the whey can rightly be described as a juice of herbs, which is purified and sweetened by the animal body.[70] Heim, too, repeated perceptions of the mountainous landscape and nature that had been introduced by early naturalists like Conrad Gessner, Johann Jakob Scheuchzer, or Albrecht von Haller.[71] The fame of mountain products was such a cliché that travellers were surprised not to see cows "walking up to their bellies in grass." Ludwig Wallrath Medicus, a German author who in 1795 wrote his report on Alpine farming, was therefore disappointed to find "just short and rather low grass".[72] Once doctors had achieved such meanings about the health of the landscape, it was just a small step to argue: "As in summer the fresh herbal juices are already mixed in the milk, and since the milk of mountain plants is always slightly astringent balsamic, it is undoubtedly the most important aid in the medical treatment of the emaciation; it is an

68 See Markus Arbenz et al., *Schweizer Ziegen* (Winterthur: Birkenhalde-Verlag, 2005), 52–57.

69 Heim, *Die Heilkräfte*, 3–7.

70 Ernst, "Nachricht von Gaiss", 198.

71 On these perceptions, see Jon Mathieu and Simona Boscani Leoni, eds., *Die Alpen! Zur europäischen Wahrnehmungsgeschichte seit der Renaissance – Les Alpes! Pour une histoire de la perception européenne depuis la Renaissance* (Bern, Berlin, Bruxelles: Lang, 2005).

72 Ludwig Wallrath Medicus, *Bemerkungen über die Alpen-Wirthschaft auf einer Reise durch die Schweiz* (Leipzig: Gräff, 1795), 24.

excellent replacement for power and juice."[73] To Heim, the sweetest and best whey, logically, was the one processed by the herdsmen themselves. It seemed not odd to him that the reader should learn in all detail how whey was produced in the Alpine huts of Appenzell,[74] while, only a few sentences later, he introduced famous names of physicians, starting with Hippocrates, who were said to have prescribed whey as a medicine since the earliest days of humankind. The experience of the past, such was the message, demonstrated that whey was particularly suitable for certain diseases.[75]

It was advantageous that sweet whey was a much tastier drink compared to the product of the earlier dairy tradition, because people were advised to drink a large amount of it. "A whey cure is a long-lasting, gradually increasing and gradually decreasing drinking of the whey towards the end of the treatment," Heim argued.[76] The drinking was integrated into a rigid daily routine that should underline its therapeutic character. Heim advised his guests to drink about 5–7 glasses of whey every quarter of an hour, at least 9–12 glasses a day. In between they should stroll over the square in front of the inn, and after this procedure enjoy the rest of the day with relaxation, small walks, and the appreciation of the "good food" of the region. The whey cure lasted a minimum of two to three weeks (for the purification of the body) and six to eight weeks in case of chronic illnesses. For offering warm whey in the morning, the dairyman had to start at midnight with the processing of cheese, which took about two hours. When the whey had reached the boiling point, it was filled in hermetically sealed wooden butts, the so-called *Schottatasen*, wrapped in cloths, and carried down from the mountain. Leaving the Alp between two and three o'clock in the morning, the shepherd should arrive Gais at six o'clock, where the spa guests gathered to the sound of a bell.[77]

73 "Da nämlich die frischen Kräutersäfte im Sommer in der Milch schon gemischt darin enthalten, und zumal die Milch von Bergpflanzen immer etwas vorstechend balsamisches führt, so ist sie ohne Zweifel das wichtigste Hilfsmittel bei der ärztlichen Behandlung der Auszehrungen; sie leistet ganz vorzüglich Ersatz an Kraft und Saft." J. C. Kottmann, *Der Weissenstein. Die Milch- und Molkenkuren, auch Molkenbäder auf dem Jura bei Solothurn* (Solothurn: Amiet Lutiger, 1829), 81.

74 Heim, *Die Heilkräfte*, 4–7.

75 Heim offered a series of indications, but, in the course of time, indications for whey cure treatments in clinical and pharmaceutical handbooks were variable. See e.g. W.[ilhelm] J.[oseph] A.[nton] Werber, *Specielle Heilmittellehre: Physikalisch, chemisch, physiologisch, diätetisch und klinisch dargestellt*, vol. 1 (Erlangen: Ferdinand Enke, 1859), 93–96.

76 "Eine Molkenkur ist ein längere Zeit anhaltendes, stufenweise steigendes und gegen das Ende der Kur stufenweise fallendes Trinken der Molken." Heim, *Die Heilkräfte*, 78.

77 Ibid., 103.

The whey cure was more than just drinking an Alpine food; descriptions of Gais included leisure activities: "The spa town of Gais is located in a lovely, graceful high mountain valley [...] of that canton, of which Ebel says, 'it resembles an immense English garden, where you find the most beautiful and varied mountain views alternating with the loveliest rural scenes.'"[78] Heim also recommended shorter walks or hikes to the surrounding mountains. Though one should avoid the highest mountains and those directed to the north, because they provide a disposition to vomit blood.[79] Heim also dealt with the health value of the atmosphere surrounding Gais. The guest need not worry about too much humidity or fog, he wrote, because "our pure and dry high air absorbs the possible exhalations of our meadows and – because of the few forests in our area and the free entrance and exit of the sunlight and the air – offers a most beautiful level of moisture and dryness of our soil and the atmosphere."[80]

Over the years, and quite similarly to other tourism initiatives, the Swiss whey cure business brought urban infrastructure to the local population. Bourgeois styles of comfort found their way into rural villages, e.g. on the open square in Gais, where the milk was distributed to the guests, lime trees were planted to provide shadow. Where it was possible (e.g. in Heinrichsbad) local authorities combined mineral baths with milk cures and whey bathing, or they extended their therapeutic offers with herb treatments, mudpacks etc.[81] A good example is the so-called cowshed-cure. Because most bath buildings were situated near farmhouses, guests were advised to stay for hours in cowshed rooms. In a cowshed, for several hours the patient should inhale the animal breath and body vapour of the cattle. The ammonia gases from urine were advertised as a balsam for infected lungs.[82]

Further on, the *Molken-Gewerbler* of Appenzell sold a plethora of medical or health products from whey, among them *serum lactis dulce* (sweet whey for children, produced with rennet and beaten egg white), *serum lactis acidulum* (sour whey, coagulated with citron or *creme tartaris*), *serum lactis vinosum* (with white wine), *serum lactis aluminosum* (coagulated by alum, prescribed

78 "Der Kurort Gais liegt in einem lieblichen, anmutigen, mit grünen Hügeln umgebenen, vom Rothbach durchflossenen, hohen Bergtale jenes Kantons, von dem Ebel sagt: 'er gleichen einem unermesslichen englischen Garten, wo man die schönsten und mannigfaltigsten Gebirgsaussichten mit den lieblichsten ländlichen Szenen abwechseln sieht.'" Ibid., 94. Heim refers to Johann Gottfried Ebel (1764–1830), a physician who travelled extensively through the Alps and published geological treatises.

79 Kottmann, *Der Weissenstein*, 129.

80 Heim, *Die Heilkräfte*, 115.

81 Rheiner, *Das Moosberger- oder Heinrichs-Bad*.

82 Ibid., 29.

in case of dysenterie, gonorrhoea, or white flux), serum lactis tamarindinatum (with tamarind), *petit-lait en poudre* as a first-aid kit (with gummi arabicum, milk sugar and milk protein), *serum lactis antiscorbuticum* (with several herbs).[83]

6 Visitors' Judgement

How did the travellers, who underwent extensive efforts and spent large amounts of money to frequent spas for medical purposes, respond? What justified the success of mountainous whey producers? In the words of the early 19th century novelist Ulrich Hegner, "Whey could have been processed at home, where we have enough cows and fatty food."[84] What then was the advantage of visiting Gais?

Ulrich Hegner (1759–1840), who had studied medicine in Strasbourg from 1776–1781, lived in Winterthur and Zurich, when he wrote and published his popular novel about the Swiss whey cure in the year 1812. The main figure was a colonel from the northern part of Germany, a hypochondriac and sceptic, who undertook the long journey to Switzerland with his niece. The young woman is characterized as a romantic person, who has fallen in love with the people and nature of the host country. In the form of letters or talks with other visitors of Gais (one French noblemen, an Englishman, a Swiss vicar) the author lets his readers recognize how mountains and people rejuvenated him. The whey cure itself, however, is partly ironically exaggerated, partly questioned in its medical effect. Obviously, the colonel doubted the effect of the health food, instead was touched by the social and natural environment. The assumed medicinal properties of the whey were insignificant, and in the end, the narrative judged social contacts more beneficial to body and mind than air and food.

Evaluation thus went further than medical reviews. Hegner offered a sophisticated social analysis, arguing that most travellers would regard the purpose of wellbeing and social contact in the same way as medical claims or discussions about physical conditions and diseases.[85] To physicians, in contrast, the virtues of goat's whey were debated by an increasing number of chemical and medical

83 See Garlichs, *Über den medizinischen Gebrauch*, 21–24.

84 Ulrich Hegner, *Die Molkenkur* (Zurich: Orell Füssli, 1812), 9.

85 Although there existed many differences among spas, this is one of the general results of the examination of spa histories. See John K. Walton, "Health, Sociability, Politics and Culture. Spas and History: An Overview," in *Mineral Springs Resorts in Global Perspective. Spa Histories*, ed. John K. Walton (London: Routledge 2014).

experiments undertaken since about 1830.[86] Scientific discourses dealt with issues like whey drinking as a treatment against chest affections, lung diseases or consumption.[87] Was whey itself the active substance or the spicy herbs of the Alps? Was living in the "Swiss" Alps healthier than living in other landscapes or mountains? Did residents in hill countries get the same diseases as people, plants, or animals in the lowlands? No easy questions, argued the British doctor John Macpherson (1817–1890), because scientific parameters do not evaluate national differences. "Alpine climates have been divided somewhat artificially into various zones, and their effects have been characterized as tonic and vivifying, tonic and exciting, &c.; but such divisions are at best dependent on latitude and on local influences."[88]

Travellers often described certain effects of the air at higher elevations, such as accelerated action of the heart, nose or ear bleeding, headaches, and "incapacity for exertion."[89] Nervous excitement or sleeplessness and other unpleasant effects made the stay in Switzerland a contra-indication. But although this might happen particularly in the beginning of a stay in the mountains, in most cases, however, the body soon gets accustomed to the change, Macpherson argued. Beside this, changes of barometric pressure and of the moisture of the atmosphere "are beneficial in giving the system alternate periods of excitement and of rest."[90] From a physiological point of view, he summarized his results, there was no clear indication that a stay in the mountains was beneficial or harmful.

Contrary to all proverbial shepherd natures, travellers confessed that the health conditions of the local population had to be assessed differently. "It is by no means easy to get very satisfactory evidence as to the health of residents in hill countries," Macpherson further reflected. In the lower parts of the Swiss mountains, dysentery, diarrhoea and low fevers seemed to be quite common. On the other hand, there had to be no doubt "that certain elevated spots enjoy a considerable amount of immunity from lung tuberculosis,"[91] while at other places one observed a good deal of phthisis and rheumatism, particularly in

86 See e.g. Christian Heinrich Ernst Bischoff, *Die Lehre von den chemischen Heilmitteln oder Handbuch der Arzneimittellehre: als Grundlage für Vorlesungen und zum Gebrauche praktischer Aerzte und Wundärzte. Enthaltend des Verfassers fernere wissenschafftliche Beyträge nebst den neueren Erwerbnissen und materiellen Bereicherungen der Arzneimittel-Lehre, auch das vollständige Register über das ganze Werk*, vol. 4 (Bonn: Eduard Weber, 1834), 490.

87 Dr. Mich. Karner, "Die Molkenkur in ihrer Beziehung zur Lungentuberkulose," *Froriep's Notizen aus dem Gebiete der Natur- und Heilkunde* 2 (1858).

88 Macpherson, *The Baths and Wells*, 36.

89 Ibid., 33.

90 Ibid., 35.

91 Ibid., 37.

Gais. From all this, he did not want to conclude that it is an unsuitable place to any consumptive patient. "I would only say that mere elevation is no sure test of the healthiness of a place."[92] Similar critical judgments about the health value of whey accumulated over the years. Many more voices than at the beginning of the whey boom mentioned the unpleasant taste of goat's whey, a fluid "with a mawkish taste"[93] or considered it an unhealthy therapy anyway.

7 Chemical Analysis and Health

Needless to say, chemists of the late 18th and early 19th century supported such views by their analysis of the specific ingredients of milk in general, and whey in particular. Their key objective was the improvement of cheese making, yet with respect to waters and other health drinks, it very soon became most common to class them according to their chemical composition. "The geo-gnostic character of the soil, from which the mineral springs as well as the diets owe their origin" can be either analyzed by their sensory properties. Even better is chemical analysis, because it "permits as a rule correct conclusions as to content and effects" of any kind of substance.[94]

However, the more spa doctors asked for scientific investigations, the more they undermined the environmental health claims. One recurrent argument runs through the medical literature of the 19th century, and this is the accusation of a pseudo or non-scientific approach of the health resorts. Rules are required for the evaluation of the milk or whey cure claimed Herrmann Lebert, Professor of Medicine at the University of Breslau, in 1869. There are too many apodictic and not proven opinions about the properties of whey.[95] Whether the cure is effective or not, is just a question of empirical experience, he criticized. The supposedly useful and beneficial influence of the aromatic Alpine herbs on the healing power of the whey he could not validate a scientific knowledge. Most practitioners in health resorts were not even trained physicians, and the "lack of knowledge about the chemical-physiological aspects of the whey is disturbing."[96]

92 Ibid.

93 Ibid., 346.

94 Gabriel Rüsch, *Anleitung zu dem richtigen Gebrauche der Bade- und Trinkcuren ueberhaupt, mit besonderer Betrachtung der schweizerischen Mineralwasser und Badeanstalten*, vol. 1 (Ebnat, St.Gallen: Abraham Keller, 1825), 270–79.

95 Herrmann Lebert, *Ueber Milch- und Molkenkuren und über ländliche Kurorte für unbemittelte Brustkranke* (Berlin: August Hirschwald, 1869).

96 Ibid., 11.

But what kind of knowledge had chemistry to offer? A chemically informed explanation of the whey cure went like this:

> Whey [...] is recommended theoretically, as supplying to the blood only non-nitrogenous elements, the nitrogenous casein and the fat being excluded; the notion is, that the constitution of the fluids and tissues of the body is altered and improved by the salts and milk-sugar which it contains, while the nitrogenous elements are withheld. But these theoretical ideas are of no real importance, as long as all patients drinking whey at the same time use a diet in which there is an abundance of nitrogenous or protein substances; some have attributed special virtues to the small quantity of the salts of potass present in whey.[97]

Reflecting the research so far, Macpherson concluded that milk and whey cures could be obtained almost everywhere. From a chemical point of view, there existed no serious difference between the lowland and mountainous products.

Scientific knowledge, in fact, delivered no clear data that would meet the practical needs of the visitors of whey spas. One experienced doctor concluded from this "that, by the same token, the educated reader will always give preference to a place which gives him pleasant memories or presents ample material for comparisons of the old and new times, and for pleasant conversation."[98] Science so to say complicated existing ways of evaluation, or the muddling towards understandings. Medico-chemical explanations confronted entrepreneurs with a completely new style of thinking, because the health value of milk like of any other organic substance changed its meaning. The technological epistemology of the modern sciences presupposed a scientifically standardized fluid. The then dominating environmental perspective, in contrast, valued the local impact of flora and fauna.

97 Macpherson, *The Baths and Wells*, 321.

98 "dass der gebildete Leser unter übrigens gleichen Umständen immer einem Orte den Vorzug geben wird, der ihm angenehme Erinnerungen erregt oder reichlichen Stoff zu Vergleichungen aus der alten und neuen Zeit und zu angenehmer Unterhaltung darbietet." Rüsch, *Anleitung*, XII.

8 A Fading Business

Almost every spa promotion of the 19th century was accompanied by an appeal to this scientific style of thinking. At first it wasn't harmful to business either. The Swiss whey cure morphed from a local to a nationwide health business, from a place that had begun its business with an environmental advantage. And yet, around the 1840s, the reference to alpine nature slowly lost its power of conviction. To Hermann Lebert it was not only the lack of science but sometimes the scientific investigation itself that damaged the business. Asserted medicinal benefits alternated too often and in the worst case even contradicted each other. Over time, Lebert observed, the Appenzeller had varied their arguments substantially. At the beginning, the effluent effect of whey was decisive, which was easily attainable by the large quantities people had to drink. Later, a resolving effect was emphasized, which, however, was a very vague and elastic concept. And when the whole world spoke of cells, whey was supposed to dissolve and remove abnormal cell formation.[99] Some praised the mild-nourishing effect of whey and were happy, if a sick patient during the cure gained a bit weight. Others, in turn, emphasized the secretion-promoting effect of whey. The consequence of this rebuttal of the popular agency of whey was that it invalidated the notion of whey as a medicine. Moreover, it questioned the credibility of the public understanding of nature insofar as it reduced the health benefits to a negotiable commodity.

Whey lost its medical importance and to Lebert the whey cure remained nothing more than a fashion that sooner or later would lose its lovers. It was highly striking, he went on, that only German-speaking regions established a whey cure business, not France, Italy, England, or the United States. Ignorance could not have been the cause, he grumbled.[100] The decisive reason why Switzerland was so successful, he concluded, must have been the climatic element, while almost all other German health resorts mixed whey with mineral water, and sold it together with baths, inhalations and other health care offers.

There is no solid data on the development of the whey cure business from the late 18th century to the end of the 19th century. It is, moreover, difficult to answer the question when exactly the Swiss whey cure reached its high point in Appenzell and beyond. The general perception is that until the 1860s, businesses went quite well and thereafter went into decline, or at least stagnated.

---

99 Lebert, *Ueber Milch- und Molkenkuren*, 66.
100 Ibid., 68.

At that time, the increasing railway infrastructure accelerated structural change, but so far no study has quantified these effects in the Swiss context. Certainly, the railway expansion promoted tourism and leisure activities, but it also increased competition between the regions and locations. For instance, with respect to pulmonary diseases locations like Davos and St. Moritz and Grison in general became much more fashionable, particularly because of their higher altitudes.[101]

Changes in curative regimes might also have been responsible for a slow decline of the whey cure. In the early days of the late 18th century, when people went to Gais, it was for treatment. Around the 1830s patients and invalids were already in a minority; Gais and its neighbouring villages had become locations of health tourism, with sociability and respectability, advertising the term *Geiß-Schotten-Kur* as a Swiss symbol. Fifty years later, the term had disappeared, displaced by the general treatment named whey cure (with only seldom mentioning goat's whey).[102] Recognizably the business had expanded, fragmented and diversified. The famous traveller's guide by Karl Baedeker emphasized in its edition of 1873 that the Weissenstein nearby Soleure had become one of the most frequented mountains for whey cures, but one could also find establishments near Zurich in Rapperswil, Einsiedeln, at the Rigi, in Interlaken or Churwalden.[103] The Appenzeller whey cure, meanwhile also whey bath, had become much less distinctive; it was just one spa among a range of many other. And not to forget, it remained a seasonal business of at least three months in the summer. Unlike spa waters whey could not be bottled for home use; the whey cure required a residential stay.

To sum up, while during the second half of the 18th century the procedure of the whey cure became an emblematic marker of the healthy environment of the Alps, during the 19th century the cure symbolized a new direction that medical practice was pursuing. It changed from its task of healing the ill to becoming a tool of prevention, wellbeing, and sociability. Although health was no longer only perceived as a day-to-day experience but as a merit "good" that must and could be processed and purchased, the place where to buy the

101 Sigismond Jaccoud, *The Curability and Treatment of Pulmonary Phthisis*, trans. Montagu Lubbock (New York: Appleton and Company, 1885), 334.

102 Spa handbooks of the 1870s illuminate that there was scarcely a bathing resort or a spa, which did not provide the possibility of drinking whey. Julius Braun, *On the Curative Effects of Baths and Waters Being a Handbook to the Spas of Europe, an Abridged Translation with Notes by Hermann Weber* (London: Smith, Elder, & Co, 1875), 439.

103 Karl Baedeker, *Switzerland: And the Adjacent Portions of Italy, Savoy, and the Tyrol. Handbook for Travellers. Sixth Edition, Revised and Augmented* (Coblenz, Leipzig: Karl Baedeker, 1873), 11.

treatment started to play a minor role. The close link between landscape and health became looser, or better to say, it had to be refashioned again. The story of Gais and its invention of the *Gaisschottenkur* were transmuted into a marketing concept that used inherited stories and tales, which people told about successful therapies. The healthy character of the mountains milk remained slightly more than symbolic capital.

Acknowledgement

This paper is based on research supported by the Family Larsson-Rosenquist Foundation.

Bibliography

Printed Sources

"Altdistriktstatthalter Samuel Heim von Gais, der älteste Landmann." *Appenzellische Jahrbücher* 4 (1860): 69–80.

Baedeker, Karl. *Switzerland: And the Adjacent Portions of Italy, Savoy, and the Tyrol. Handbook for Travellers. Sixth Edition, Revised and Augmented.* Coblenz, Leipzig: Karl Baedeker, 1873.

Behr, Georg Heinrich. *Zwey Bücher von der Materia Medica, oder vollständige Beschreibung aller und jeder Arzeney-Mittel: Samt beygefügter wohl-eingerichteter und höchst-nutzbarer Therapie.* Straßburg: Johannes Beck, 1748.

Bischoff, Christian Heinrich Ernst. *Die Lehre von den chemischen Heilmitteln oder Handbuch der Arzneimittellehre: als Grundlage für Vorlesungen und zum Gebrauche praktischer Aerzte und Wundärzte. Enthaltend des Verfassers fernere wissenschafftliche Beyträge nebst den neueren Erwerbnissen und materiellen Bereicherungen der Arzneimittel-Lehre, auch das vollständige Register über das ganze Werk*, vol. 4. Bonn: Eduard Weber, 1834.

Braun, Julius. *On the Curative Effects of Baths and Waters Being a Handbook to the Spas of Europe, an Abridged Translation with Notes by Hermann Weber.* London: Smith, Elder, & Co, 1875.

Ernst, [Johann Heinrich]. "Nachricht von Gaiss, und von dem daselbst üblichen Gebrauch der Ziegenmolken." *Museum der Heilkunde* 3 (1795): 196–208.

Garlichs, Friedrich. *Über den medizinischen Gebrauch der Milch und der Molke: Eine Inaugural-Abhandlung.* Würzburg: Becker, 1837.

Hegner, Ulrich. *Die Molkenkur.* Zurich: Orell Füssli, 1812.

Heim, Johann Heinrich. *Die Heilkräfte der Alpenziegen-Molken und der Molkenkurort Gais*. Zurich: Schultheß, 1844.

Heim, Johann Heinrich. *Ueber den medicinischen Gebrauch der Molken. Eine Inaugural-Abhandlung*. St. Gallen: Wegelin und Rätzer, 1824.

Hippocrates. *On Airs, Waters, and Places*. Translated by Francis Adams. Last modified July 3, 2018. http://classics.mit.edu//Hippocrates/airwatpl.html, accessed on 3 July 2018.

Jaccoud, Sigismond. *The Curability and Treatment of Pulmonary Phthisis*. Translated by Montagu Lubbock, New York: Appleton and Company, 1885.

Karner, Dr. Mich. "Die Molkenkur in ihrer Beziehung zur Lungentuberkulose." *Froriep's Notizen aus dem Gebiete der Natur- und Heilkunde* 2 (1858): 215–19.

Kottmann, J[ohann] C[arl]. *Der Weissenstein. Die Milch- und Molkenkuren, auch Molkenbäder auf dem Jura bei Solothurn*. Solothurn: Amiet_Lutiger, 1829.

Kronfels, Fr.[iedrich] K. v. *Gais, Weisbad und die Molkenkuren im Canton Appenzell*. Constance: W. Wallis, 1826.

Lebert, Herrmann. *Ueber Milch- und Molkenkuren und über ländliche Kurorte für unbemittelte Brustkranke*. Berlin: August Hirschwald, 1869.

Macpherson, John. *The Baths and Wells of Europe: Their Action and Uses, with Hints on Change of Air and Diet Cures*. London: Macmillan and Co, 1869.

Malten, Heinrich von. *Beschreibung aller berühmten Bäder in der Schweiz. Nebst einer allgemeinen Uebersicht der Bäder zweiten Ranges und der unbenutzten Heilquellen*. Aarau: Sauerländer, 1830.

Medicus, Ludwig Wallrath. *Bemerkungen über die Alpen-Wirthschaft auf einer Reise durch die Schweiz*. Leipzig: Gräff, 1795.

Morgenblatt für gebildete Leser, no. 3, January 3, 1849, 5.

Müller, Vincenz. *Specielle Beschreibung der Molkenkur-Anstalten des Königreichs Bayern*. Munich: Eigenverlag, 1843.

Roubaud, Félix. *Les Cures de Petit-Lait en Suisse, en Allemagne, dans le Tyrol et la Styrie*. Paris: Adrien Delahaye, 1867.

Rheiner, H. *Das Moosberger oder Heinrichs-Bad im Kanton Appenzell, historisch, chemisch und topographisch beschrieben*. St. Gallen: Huber und Compagnie, 1833.

Rüsch, Gabriel. *Anleitung zu dem richtigen Gebrauche der Bade- und Trinkcuren ueberhaupt, mit besonderer Betrachtung der schweizerischen Mineralwasser und Badeanstalten*, 2 vols. Ebnat, St.Gallen: Abraham Keller, 1825, 1826.

Rüsch, Gabriel. *Heiden und seine Molkenkuranstalt im Freihof*. Trogen: Schläpfer, 1854.

Scheuchzer, Johann Jakob. *Beschreibung der Natur-Geschichten des Schweizerlands*, vol 1, no. 8. Zurich: Michael Schaufelb.[erger]s. E.[rben] und Christoff Hardmeier, 1706.

Shoberl, Frederic. *The World in Miniature. Switzerland, Containing a Description of the Characters, Manners, Customs, Diversions, Dress &c. of the People of that Country in General, and of the Inhabitants of the twenty-two Cantons in particular*. London: R. Ackermann, 1827.

Strasser, J. J. *Interlaken im Berner Oberlande als Luft- und Molkenkur-Ort, vom therapeutischen Standpunkte aus betrachtet*. Interlaken: Kurhaus-Verwaltung, 1863.

Weber, Herman. "Notes on the Climate of the Swiss Alps, and on some of their Health Resorts and Spas." *The Dublin Quarterly Journal of Medical Science* 37, no. 2 (1864): 333–64.

Werber, W.[ilhelm] J.[oseph] A.[nton]. *Specielle Heilmittellehre: Physikalisch, chemisch, physiologisch, diätetisch und klinisch dargestellt*, vol. 1. Erlangen: Ferdinand Enke, 1859.

Zeller, F. B. *Die Molkenkur in Verbindung der Mineral-Bronnenkur*. Würzburg: Etlinger, 1826.

Zellweger, Laurenz. "Versuch einiger physicalisch und medicinischer Betrachtungen." In *Abhandlungen der Naturforschenden Gesellschaft in Zürich* 2, 309–60. Zurich: Heidegger und Compagnie, 1764.

Literature

Association Internationale pour l'Histoire des Alpes, ed. *Histoire des Alpes*. Zurich: Chronos, 1996–2018.

Arbenz, Markus et al. *Schweizer Ziegen*. Winterthur: Birkenhalde-Verlag, 2005.

Bätzing, Werner. *Die Alpen. Geschichte und Zukunft einer europäischen Kulturlandschaft*, 4th. rev. ed. Munich: Beck, 2015.

Barton, Susan R. *Healthy Living in the Alps: The Origins of Winter Tourism in Switzerland, 1860–1914*. Manchester: Manchester University Press, 2008.

Beattie, Andrew. *The Alps: A Cultural History*. Oxford: Oxford University Press, 2006.

Bircher, Eugen. "Die Molkenkur, ihre Geschichte und geographische Verbreitung. Eine geschichtliche Studie." *Schweizerische Medizinische Wochenschrift* 39 (1953): 937–53.

Boscani Leoni, Simona. "Tra Zurigo e le Alpi: le 'Lettres des Grisons' di Johann Jakob Scheuchzer (1672–1733). Dinamiche della comunicazione erudita all'inizio del Settecento." In *Die Alpen! Zur europäischen Wahrnehmungsgeschichte seit der Renaissance – Les Alpes! Pour une histoire de la perception européenne depuis la Renaissance*, edited by Jon Mathieu and Simona Boscani Leoni, 157–171. Bern: Peter Lang, 2005.

Boscani Leoni, Simona, ed. *Wissenschaft – Berge – Ideologien. Johann Jakob Scheuchzer (1672–1733) und die frühneuzeitliche Naturforschung – Scienza – montagna – ideologie. Johann Jakob Scheuchzer (1672–1733) e la ricerca naturalistica in epoca moderna*. Basel: Schwabe, 2010.

Cantor, David, ed. *Reinventing Hippocrates*. Ashgate: Aldershot, 2002.

Cavallo, Sandra and Tessa Storey, eds. *Conserving Health in Early Modern Culture. Bodies and Environments in Italy and England*. Manchester: Manchester University Press, 2017.

Classen, Albrecht, ed. *Bodily and Spiritual Hygiene in Medieval and Early Modern Literature: Explorations of Textual Presentations of Filth and Water*. Berlin, Boston: De Gruyter, 2017.

Classen, Albrecht. "Introduction: Bathing, Health Care, Medicine, and Water in the Middle Ages and Early Modern Age." In *Bodily and Spiritual Hygiene in Medieval and Early Modern Literature: Explorations of Textual Presentations of Filth and Water*, edited by Albrecht Classen, 1–87. Berlin, Boston: De Gruyter, 2017.

Coley, Noel G. "Physicians, Chemists and the Analysis of Mineral Waters: 'The most difficult part of Chemistry'." In *The Medical History of Water and Spas*, edited by Roy Porter, 56–66. London: Wellcome Institute for the History of Medicine, 1990.

Eddy, Matthew Daniel. "The Sparkling Nectar of Spas, or, Mineral Water as a Medically Commodifiable Material in the Province, 1770–1805." In *Materials and Expertise in Early Modern Europe. Between Market and Laboratory*, edited by Ursula Klein and Emma C. Spary, 198–224. Chicago, London: The University of Chicago Press, 2010.

Eisenhut, Heidi. "Gelehrte auf Molkenkur. Laurenz Zellweger und sein Kreis in Trogen." In *Heilkunst und schöne Künste. Wechselwirkungen von Medizin, Literatur und bildender Kunst im 18. Jahrhundert*, edited by Heidi Eisenhut, Anett Lütteken, and Carsten Zelle, 271–302. Göttingen: Wallstein, 2011.

Esser, Raingard and Thomas Fuchs, eds. *Bäder und Kuren in der Aufklärung: Medizinaldiskurs und Freizeitvergnügen*. Berlin: Berliner Wissenschafts-Verlag, 2003.

Flubacher, Silvia. "Alpen-Tiere. Lokale Wissenswelten in der schweizerischen Naturgeschichtsschreibung." In *Wissenschaftsgeschichte und Geschichte des Wissens im Dialog – Connecting Science and Knowledge: Schauplätze der Forschung*, edited by Kaspar von Greyerz, Silvia Flubacher, and Philipp Senn, 347–74. Göttingen: V&R unipress, 2013.

Fürbeth, Frank. "Adaptionen gelehrten Wissens für laikale Zwecke in der Bäderheilkunde der frühen Neuzeit." In *Wissenschaftsgeschichte und Geschichte des Wissens im Dialog – Connecting Science and Knowledge: Schauplätze der Forschung*, edited by Kaspar von Greyerz, Silvia Flubacher, and Philipp Senn, 211–30. Göttingen: V&R unipress, 2013.

Gebhard, Ulrich and Thomas Kistemann, eds. *Landschaft, Identität und Gesundheit. Zum Konzept der therapeutischen Landschaften*. Wiesbaden: Springer VS, 2016.

Grosser, Hermann. "Der Fremdenverkehr und seine Entwicklung vom 17. Jahrhundert bis zur Gegenwart im Appenzellerland: unter besonderer Berücksichtigung von Appenzell Innerrhoden." *Innerrhoder Geschichtsfreund* 22 (1977–1978): 45–60.

Habrich, Christa. "Medizin- und naturwissenschaftshistorische Aspekte der Alpen als Heilungs- und Erholungsraum." In *Die Alpen als Heilungs- und Erholungsraum. Historikertagung in Meran – Le alpi: luogo di cura e riposo: convegno storico a Merano: 19.–21.X.1988*, edited by Josef Nössing, 11–34. Bozen: Verlagsanstalt Athesia, 1994.

Hofmann, Heini. *Gesundheits-Mythos St. Moritz. Sauerwasser, Gebirgssonne, Höhenklima*. St. Moritz: Montabella-Verlag, 2011.

Hunziker, F. "Das Appenzellerland nach früheren französischen Reisebeschreibungen 1750–1840." *Appenzellische Jahrbücher* 63 (1936): 1–36.

Jankovic, Vladimir. "The Last Resort: A British Perspective on the Medical South, 1815–1870." *Journal of Intercultural Studies* 27, no. 3 (2006), 271–98.

Krämer, Daniel. *"Menschen grasten nun mit dem Vieh.": Die letzte grosse Hungerkrise der Schweiz 1816/17*. Basel: Schwabe 2015.

Lotz-Heumann, Ute. "Finding a Cure. Representations of Holy Wells and Healing Waters in Early Modern Germany." In *Wissenschaftsgeschichte und Geschichte des Wissens im Dialog – Connecting Science and Knowledge: Schauplätze der Forschung*, edited by Kaspar von Greyerz, Silvia Flubacher, and Philipp Senn, 233–54. Göttingen: V&R unipress, 2013.

Mathieu, Jon and Simona Boscani Leoni, eds. *Die Alpen! Zur europäischen Wahrnehmungsgeschichte seit der Renaissance – Les Alpes! Pour une histoire de la perception européenne depuis la Renaissance*. Bern, Berlin, Bruxelles: Lang, 2005.

Mathieu, Jon. *Eine Agrargeschichte der inneren Alpen. Graubünden, Tessin, Wallis 1500–1800*. Zurich: Chronos-Verlag, 1992.

Mathieu, Jon, Norman Backhaus, Katja Hürlimann, and Matthias Bürgi, eds. *Geschichte der Landschaft in der Schweiz. Von der Eiszeit bis zur Gegenwart*. Zurich: Orell Füssli, 2016.

Nägeli, Albert. "Die Molkenkuren im Appenzellerland." *Appenzeller Kalender* 220 (1941): n.p.

Neswald, Elizabeth. "Asserting Medical Identities in Mid-Nineteenth Century Ireland: the Case of the Water Cure in Cork." In *Science and Technology in Nineteenth-Century Ireland*, edited by Janina Adelman and Eadaoin Agnew, 32–47. Dublin: Four Courts Press, 2010.

Orland, Barbara. "Alpine Milk: Dairy Farming as a Pre-modern Strategy of Land Use." *Environment and History* 3 (2004): 327–64.

Orland, Barbara. "Le régime suisse du petit lait (18/19ème siècle)," in *Allaiter. Histoire(s) et cultures d'une pratique*, edited by Yasmina Foehr-Janssens and Daniela Solfaroli Camillocci. Turnhout: Brepols Publisher 2022 (forthcoming).

Palmer, Richard. "'In this our lightye and learned tyme': Italian Baths in the Era of the Renaissance." In *The Medical History of Water and Spas*, edited by Roy Porter, 14–22. London: Wellcome Institute for the History of Medicine, 1990.

Riley, James C. *The eighteenth-century campaign to avoid disease*. New York: Knopf, 1987.

Reichler, Claude. *La montagne réinventée. Géographes, naturalistes et sociétés (XVIIIe–XXe siècles) – Reinventing Mountain Areas: Geographers, Naturalists and Society (18th–20th centuries)*. Grenoble: Revue de géographie alpine (Institut de Géographie Alpine), 1994.

Reichler, Claude. *La découverte des Alpes et la question du paysage*. Chêne-Bourg: Georg, 2002.

Ruesch, Hanspeter. *Lebensverhältnisse in einem frühen schweizerischen Industriegebiet. Sozialgeschichtliche Studie über die Gemeinden Trogen, Rehetobel, Wald, Gais, Speicher und Wolfhalden des Kantons Appenzell Ausserrhoden im 18. und frühen 19. Jahrhundert*, Basler Beiträge zur Geschichtswissenschaft, vol. 139/140. Basel, Stuttgart: Helbing & Lichtenhahn, 1979.

Ruesch, Hanspeter. "Medizinhistorisches aus Appenzell Ausserrhoden (1800–1830)." *Gesnerus* 36 (1979): 21–34.

Sargent, Frederick II. *Hippocratic Heritage: a History of Ideas about Weather and Human Health*. New York: Pergamon Press, 1982.

Schläpfer, Walter. *Wirtschaftsgeschichte des Kantons Appenzell Ausserrhoden bis 1939*. Herisau: Appenzell-Ausserrhodische Kantonalbank, 1984.

Schürer, Christian. *Der Traum von Heilung. Eine Geschichte der Höhenkur zur Behandlung der Lungentuberkulose*. Baden: hier + jetzt, 2017.

Senn, Philipp. "Forscher vor Ort. Johann Jakob Scheuchzer (1672–1733), Bündner Gönner und die Balneologie." In *Wissenschaftsgeschichte und Geschichte des Wissens im Dialog – Connecting Science and Knowledge: Schauplätze der Forschung*, edited by Kaspar von Greyerz, Silvia Flubacher, and Philipp Senn, 255–92. Göttingen: V&R unipress, 2013.

Sperling, Jutta Gisela, ed. *Medieval and Renaissance Lactations. Images, Rhetorics, Practices*. Burlington: Ashgate, 2013.

Tissot, Laurent. "From Alpine Tourism to the 'Alpinization' of Tourism." In *Touring Beyond the Nation: A Transnational Approach to European Tourism History*, edited by Eric G. E. Zuelow, 59–78. Farnham, Uk: Ashgate, 2011.

Treichler, Hans Peter. "Molken, Milch und Traubenberge. Neue Ansätze im Kurwesen." In *Zauberberge. Die Schweiz als Kraftraum und Sanatorium*, edited by Felix Graf and Eberhard Wolff, 49–52. Baden: hier + jetzt, 2010.

Vaj, Daniela. "La géographie médicale et l'immunité phtisique des altitudes aux sources d'une hypothèse thérapeutique = [Medical Geography and Phthisic Immunity in the High Altitudes : the Origins of a Therapeutic Hypothesis]." *Revue de géographie alpine – Journal of Alpine Research* 93, no. 1 (2005): 21–33.

Vaj, Daniela. "Le Bon air des Alpes et le développement des stations climatiques d'altitude." In *Bon air des Alpes: des stations climatiques au tourisme de bien-être: actes du colloque, Sierre, 7–8.10.2004 – Die gute Alpenluft: von den Luftkurorten zum Wellnesstourismus: Tagungsband, Siders, 7.–8.10.2004*, edited by HES-SO Valais, 53–65. Sierre: HES-SO Valais, 2007.

Vogler, Werner. "Heilbäder in den Alpen im Spätmittelalter und in der Renaissance am Beispiel von Pfäfers." In *Die Alpen als Heilungs- und Erholungsraum. Historikertagung in Meran – Le alpi: luogo di cura e riposo: convegno storico a Merano: 19.–21.X.1988*, edited by Josef Nössing, 137–56. Bozen: Verlagsanstalt Athesia, 1994.

Walsham, Alexandra. *The Reformation of the Landscape. Religion, Identity, and Memory in Early Modern Britain and Ireland.* Oxford: Oxford University Press, 2011.

Walton, John K. "Health, Sociability, Politics and Culture. Spas and History: An Overview." In *Mineral Springs Resorts in Global Perspective. Spa Histories*, edited by John K. Walton, 1–14.

Walton, John K., ed. *Mineral Springs Resorts in Global Perspective. Spa Histories.* London, New York: Routledge 2014.

Weishaupt, Matthias. "'Viehveredelung' und 'Rassenzucht'. Die Anfänge der Appenzellischen Viehschauen im 19. Jahrhundert." In *Appenzeller Viehschauen*, edited by Mäddel Fuchs, 11–50. St. Gallen: Typotron AG, 1998.

Wood, Karl E. *Health and Hazard: Spa Culture and the Social History of Medicine in the Nineteenth Century.* Newcastle upon Tyne: Cambridge Scholars Publishing, 2012.

Wyder, Margrit. *Kräuter, Kröpfe, Höhenkuren: die Alpen in der Medizin – die Medizin in den Alpen; Texte aus zehn Jahrhunderten.* Zurich: Verlag NZZ, 2003.

Zimmer, Oliver. "In Search of Natural Identity: Alpine Landscape and the Reconstruction of the Swiss Nation." *Comparative Studies in Society and History* 40, no. 4 (1998): 637–65.

CHAPTER 11

Creation of "Scientific" Knowledge: The Asiatick Society and Exploration of the Himalaya, 1784–1850

Chetan Singh

1 Background and Context

The English East India Company started as a trading concern in 1600 CE, but by the early 19th century it ruled large parts of South Asia. This drew it deeper into matters of governance. It was, therefore, compelled to assess the resources of Company-held territories and understand their very diverse people. A trading company was drawn into the difficult task of governing an extensive area and unfamiliar people.

Early European accounts were its initial source of information. Many of these described markets and commodities. Others noted the "strangeness" of Indian practices and beliefs. These initial, sometimes misleading, descriptions attracted European curiosity. However, methodically collected data was needed for a better understanding. Increasing contact with dissimilar cultures prompted several "scientific explorations" to different parts of the globe. Information was gathered, examined and classified in accordance with European methods. Gradually, a systematic description of the world emerged. A European idea of modern science was gradually dispersed to other regions. For non-Europeans, this encounter with western science began an unimagined transformation. Their world was practically and conceptually re-ordered.

2 The Asiatick Society

In India, British bureaucrats, army officers and explorers personally initiated the study of various aspects of the country. Their work was better organised than the impressionistic travel writing that preceded. However, a forum for critical deliberation to establish the scientific veracity of this work was still lacking. William Jones (1746–1794), a linguist and scholar – also a judge of the Supreme Court at Calcutta – realised the importance of such a forum. Through his efforts, an organisation called "The Asiatick Society" was founded on 15 January 1784. Its name, though, went through several changes in the ensuing years.[1] William Jones explicitly stated the objectives of the Society in the inaugural address. They were to explore what lay within the "geographical limits of Asia" and to enquire into "whatever is performed by man or produced by nature."[2]

Jones remained the President of the Society from its inception till his death in 1794. His annual addresses to the Society articulated the grand issues, which spawned numerous smaller questions. They foretold the path that organized research in colonial South Asia would follow. The diversity of interests is revealed by the Society's appeal to its members for communications on "religion, policy, jurisprudence, manners and customs, geography (towns, countries, provinces, rivers, mountains), biography (Buddha, history of saints); commerce, natural history, *materia medica*, medicine and surgery, language and literature."[3]

In 1829, the Society began supporting a publication called *Gleanings in Science* that focused on more "scientific" subjects. It also sought to collect as much geographical and geological information as was possible.[4] The Society was gradually able to source information even from areas that lay beyond British influence. For example, Alexander Csomo Korosi a widely travelled Hungarian, requested the Society's support in 1834 to research the religious literature of Tibet which till then was "a vast *terra incognita* of Oriental literature."[5] The proceedings of Society meetings reveal the great diversity of issues being explored by researchers in regions unknown to Europeans.

George William Traill's, *Statistical Sketch of Kamaon* typically illustrates the accumulation of such comprehensive data by 1828. The geographical features and the flow of rivers and streams were described in detail. Footpaths, mule-tracks and trade-routes, passes, bridges of different kinds, buildings, temples, water springs and a host of other details were meticulously noted. Villages, hamlets and total number of houses were enumerated. The minerals, flora and fauna of Kumaon also found mention.

---

1 L. L. F., "The Bicentenary of the Birth of Sir William Jones, F.R.S., Founder of the Royal Asiatic Society of Bengal," *Notes and Records of the Royal Society of London* 4, no. 1 (April 1946): 58; http://www.jstor.org/stable/531239, accessed on 20 July 2019.

2 Anonymous, "Appendix: Rules of the Asiatick Society," *Asiatick Researches* 12 (1818).

3 Anonymous, "Desiderata," *Asiatick Researches* 6 (1801): iii–vi (passim).

4 Anonymous, "Preface," *Gleanings in Science* 2 (January to December 1830): viii.

5 M. Alexander Csomo Korosi, "Biographical Sketch of M. Alexander Csomo Korosi, the Hungarian Traveller; Extracted from a Letter Addressed by the Gentleman to Captain C.P. Kennedy, of the Hon. East-India Company's Service, Assistant to the Political Resident at Delhi, &c. Communicated by Charles Elliot, of the Hon. East-India Company's Civil Service, late Political Resident at Delhi, &c, &c," *The Journal of the Asiatic Society of Great Britain and Ireland* 1 (1834): 133.

3 Understanding the Other

An explicit notion of race grew as colonialism expanded. This notion seemed to geographically overlap with the concepts of "people" and "nation." These ideas intersected in the real world and became identifiable with one or the other region. China, Central Asia, Iran, Arabia, etc., were seen as peopled by nations or races that could be understood and classified as were other creatures of the natural world. This seemed in accordance with a world becoming increasingly "scientific." By the 18th century the idea of race emerged as a new and influential classificatory category.[6]

William Jones had a keen interest in the origin of diverse populations. The people of Asia particularly fascinated him. He believed they were descended from three populous "nations" called: the Hindu, Arabs and Tartars. These he divided into sub-groups that differed markedly from each other in appearance, language, religion, etc. Notwithstanding the diversity, he felt they had "a common root." William Jones argued that the Chinese, Japanese and Persians, too, descended from "stocks" that intermingled over a long period of time.[7] The lack of information on China which adjoined the British Indian empire, was of special concern to him. He saw the study of Chinese "laws, politics and morals" (instead from its "manufactures and trade") as a means of appreciating Asian thinking. Translations of important Chinese texts were valued as sources of Asiatic philosophy and laws.[8] The British hoped their location in India at "the utmost verge of the habitable globe known to the best geographers of Old Greece and Egypt" would allow them a glimpse into the extensive Chinese empire.[9]

Beyond the military might and high culture of grand empires were societies inhabiting the political and cultural fringes. Jones called them "races of borderers" located "on the limits of Arabia, Persia, India, China, and Tartary" and "wild tribes in the mountainous parts of those extensive regions."[10] Though peripherally situated, these peoples were integral to the political economy of

6 Snait B. Gissis, "'Race' in the Eighteenth Century," *Historical Studies in the Natural Sciences* 41, no. 1 (Winter 2010).

7 William Jones, "Fifth Anniversary Discourse on the Tartars. Delivered 21 February, 1788 by the President," *Asiatick Researches* 2 (1807): 40–41.

8 William Jones, "On the Second Classical Book of the Chinese," *Asiatick Researches* 2 (1807): 195. See also Jones, "Seventh Anniversary Discourse on the Chinese. Delivered 25 February, 1790 by the President," *Asiatick Researches* 2 (1807): 365.

9 Jones, "Seventh Anniversary Discourse on the Chinese," 365.

10 William Jones, "The Eight Anniversary Discourse on the Borderers, Mountaineers and Islanders of Asia. Delivered 24 February, 1791 by the President," in *Dissertations and Miscellaneous Pieces Relating to the History and Antiquities, the Arts, Sciences and Literature*

Asia. Attempts to explain social phenomena of this kind in India were part of the early colonial knowledge creation enterprise.[11] At an annual meeting of the Asiatick Society, William Jones argued that in the mountains of north India were "many races of wild people with more or less of that pristine ferocity, which induced their ancestors to secede from the civilized inhabitants of the plains and valleys."[12] The relatively isolated people of Tibet, too, were believed to have drawn upon old Sanskritic traditions of India.[13] Jones understood how fields of functional knowledge in India were overlaid by mythical belief. Therefore, he insisted that "accurate knowledge of Sanscrit and a confidential intercourse with learned Brahmins, are the only means of separating truth from fable."[14]

4 Himalayan Societies: Categorizing People

It was the creation of a vast Himalayan empire through conquests by the kingdom of Nepal – in the late 18th and early 19th centuries – that first attracted the British. The empire stretched from Sikkim in the east to the outer defences of the besieged Kangra fort in the far west. Its economic core was Kathmandu valley that lay on a trans-Himalayan trade route. The valley nurtured a rich culture and was economically connected with distant people and places. In 1790, Father Giuseppe a missionary in Kathmandu, was particularly impressed by the many rich and well-built temples of the towns in the valley.[15]

In much of the Himalaya, however, mountain communities were not particularly prosperous. They had a transhumant, agro-pastoral economy that was different from the agrarian system of the Gangetic plains. Historical texts repeatedly mention Khasa mountain tribes. Zahiruddin Babar, the sixteenth

of Asia, vol. 3, ed. William Jones et al. 1–20 (London: Vernor and Hood, Birchin-Lane, and Darton and Harvey 1796), 1.

11 Bhargya Bhukya, "The Mapping of the Adivasi Social: Colonial Anthropology and Adivasis," *Economic and Political Weekly* 43, no. 39 (Sept. 27–Oct. 3, 2008): 103.

12 Jones, "The Eight Anniversary Discourse on the Borderers, Mountaineers and Islanders of Asia," 10. See also Bhukya, "The Mapping of the Adivasi Social," 105, 106, 107. The idea of tribal communities abandoning the mainstream was romanticized in colonial anthropology after the first violent resistance to the British by the Pahariya Sirdars of Bihar in 1778.

13 Jones, "The Eight Anniversary Discourse on the Borderers, Mountaineers and Islanders of Asia," 13.

14 William Jones, "The Tenth Anniversary Discourse on Asiatic History: Civil and Natural, Delivered 28 February, 1793 by the President," *Asiatic Researches* 4 (1807): xxiv.

15 Father Giuseppe, "Account of the Kingdom of Nepal by Father Giuseppe, Prefect of the Roman Mission. Communicated by John Shore," *Asiatick Researches* 2 (1807): 310.

century founder of the Mughal empire, wrote in his memoirs that Khasas inhabited the extensive mountain range. Three centuries later, Captain Francis Wilford's report to the Asiatic Society again referred to the Khasas as a tribe that inhabited the mountain range from the 'eastern limits of India to the confines of Persia' since very ancient times.[16] Wilford was pursuing the ideas of William Jones to understand peoples in a more "scientific" manner. In this colonial endeavour, the classification and characterisation of Himalayan people was part of this new anthropological enterprise.

European explorers saw the Himalaya as a meeting point of two "nations": the "Hindu" and "Tartar." Each of these "nations" were believed to have distinguishing characteristics. The degree to which these were present or absent in inhabitants also defined the territory. Internal distinctions within different mountain tribes and communities were similarly made. The popular myth of martial hill communities, that provided recruits to the British Indian army, originated in such distinctions.

During his search for the source of river Ganges, Raper became familiar with the people of Garhwal. He was surprised that Manah, a town bordering Tibet, had inhabitants of a "different race" with "features of the Tartars or Butias," but who professed Hinduism and called themselves Rajputs.[17] Raper had till then not noticed the lack of religious and caste rigidity in Himalayan societies.[18] Four years later (1812), William Moorcroft (1765–1825) a veterinarian turned trader-explorer, passed through another part of Garhwal on his way to Mansarovar lake (Tibet) through the Niti pass. His observations echo those of Raper. At Malari village, the people had a "Tartar countenance mixed with that of the Hindu."[19] The transhumant agro-pastoral nature of the higher Himalayan economy was noticed by all travellers. The inhabitants of Malari, too, migrated in winter to villages in lower areas. Moorcroft also noticed that the Malaris functioned as intermediaries in the trade between trans-Himalayan Tibet and the lower hills of India.[20]

Sheep-raiding was practised against rival villages. Fraser and Hodgson (passing through Garhwal in 1820 and 1822 respectively) were astonished at the brazenness of the robbers. At Durali village, Fraser learnt that all the men had "gone to buy corn, or to steal sheep," seemingly part of normal "business,

16 Francis Wilford, "On Mount Caucacus," *Asiatick Researches* 6 (1801): 455–56.

17 F. V. Raper, "Narrative of a Survey for the Purpose of Discovering the Sources of the Ganges," *Asiatic Researches* 11 (1812): 525–26.

18 Ibid., 527–28.

19 William Moorcroft, "A Journey to Lake Manasarovara in Un-des, a Province of Little Tibet," *Asiatick Researches* 12 (1818): 398–99.

20 Ibid., 399.

too ordinary and common to conceal."[21] He later met these men returning with about 500 stolen sheep and goats. They were pursued by another 100 or so armed men – not owners of the sheep – but themselves intending to rob the robbers. Fraser was surprised that they saw nothing illegal in their acts. His views and those of the hill-men represented opposing perspectives on legality and justice.[22] Similar transhumant practices existed – with some variations – in eastern Himalayan Bhutan and other areas of north-eastern India.[23] Here shifting cultivation (*jhoom*) was common. This required the slashing and burning of thick forests for cultivation and then shifting to another place every few years.

The delineation of Himalayan people, their peculiarities and their territories remained an integral part of British interest. An example, is a study of three tribes (Lepchas, Limbos and Bhotias) resident in Sikkim wherein their social ranking and the areas that they dominated are demarcated.[24] In parts of Garhwal, it was the unhygienic state of houses and the lack of material possessions that struck the traveller.[25] People were often judged and identified by physical appearance. For instance, Bashahris (inhabitants of Kinnaur) were described as:

> stout well-built figures, are frequently fair [...]; their eyes often blue, and their hair and beards curled, and of a light or red colour. They seem admirably calculated to be formed into soldiers for a hilly region. Here and there traces may be detected of the Tartar features, the small eye, high cheekbones and meagre moustaches [...][26]

---

21 James B. Fraser, "An Account of a Journey to the Sources of the Jumna and Bhagirathi Rivers," *Asiatick Researches* 13 (1820): 215.

22 Ibid., 235; J. A. Hodgson, "Journal of a Survey to the Heads of the Rivers, Ganges and Jumna," *Asiatick Researches* 14 (1822). See 126 for customary sheep-raiding practices of people of Rawain and Bashahr.

23 Kishen Kant Bose, "Some Account of the Country of Bhutan," *Asiatick Researches* 15 (1825): 141, which mentions that cold winters compelled peasants to shift down to lower areas. They maintained two houses in separate areas to cultivate crops in different seasons.

24 J J. T., "Some further particulars of the Country of Siccim, and of its Inhabitants, the Lepchas and Bhotias," *Gleanings in Science* 2 (1830): 180.

25 Moorcroft, "A Journey to Lake Manasarovara in Un-des, a Province of Little Tibet," 399; Hodgson, "Journal of a Survey to the Heads of the Rivers, Ganges and Jumna," 65 mentions that though the houses are well built with wood and stone, they are "exceedingly filthy within, and full of vermin."

26 Fraser, "An Account of a Journey to the Sources of the Jumna and Bhagirathi Rivers," 192.

In terms of cultural characteristics, Raper regarded the lack of "bashfulness and reserve" amongst women of the higher mountains as a mark of being "more civilized."[27] "Racial" traits and behavioural habits were seen to overlap. Herbert noted that "Tartar" women (referring to Kinnaur women) had fair complexions that "emulate those of Europe." He reiterated Raper's observation that these women "exhibited a variety of character and expression" different from the women of Hindustan.[28] Herbert delighted at seeing them fetch water as they did in Europe: "with ruddy cheerful countenances, unconcealed and unsuspicious, and a wooden pail under the arm."[29]

Colonial administrators soon discovered that the Himalaya encompassed considerable socio-cultural diversity. At the Shipki pass – between Kinnaur with Tibet – Herbert saw an "assemblage of Hindustanis, Kanawaris, and Tartars seated in groups; the contrast of whose dresses was scarcely less striking than that of their features and of their speech."[30] This made it abundantly clear that knowledge about Hindu society of the Gangetic doab was not of great help in understanding the religion and society of the mountains. Fraser's description of a religious celebration at the Garhwali village of Cursall further reinforces this. At the celebration, liquor was consumed in huge quantities, loud music was played and everyone danced together. Fraser called these "savage and inconsistent" practices enough to shock even "Hindus who are less bigoted."[31] The heterodoxy that marks Himalayan Hinduism, was fully appreciated only much later.

The presence of a large Buddhist population further differentiated the Himalaya and trans-Himalaya from the plains. The value of ancient Buddhist religious texts preserved in Himalayan monasteries was gradually appreciated by European explorers, scholars and administrators with the help of Indian specialists. These were later translated and helped retrieve a substantial part of Buddhist philosophy. However, some early 18th century European explorers regarded literature of this kind as impractical for impoverished Himalayan societies. Hodgson lived many years in Nepal where books on Buddhism

27 Raper, "Narrative of a Survey for the Purpose of Discovering the Sources of the Ganges," 472.

28 J. D. Herbert, "An Account of a Tour made to lay down the Course and Levels of the River Setlej or Satudra, as far as Traceable within the Limits of the British Authority, performed in 1819," *Asiatic Researches* 15 (1825): 372.

29 Ibid., 392.

30 Ibid., 374. On his way up to Shipkila, Herbert noticed that he had entered a zone of cultural transition. He wrote: "Hitherto we had been accustomed to brahmins, (of a degenerate race, no doubt) but still Hindus, but here we had the worship of Budh fairly established […]".

31 Fraser, "An Account of a Journey to the Sources of the Jumna and Bhagirathi Rivers," 192.

proliferated through printing technology introduced from China. Yet, he was surprised that "literature of any kind should be so widely diffused as to reach persons covered with filth."[32] Furthermore, he felt that literature should bring about change as it did in Europe.[33] European interest in Himalayan Buddhist literature and the "moral feelings and religious principles of the people" of that region, nevertheless, continued to grow.[34]

5 The Geography of Disease

There were some diseases and physiological problems that particularly afflicted people in the mountains. The journals of European explorers of the early 19th century specifically mention these.

5.1 *Goitre*

Goitre was common in areas where iodine had leached out from the soil through the effects of snow, water flow and heavy rainfall. Among these were the mountain areas of Europe, the Andes of South America, highlands of China, the uplands of Africa as well as North India, especially the Himalayan region. Thomas Hardwicke, who travelled through the Garhwal-Himalaya in 1799, saw many people affected by this deficiency in Srinagar. But they seemed to suffer no pain or discomfort from it. After discussions with local residents he concluded that goitre affected "natives who reside near rivers which receive increase from the melting snows."[35] This explanation provided by locals was widely accepted even by Europeans who came later. Raper (1808) met people with large goitres near Chamoli village and initially concluded that those who lived 'in the vicinity of snow' were especially prone to this.[36] But he began doubting this local wisdom when he met many goitre afflicted people at

32 Brian Houghton Hodgson, "Notices of the Language and Literature, and Religion of the Bauddhas of Nepal and Bhot," *Asiatic Researches* 16 (1828): 419.

33 Ibid., 421. He wrote dismissively: "[…] the general diffusion of books (that most potent of spurs to improvement is our ideas) becomes, in Bhot," an "agreeable means of filling up the tedious hours of the twilight of civilization."

34 Horace Hayman Wilson, "Note on the Literature of Thibet," *Gleanings in Science* 3 (January to December 1831): 243–44. The Asiatic Society successfully acquired the multi-volume texts (the complete sets being a total of 325 volumes) of two major Tibetan Buddhist works.

35 Thomas Hardwicke, "Narrative of a Journey to Sirinagur," *Asiatick Researches* 6 (1801): 344.

36 Raper, "Narrative of a Survey for the Purpose of Discovering the Sources of the Ganges," 473.

Sancot village in the lower hills where the water was not drawn from a snow-melt source.[37]

Traill, who lived for long in the region, knew how widespread the goitre was. Yet, he failed to understand its cause. He realized that water from snow-melt did not cause the problem but felt that: "Goitre is in part produced by the effects of the keen mountain air acting on the exposed throat."[38] By the mid-19th century, iodine was being used for treating goitre. But it was not till the end of the century that the nature and cause of the disease came to be fully explained.

5.2 *High Altitude Sickness*

Explorers often experienced the debilitating effects of rarefied air at high altitudes. Local people too suffered whenever they ascended to higher heights. The latter believed that travellers became senseless because certain flowers in high mountains released "poisonous perfumes."[39] Around 1822, Hodgson's search for the source of the Ganges and Yamuna took him to the confluence of the Bhagirathi and Jahnavi rivers. To the north of this confluence, near Nilang village, was a difficult pass through which Indian traders crossed into Tibet. A local Brahmin informed Hodgson that travellers had difficulty in breathing while traversing this pass because of "the *bi'ch* or *bish*, i.e. exhalations from poisonous herbs which grow on the bare high knolls."[40]

Fraser experienced the debilitating effect of high altitude sickness on his trek across an elevated ridge between the Yamuna catchment area and the Ganges valley. His exhausted porters complained of "poisoned wind." He too "experienced much difficulty and oppression, as if there were an insufficiency of air [...]." Fraser realised that "this supposed poison was nothing more than the effect, which the rarefied state of the air, from the great height we have reached, has on the lungs [...]."[41] A complete explanation still eluded him. Herbert's experience at Gunas pass (4775 m) above Sangla village in Kinnaur was similar. He refers to the "fatigue" and "hurried breathing" he suffered and concluded like Fraser that these were caused by "a want of sufficient air".[42]

---

37 Ibid., 551.

38 George William Traill, "Statistical Sketch of Kamaon," *Asiatic Researches* 16 (1828): 215–16.

39 Fraser, "An Account of a Journey to the Sources of the Jumna and Bhagirathi Rivers," 200–01.

40 Hodgson, "Journal of a Survey to the Heads of the Rivers, Ganges and Jumna," 92.

41 Fraser, "An Account of a Journey to the Sources of the Jumna and Bhagirathi Rivers," 201.

42 Herbert, "An Account of a Tour made to lay down the Course and Levels of the River Setlej or Satudra, as far as Traceable within the Limits of the British Authority, performed in 1819," 347.

During his second attempt on Mt. Blanc in 1787, Horace-Bénédict de Saussure (1740–1799) experienced severe mountain sickness caused by "the rarity of the air" as it had "hardly more than half of its density."[43] Many years after de Saussure, Humboldt recounted his experience of climbing Chimborazo. He argued that oxygen in the air at higher altitudes was the same as at lower elevations, but the lower barometric pressure reduced inhalation of adequate amount of oxygen. Despite his flawed argument, Humboldt came nearest to providing an explanation for high altitude sickness by placing oxygen at the centre.[44] Only years later was it understood that insufficiency of oxygen in the blood stream at high altitudes critically affected the functioning of body organs.[45]

6 Navigating the Wilderness: Resources, Trade and Traders

Early surveys of the Himalaya had little prior information to rely upon. Descriptions by preceding travellers evoked inquisitiveness amongst British administrators. This prompted further studies. Interests of the East India Company were the prime concern. In 1829, a contributor to *Gleanings in Science* felt that the Himalayan possessions of the Company offered enormous unexploited wealth.[46] On the contrary, British reports argued that the region suffered extreme impoverishment under Nepalese rule (late 18th century to 1815). When the British annexed some parts of it in 1816, the region's agriculture and trade were reportedly in decline.[47]

However, networks created over centuries do not disappear easily. Himalayan societies survived through interdependence. William Moorcroft, who traversed extremely difficult trans-Himalayan areas in 1812, illustrates this point. His notes reveal the complex socio-economic (even political) network

43 As quoted in Charles S. Houston, David E. Harris, and Ellen J. Zeman, *Going Higher: Oxygen, Man, and Mountains*, 5th ed. (Seattle: The Mountaineers Books, 2005), 85.

44 Ibid., 86.

45 In 1862, Denis Jordanet, a French physician discovered that low levels of oxygen in the blood caused high altitude sickness (hypoxia). Paul Bert is also sometimes credited with being the first person to make this discovery in 1878. The two knew each other and were familiar with each other's work.

46 Anonymous, "On the Introduction of the Iron Chain Suspension Bridge into the Himmalaya Mountains," *Gleanings in Science* 1, no. 12 (12. December 1829): 349–50.

47 Raper, "Narrative of a Survey for the Purpose of Discovering the Sources of the Ganges," 495, 496, 499. Raper argued that heavy exactions of the Nepali rulers caused a decline of trade. Garhwal was compelled to pay an annual sum of Rs. 3000 to avoid annexation. In 1803, Garhwal was nevertheless annexed.

that endured into modern times in the most challenging landscapes. Moorcroft had to skilfully negotiate his passage through the intricate web that knit the region together. At Niti, the last Indian village on the Tibet border, he had to persuade reluctant local people to help him cross into the Tibetan plateau. He sighted Gartok (a collection of "black tents") in the distance as he descended into the plateau. The intervening plain was "covered with prodigious bodies of sheep, goats and yaks" estimated at not "less than 40,000".[48]

This nondescript trading post on a desolate plain was a busy summer market for traders from distant places. Gartok was connected – directly or indirectly – with Ladakh, Kashmir, Amritsar, Delhi, Yarkand, Bokhara, Bashahr and numerous other centres of trade. Even Russian traders came to Gartok with large caravans.[49] Moorcroft learnt about the Bashahr route from traders at Gartok and wished to explore it on his return. The Tibetan administration, however, refused permission and he reluctantly returned by the route he had entered.

Herbert, who visited Kinnaur in 1819, heard about the commercial importance of Gartok (also known as Garo) and its rich and extensive pastures to which large Kinnaur flocks were herded in summer.[50] A Tibetan trader reconfirmed to Herbert that Yarkandi traders visited Gartok.[51] The extensive economic linkages of Himalayan societies with trans-Himalayan Tibet and through it to Central Asia were slowly being revealed.[52] Rampur – the capital of Bashahr state – was the most important Indian trading town corresponding to Gartok. Herbert, who witnessed the annual trade fair at Rampur, has described how transactions were conducted by traders from different countries.[53]

Trade was crucial for the survival of communities in the Himalaya. So was the supervisory influence exercised by political structures. In the hostile, barren terrain of the greater Himalaya, regulation of *points of crossing* and *locations*

48 Moorcroft, "A Journey to Lake Manasarovara in Un-des, a Province of Little Tibet," 444. He describes Gartok (Ghertope) as an "assemblage of tents in clusters made of blankets surrounded by hair ropes fixed to stakes."

49 Ibid., 452–54.

50 Herbert, "An Account of a Tour made to lay down the Course and Levels of the River Setlej or Satudra, as far as Traceable within the Limits of the British Authority, performed in 1819," 359.

51 Ibid., 378.

52 Ibid., 340, 368.

53 Ibid., 402–03; Fraser, "An Account of a Journey to the Sources of the Jumna and Bhagirathi Rivers," 340; says that the remotest parts of Satlej valley were connected with neighbouring regions. Considerable trade existed between western Tibet to the north and the Panjab to the south. Garhwal villages, on the other hand, were isolated. This explained the economic disparity between Bashahr and Garhwal.

for interaction was no less important than maintenance of *lines of separation*. Political allegiances were tempered by pragmatic necessities and mutual interest. Communities at the border were the crucial link and had the aptitude to win the trust of different sides. Typically, people of Tungsah, a frontier area, paid land revenue to Tibet. They also sent a tribute of blankets to Bashahr every third year and a complimentary tribute of raisins to Garhwal. These payments kept the trade flowing while the linguistic skills of the Tungsah people kept the traders' conversation going.[54]

7 Where Do the Rivers Come From?

Hindu pilgrimage centres are often located along rivers. Almost by default, itinerant holy men and mendicants become the most knowledgeable about them. Param Swantantra Prakashanand Brahmachari, a religious ascetic and pilgrim whose family was settled in eastern India, was one such person. He had travelled to Mansarovar lake in Tibet through Nepal and returned to India by the Kashmir route.[55] Jonathan Duncan (1797) mentions two other *fakirs* (ascetics) who were great travellers. They journeyed through the Himalaya and Tibet to Mansarovar as well as to distant places in the Caspian, Moscow, Central Asia and the Middle East.[56] Little was, however, obtained from their testimonies. Suspicions that "natives" tended to hide the truth and the conviction that western science was vastly superior to indigenous knowledge made European explorers sceptical about information provided by Indians. Contrary to information provided by knowledgeable Indians, Major James Rennell – the first Surveyor-General of India – had argued that the Ganges originated near Mansarovar in Tibet, and then cut across the Himalaya into India. As part of the colonial policy to verify and record, therefore, an 'actual survey of the Ganges, above Haridwar to the farthest point to which it had been traced by Hindu pilgrims' was ordered.[57]

Lieutenant Webb, assistant to the Surveyor-General of India, was placed in charge and was assisted by Captain F. V. Raper and Captain Hearsay. In 1808,

54 Hodgson, "Journal of a Survey to the Heads of the Rivers, Ganges and Jumna," 92.

55 Param Swatantra Prakashanand Brahmachari, "Translation of the Relation Delivered by Purrum Soantuntre Purkasanund Brehmchary of his Travels and Life: Delivered on the 14th of August 1792," *Asiatick Researches* 5 (1807).

56 Jonathan Duncan, "An Account of Two Fakeers with their Portraits," *Asiatick Researches* 5 (1807).

57 Henry Thomas Colebrooke, "On the Sources of the Ganges in the Himadri or Emodus," *Asiatic Researches* 11 (1812): 441.

Raper submitted a report on this survey clearly stating that the source of the Ganges lay to the south of the Himalaya. Webb, reiterated this fact in a letter to the Society in 1810, and called it the "most important information gained."[58] In 1812, Moorcroft's report on his own expedition to Mansarovar in Tibet was published in the Society's journal. Herein he correctly asserted that the Ganges did not originate in Tibet. But he also erred significantly in another way. Henry Thomas Colebrooke, (1765–1837) an Orientalist, Sanskrit scholar and President of the Asiatic Society of Calcutta highlighted three important achievements of Moorcroft's expedition in his introduction to the Report. He wrote it: (1) "ascertained, and approximately determined the situation of Manasarovara, verifying at the same time the fact, that"; (2) "it gives origin neither to the Ganges," (3) "nor any other of the rivers reputed to flow from it."[59] While the first two points were correct, the last of these claims was entirely false. Despite reaching the lake, Moorcroft had failed to discover the surrounding glaciers from which four major Himalayan rivers originate. Yet, his erroneous assertion that no river originated from the lake was readily accepted by those in authority.

British occupation of Kumaon and Garhwal after 1816 made it easier for "scientific" explorations to be conducted in the area. James Fraser volunteered to explore the higher valleys of the Ganges and Yamuna.[60] He first went up to Jamnotri, very near the source of the Yamuna. His account provides details of the temple, its physical surroundings and the vegetation around.[61] At Gangotri, however, the actual source of the Ganges river was much higher up. His exploration around Gangotri reconfirmed that unlike the Satlej, the Ganges originated south of the Himalaya and did not cut across.[62] Having failed to reach the actual source, Fraser (unlike some other Europeans) accepted the testimony of local guides and traders who were well informed.

In 1822, J. A. Hodgson trekked up the Ganges valley beyond the Gangotri. Raper and Webb had already explored the lower portion in 1808 and 1810 respectively.[63] Hodgson hoped to locate the exact source of the Ganges; to measure the heights and locations of villages and also to mark the course of the rivers. His equipment was probably superior to that used in earlier explorations.

---

58 Ibid., 442.

59 Moorcroft, "A Journey to Lake Manasarovara in Un-des, a Province of Little Tibet," 473.

60 James Fraser was the brother of William Fraser, First Assistant, Delhi Residency who was at this time assigned the task of administering the Garhwal territory recently freed from Nepali control. See Fraser, "An Account of a Journey to the Sources of the Jumna and Bhagirathi Rivers," 171.

61 Ibid., 195–97.

62 Ibid., 224–27.

63 Hodgson, "Journal of a Survey to the Heads of the Rivers, Ganges and Jumna," 118.

On proceeding beyond Gangotri he discovered that the river disappeared beneath the snow for about 4 or 5 miles.[64] Regarding the Yamuna, Hodgson realised that the maps he had were inaccurate. The Tons, a tributary of the Yamuna, was for instance, about thrice the size of the latter at their junction.[65]

The territories to the east of the Satlej river came under British control in 1816. Soon thereafter in 1819, J. D. Herbert surveyed the Satlej watershed area.[66] He travelled up to Shipki pass where Tibetan authorities blocked his intended passage to the source of the Satlej. However, at Dubling village near the Tibetan border, Herbert met its headman Lama Ring Jing [sic] (probably Rinzen) who had travelled extensively in Tibet. Rinzen revealed that the Satlej flowed past Chaprang in Tibet that was nine days journey from Dubling. At Chaprang, too, the river was large and unfordable. Mansarovar (Mapang) was 18 days' journey from Shipki pass. More interestingly, Lama Rinzen told Herbert that Mansarovar (Mapang) lake was the source of four major rivers that flowed into India from different directions. Unfortunately, Moorcroft's mistaken information had already found wide acceptance. Herbert refused to accept the lama's statement. His disbelief perhaps indicates initial European scepticism towards information provided by "natives." Herbert was intrigued why Rinzen insisted on something "so completely contradicted by Mr. Moorcroft's journey, and which no one can believe to be other than some legend of their sacred books."[67]

Mansarovar lake held the answers to several questions agitating 19th century British explorers. One of these was the mysterious source of the Brahmaputra that flowed into India's far eastern region of Assam. A brief report about Assam was sent by John Neufville in 1828 to the Asiatic Society.[68] He described different tribes: some of whom were nominally "Hindu" while others he saw as clearly "wild."[69] He encountered many of these tribes in his attempt to trace the course of the Brahmaputra. Tribal resistance and the difficulty of

64 Ibid., 118–19.

65 Ibid., 129.

66 Captain Alexander Gerard and Dr. Govan had travelled along the river in 1817. Again in 1818, Alexander was accompanied by his brother Dr. J. G. Gerard. Their journals remained inaccessible till they were later edited by George Lloyd and published by James Madden & Co., London, 1841.

67 Herbert, "An Account of a Tour made to lay down the Course and Levels of the River Setlej or Satudra, as far as Traceable within the Limits of the British Authority, performed in 1819," 369: "[…] he maintains, how much soever questioned that four rivers originate from it: – 1. Tamja Kampa flows through Usang; 2. Mamja Kampa through Purang; 3. Lang Jing Kampa through Kanawer; and 4. Sin Jing Kampa through Ladak."

68 John Bryan Neufville, "On the Geography and Population of Asam," *Asiatic Researches* 16 (1828).

69 Ibid., 334.

navigating turbulent streams, prevented a definite conclusion. Predominant thinking was that the Brahmaputra – like the Ganges – originated in the Himalaya. Neufville, however, regarded this as "far from probable." But he mistakenly thought the Brahmaputra had the same source as the Irrawaddy that turned towards Burma.[70] Neufville guessed correctly that the Dihong (called the Siang further north), which flowed down from the north and joined the "Brahmaputra" in Assam, was a river that cut across the Himalaya. Its heavy flooding and strong current indicated its elevated trans-Himalayan origins. But he wrongly suggested that the Dihong also originated in the glaciers of eastern Tibet.[71] European explorers till then were unaware that the Tsangpo (Tibetan name of Brahmaputra) arose in the far west, flowed into eastern Tibet and then turned sharply southwards. It cut across the Himalaya through deep and narrow gorges and descended to Assam as the Siang (later Dihong) that then became the Brahmaputra.

In 1830, Lieutenant R. Wilcox, was deputed to survey Assam. He chanced upon a Chinese lama-pilgrim who had travelled through the area for the last ten years. To illustrate the course of his journey, Kho-shang Lama drew a sketch for Wilcox that solved the puzzle troubling the British. It revealed that the Tsangpo did not flow into Burma to become the Irrawaddy but penetrated the towering Himalaya into Assam to became the Brahmaputra.[72]

8 How High Are the Mountains?

It was the enormity and ruggedness of the Himalaya that first struck Europeans visitors. A description of the Garhwal Himalaya illustrates this fact:

> From the edge of the scarp, the eye extended over seven of eight distinct chains of hills, one rising above the other, till the view was terminated by the *Himalaya*, or snowy mountains. It is necessary for a person to place himself in our situation before he can form a just conception of the scene. The depth of the valley below, the progressive elevation of the intermediate hills, and the majestic splendour of the "cloud-capt"

70 Ibid., 336.

71 Ibid., 335, 336.

72 R. Wilcox, "On the Identity of the Sampu and Irawadi Rivers," *Gleanings in Science* 2 (1830): 66–67. To quote Wilcox, the lama: "travelled from Lhassa to Koombo, from Koombo to Deva Dharma, from Deva Dharma to Assam, from Assam to Bengal. Says the road from Lhassa to Koombo lies occasionally on bank of river, occasionally over mountains."

> Himalaya, formed so grand a picture, that the mind was impressed with a sensation of dread rather than of pleasure.

Fraser described the "scenery" of the Yamuna valley becoming increasingly "savage" as he went higher. Upon his departure from upper Satlej valley in Kinnaur, Captain Herbert heaved a sigh of relief. His journey through "naked and lofty precipices" and "broken declivities" had numbed the senses to a point where they "had, in Macbeth's words, 'supped full with horror'."[73]

Thomas Hardwicke who was in Srinagar (Garhwal) in 1796, also mentions the ranges that were visible from Srinagar. His visual calculation of the distance to the snow range prompted him to dispute the estimate made by the well-known landscape artist Thomas Daniell (1749–1840) who – along with his nephew William Daniell – visited Srinagar in 1789. On the basis of information provided by local people, Daniell had concluded that the base of the snow-covered mountains was only 14 or 15 "geographic miles" from where he stood. Hardwicke, however, calculated the snow range at no less than "eighty *English* miles" distant.[74] The difference in these estimates was important because it affected calculations pertaining to the heights of peaks.

Early estimates of the height of the Himalaya were met with scepticism. In the absence of proper instruments, a definite conclusion was difficult. Yet the towering presence of the Himalaya was obvious. Henry Thomas Colebrooke was reluctant to accept the initial findings. Rather grudgingly he suggested that "Without however supposing the Himalaya to exceed the Andes, there is still room to argue, that an extensive range of mountains, which rears, high above the line of perpetual snow, in an almost tropical latitude, an uninterrupted chain of lofty peaks, is neither surpassed nor rivalled by any other chain of mountains but the Cordilleras of the Andes."[75]

The British government in Bengal had authorized a systematic survey in 1807 to determine the sources of important rivers and location of important places. It further directed the height of every halting place to be documented. The latter was to enable trigonometrical calculations for determining the heights

73 Herbert, "An Account of a Tour made to lay down the Course and Levels of the River Setlej or Satudra, as far as Traceable within the Limits of the British Authority, performed in 1819," 399.

74 Hardwicke, "Narrative of a Journey to Sirinagur," 345. [The geographical mile is a unit of length determined by 1 minute of the arc along the Earth's equator. For the 1924 International Spheroid this equalled 1855.4 metres. The English mile at that point of time would have been about 1609.3 metres.]

75 Colebrooke, "On the Sources of the Ganges in the Himadri or Emodus," 445.

of Himalayan peaks. Raper's tour, mentioned above, was part of this effort.[76] Colebrooke now re-examined the supplementary material collected by Webb and the information provided by Lieutenant Colonel Crawford's survey of 1805. Crawford had estimated the heights of some peaks around Kathmandu when he was there in 1802.[77] These mountains were visible from as far as Patna in Bihar. Colebrooke was thereafter compelled to revise his views. He stated explicitly: "I consider the evidence to be now sufficient to authorize an unreserved declaration of the opinion, that the Himalaya is the loftiest range of Alpine mountains which has been yet noticed, its most elevated peaks greatly exceeding the highest of the Andes."[78]

Several Himalayan peaks were visible from certain locations even in distant Bengal. This implied that these peaks were higher than earlier believed.[79] On his journey through the Gangetic plain, Webb halted at Gorakhpur from where Dhaulagiri peak was clearly visible. He noted measurements from four different stations to assess its height. His estimate of the peak at 26,862 feet (8,187 metres) above sea level was fairly accurate despite the lack of sophisticated instruments.[80] Measurements pertaining to several other mountain peaks were also taken: many of these located in Nepal. Their heights were calculated and listed in the journal of the Asiatic Society.[81]

Systematic surveys were conducted across the Himalaya as part of colonial policy. In 1820, Captain W. S. Webb recorded the latitudes, longitudes and elevations of important places in Kumaon.[82] Apart from being an "extension of geographical knowledge," the measuring of heights and positions of the important mountain peaks was important for accurately determining "the geographical position of any place, from whence one, or more of them are visible."[83] In these

76 Raper, "Narrative of a Survey for the Purpose of Discovering the Sources of the Ganges," 447.

77 Henry Thomas Colebrooke, "On the Heights of the Himalaya Mountains," *Asiatick Researches* 12 (1818): 263–64.

78 Ibid., 253.

79 Ibid., 256. Colebrooke suggested that considering the state of the atmosphere, only an 'elevation exceeding 28,000 feet' would be visible at such a distance.

80 Colebrooke, "On the Heights of the Himalaya Mountains," 267–70. Dhaulagiri I stands at 26,795 feet (8,167 metres) above sea level. By way of comparison, the report indicates that George Shuckburg calculated the height of Mont Blanc by this method at 15,662 feet. This is a bit short of 15,780 feet (4810 metres) estimated today.

81 Ibid., 276.

82 W. S. Webb, "Memoir Relative to a Survey of Kemaon, with some Account of the Principles, Upon Which It has been Conducted," *Asiatick Researches* 13 (1820).

83 J. A. Hodgson and J. D. Herbert, "Account of the Trigonometrical and Astronomical Operations for determining the Heights and Positions of the principal Peaks of the Himalaya Mountains," *Asiatick Researches* 14 (1822): 188.

explorations, colonial officers emulated the best European scholars and followed the methods of the most "skilful observers." Webb took pains to explain the method he had adopted. He learnt subsequently that Humboldt, too, had used the same method in his survey of Mexico and that the latter had adopted the tables prepared by Mendoza de Rios. He referred specifically to the great interest evoked by Humboldt's study on "New Spain" but argued that there was a need to provide accurate measurements for creating vertical sections. In February 1830, the author of a letter to the editor of *Gleanings in Science* wrote, "Perhaps, in England they think that officers of the army are unequal to the task, but really it is not mysterious; good instruments, time, care, perseverance, and a moderate skill in calculation are all that is required."[84] Noting the unwillingness of Europeans "to believe that the Himmalaya are higher than the Andes," the letter claimed that Humboldt himself would "fully confirm" this and other related research were he to come to India.[85]

While much was discovered about the heights of peaks and the course of rivers; there was even more that remained unknown about the Himalaya and its people. Early British explorers and contributors to the journal of the Asiatic Society had made an exciting beginning. The future promised to be even more fascinating.

Bibliography

An abbreviated title *Asiatick Researches* has been used. The full form being: *Asiatick Researches; or, Transactions of the Society Instituted in Bengal, for Inquiring into the History and Antiquities, the Arts, Sciences, and Literature of Asia*].

Anonymous. "Appendix: Rules of the Asiatick Society." *Asiatick Researches* 12 (1818): i–viii.

Anonymous. "Desiderata." *Asiatick Researches* 6 (1801): iii–vi.

Anonymous. "Letter from the Himalaya." *Gleanings in Science* 2, no. 14 (February, 1830): 48–52.

Anonymous. "On the Introduction of the Iron Chain Suspension Bridge into the Himmalaya Mountains." *Gleanings in Science* 1, no. 12 (12. December 1829): 349–56.

Anonymous. "Preface." *Gleanings in Science* 2 (January to December 1830): v–viii.

Bhukya, Bhargya. "The Mapping of the Adivasi Social: Colonial Anthropology and Adivasis." *Economic and Political Weekly* 43, no. 39 (Sept. 27–Oct. 3, 2008): 103–09.

84 Anonymous, "Letter from the Himalaya," *Gleanings in Science* 2, no. 14 (February, 1830): 49.

85 Ibid., 49.

Bose, Kishen Kant. "Some Account of the Country of Bhutan." *Asiatick Researches* 15 (1825): 128–56.

Brahmachari, Param Swatantra Prakashanand. "Translation of the Relation Delivered by Purrum Soantuntre Purkasanund Brehmchary of his Travels and Life: Delivered on the 14th of August 1792." *Asiatick Researches* 5 (1807): 49–53.

Colebrooke, Henry Thomas. "On the Sources of the Ganges in the Himadri or Emodus." *Asiatic Researches* 11 (1812): 429–45.

Colebrooke, Henry Thomas. "On the Heights of the Himalaya Mountains." *Asiatick Researches* 12 (1818): 253–93.

Csomo Korosi, Alexander. "Biographical Sketch of M. Alexander Csomo Korosi, the Hungarian Traveller; Extracted from a Letter Addressed by the Gentleman to Captain C. P. Kennedy, of the Hon. East-India Company's Service, Assistant to the Political Resident at Delhi, &c. Communicated by Charles Elliot, of the Hon. East-India Company's Civil Service, late Political Resident at Delhi, &c, &c." *The Journal of the Asiatic Society of Great Britain and Ireland* 1 (1834): 128–33.

Duncan, Jonathan. "An Account of Two Fakeers with their Portraits." *Asiatick Researches* 5 (1807): 37–48.

Fraser, James B. "An Account of a Journey to the Sources of the Jumna and Bhagirathi Rivers." *Asiatick Researches* 13 (1820): 171–249.

Gissis, Snait B. "'Race' in the Eighteenth Century." *Historical Studies in the Natural Sciences* 41, no. 1 (Winter 2010): 41–103.

Giuseppe, Father. "Account of the Kingdom of Nepal by Father Giuseppe, Prefect of the Roman Mission. Communicated by John Shore." *Asiatick Researches* 2 (1807): 307–22.

Hardwicke, Thomas. "Narrative of a Journey to Sirinagur." *Asiatick Researches* 6 (1801): 309–81.

Herbert, J. D. "An Account of a Tour made to lay down the Course and Levels of the River Setlej or Satudra, as far as Traceable within the Limits of the British Authority, Performed in 1819." *Asiatic Researches* 15 (1825): 339–428.

Hodgson, Brian Houghton. "Notices of the Language and Literature, and Religion of the Bauddhas of Nepal and Bhot." *Asiatic Researches* 16 (1828): 409–48.

Hodgson, J. A. "Journal of a Survey to the Heads of the Rivers, Ganges and Jumna." *Asiatick Researches* 14 (1822): 60–152.

Hodgson, J. A. and J. D. Herbert. "Account of the Trigonometrical and Astronomical Operations for determining the Heights and Positions of the principal Peaks of the Himalaya Mountains." *Asiatick Researches* 14 (1822): 187–372.

Houston, Charles S., David E. Harris, and Ellen J. Zeman. *Going Higher: Oxygen, Man, and Mountains*, 5th ed. Seattle: The Mountaineers Books, 2005.

J. T. “Some Further Particulars of the Country of Siccim, and of its Inhabitants, the Lepchas and Bhotias.” *Gleanings in Science* 2 (1830): 177–82.

Jones, William. “The Eight Anniversary Discourse on the Borderers, Mountaineers and Islanders of Asia. Delivered 24 February, 1791 by the President.” In: *Dissertations and Miscellaneous Pieces Relating to the History and Antiquities, the Arts, Sciences and Literature of Asia*, vol. 3, edited by Wiliam Jones et al. 1–20. London: Vernor and Hood, Birchin-Lane, and Darton and Harvey 1796.

Jones, William. “Fifth Anniversary Discourse on the Tartars. Delivered 21 February, 1788 by the President.” *Asiatick Researches* 2 (1807): 19–41.

Jones, William. “Seventh Anniversary Discourse on the Chinese. Delivered 25 February, 1790 by the President.” *Asiatick Researches* 2 (1807): 365–81.

Jones, William. “The Tenth Anniversary Discourse on Asiatic History: Civil and Natural, Delivered 28 February, 1793 by the President.” *Asiatic Researches* 4 (1807): xi – xxxv.

Jones, William. “On the Second Classical Book of the Chinese.” *Asiatick Researches* 2 (1807): 194–203.

L. L. F. “The Bicentenary of the Birth of Sir William Jones, F.R.S., Founder of the Royal Asiatic Society of Bengal.” *Notes and Records of the Royal Society of London* 4, no. 1 (April 1946): 58–62; http://www.jstor.org/stable/531239, accessed on 20 July 2019.

Moorcroft, William. “A Journey to Lake Manasarovara in Un-des, a Province of Little Tibet.” *Asiatick Researches* 12 (1818): 380–536.

Neufville, John Bryan. “On the Geography and Population of Asam.” *Asiatic Researches* 16 (1828): 331–52.

Raper, F. V. “Narrative of a Survey for the Purpose of Discovering the Sources of the Ganges.” *Asiatic Researches* 11 (1812): 446–564.

Traill, G. W. “Statistical Sketch of Kamaon.” *Asiatic Researches* 16 (1828): 137–234.

Webb, W. S. “Memoir Relative to a Survey of Kemaon, with some Account of the Principles, Upon Which It has been Conducted.” *Asiatick Researches* 13 (1820): 293–310.

Wilcox, R. “On the Identity of the Sampu and Irawadi Rivers.” *Gleanings in Science* 2 (1830): 66–67.

Wilford, Francis. “On Mount Caucacus.” *Asiatick Researches* 6 (1801): 455–539.

Wilson, Horace Hayman. “Note on the Literature of Thibet.” *Gleanings in Science* 3 (January to December 1831): 243–48.

Index of Places and Names

www.ingramcontent.com/pod-product-compliance
Lightning Source LLC
Chambersburg PA
CBHW071323230926
51849CB00002B/305
9789004412460